COLLECTED WORKS
OF
COUNT RUMFORD

VOLUME II

COLLECTED WORKS
OF
COUNT RUMFORD

EDITED BY SANBORN C. BROWN

VOLUME II

PRACTICAL APPLICATIONS
OF HEAT

THE BELKNAP PRESS OF
HARVARD UNIVERSITY PRESS
CAMBRIDGE, MASSACHUSETTS
1969

PREFACE

Count Rumford's lasting contribution to science and technology was his consistent demonstration that progress in inventions and technical improvements was basically served by understanding the fundamental physical principles before attempting to make practical applications. The nature of Rumford's belief that scientific understanding must precede technological applications can be clearly seen from studying the contents of Volumes I and II of these *Collected Works*.

The papers in Volume I deal with Rumford's theories of the nature of heat and such experiments as could not be explained by the fluid caloric theory. His experiments on radiation, and particularly on the properties of the surfaces that would enhance or diminish heat loss, led him directly to designing dishes and other containers, as well as various fabrics, which would minimize such loss. His descriptions of many of these improvements are contained in Volume II. His studies also suggested to him the proper shape and structure of chimney fireplaces by which he solved all the basic problems of a smoky chimney. His papers on this subject contained in this volume can be used to this day as proper guide lines for fireplace and chimney construction.

Rumford's remarkable discovery of convection currents, as described in Volume I, not only allowed

v

him to solve the problem of the smoky chimney by proper design of the smoke shelf, throat, and fire-place opening, but led to a long study of the insulating properties of all kinds of substances based on studies demonstrating that the trapped air was accountable for the insulation. These studies of the practical application of his discoveries are included in Volume II.

Collected here also are those applications to natural phenomena which resulted from Rumford's studies of water at maximum density given in Volume I. The experiments on the maximum density of water were made to capitalize on a series of observations that were in direct conflict with the caloric-theory statement that caloric fluid added to a body always expanded it.

In the process of worrying about heat insulation to increase the efficiency of stoves and ovens for cooking, Rumford quite naturally extended his thinking in the direction of increasing also the efficiency of fuel. To this end he first designed a combustion calorimeter and then used it in a long series of measurements, reproduced in this volume. (Appended to the last of these papers in the Academy edition of Rumford's *Works* is the note: "On pages 392, 410, and 412 there are several numerical errors and inconsistencies; but, as the original French memoirs are not accessible, no attempt has been made to reconcile them." Working through the arithmetic of Rumford's arguments shows that these errors were simply and obviously typographical in nature, and they have been corrected in this edition.)

Sanborn C. Brown

CONTENTS

Experiments on cooling Bodies (Experimental Investigations concerning Heat, Section II)   1

Experiments tending to show that Heat is communicated through solid Bodies, by a Law which is the same as that which would ensue from Radiation between the Particles (Experimental Investigations concerning Heat, Section III)   9

Description of a new Instrument of Physics   24

An Account of a curious Phenomenon observed on the Glaciers of Chamouny; together with some occasional Observations concerning the Propagation of Heat in Fluids   31

An Account of some new Experiments on the Temperature of Water at its maximum Density   39

Inquiries concerning the Mode of the Propagation of Heat in Liquids   57

Of the slow Progress of the spontaneous Mixture of Liquids disposed to unite chemically with each other   74

Researches upon the Heat developed in Combustion and in the Condensation of Vapours   80

Experiments and Observations on the Cooling of Liquids in Vessels of Porcelain, Gilded and not Gilded   135

On the Capacity for Heat or calorific Power of various Liquids   145

Observations relative to the Means of increasing the Quantities of Heat obtained in the Combustion of Fuel   ·155

Account of some new Experiments on Wood and Charcoal   162

Inquiries relative to the Structure of Wood, the specific Gravity of its solid Parts, and the Quantity of Liquids and elastic Fluids contained in it under various Circumstances; the Quantity of Charcoal to be obtained from it; and the Quantity of Heat produced by its Combustion   171

Chimney Fireplaces, with Proposals for improving them to save Fuel; to render Dwelling-houses more comfortable and salubrious, and effectually to prevent Chimneys from smoking   221

Supplementary Observations concerning Chimney Fireplaces   296

Of the Management of Fire and the Economy of Fuel   309

Experiments and Observations on the Adhesion of the Particles of Water to each other   478

Continuation of Experiments and Observations on the Adhesion of the Particles of Liquids to each other   488

References to Rumford's own Works   507

Facts of Publication   509

Index   519

COLLECTED WORKS

OF

COUNT RUMFORD

VOLUME II

EXPERIMENTAL INVESTIGATIONS CONCERNING HEAT

Experiments on cooling Bodies.

It is only by careful observation of the phenomena which accompany the heating and cooling of bodies, that we can hope to acquire exact notions of the nature of heat and its manner of acting.

Many experiments have been made by different persons, at different times, with a view to determine what has been called the conducting quality of different substances with regard to heat. I have myself made a considerable number; and it is from their results, often no less unexpected than interesting, that I have been gradually led to adopt the opinions on the nature of heat which I have presumed to submit to the judgment of this illustrious assembly. The flattering attention with which the Class has honoured the three Memoirs I have lately presented encourages me to communicate the continuation of my researches.

All philosophers are agreed in considering glass as one of the worst conductors of heat which exists; and

when it is proposed to confine the heat in a body, of
which the temperature has been raised, or to hinder its
dissipation as much as possible, care is taken to surround the heated body with substances known to be
bad conductors of heat.

The results of many of my experiments having led
me to suspect that the cooling of bodies is not effected
in the manner which is generally supposed, I made the
following experiment, with the intention of clearing up
this interesting part of the science.

I procured two bottles, nearly cylindrical, of the same
form and the same dimensions when measured externally, — one being of glass, and very thick, and the
other of tin or tinned iron, which was very thin.   Each
of them is three inches ten lines in diameter, very
nearly, and five inches in height; and each has a neck
one inch three lines in diameter, and one inch two lines
in height.   The glass bottle weighs 13 ounces, 1 gros,
and 18 grains poids de marc; and the other thin metallic vessel weighs only 5 ounces, 1 gros, and 65 grains.

Having very exactly weighed the bottle of tinned
iron, I found its exterior surface to be 54.462 inches,
which give 0.21142 of a line for the thickness of
its sides, taking the specific gravity of the metal at
7.8404.

The mean thickness of the sides of the glass bottle
is more than six times as great, as may be easily de-
duced from a calculation founded on the weight of the
bottle, the quantity of its surface, and the specific gravity of glass.

Having filled these two bottles with boiling water,
I hung them up by slender strings in the midst of
the tranquil air of a large chamber, at the height of

five feet from the floor, and at the distance of four feet asunder.

The temperature of the air of the chamber, which did not vary a quarter of a degree during the whole time of the experiment, was $9\frac{3}{4}$ degrees of Reaumur's scale.

An excellent mercurial thermometer, with a cylindrical bulb, of four inches long and two lines and a half in diameter, suspended in the axis of each of these bottles, indicated the temperature of the contained water ; and the time employed in its cooling for every five degrees of Fahrenheit's thermometer was carefully observed, during eight hours.

The glass being considered as a very bad conductor of heat, and the sides of the bottle being so thick, who would not have expected that the water in this bottle would have been more slowly cooled than that in the very thin bottle of tin ?

The contrary, however, was the event ; the bottle of glass was cooled almost twice as quickly as that of tin.

While the water included in the bottle of tinned iron employed 56 minutes to pass through a certain interval of cooling, — namely, through ten degrees, between the 50th and 40th degree of the thermometer of Fahrenheit above the temperature of the air of the chamber, — the water in the glass bottle employed only 30 minutes for the same change.

It appears to me that the result of this experiment throws great light on the mysterious operation of the communication of heat.

If we admit the hypothesis that hot bodies are cooled, not by losing or acquiring some material substance, but by the action of colder surrounding bodies, communicated by undulations or radiations excited in an ethereal fluid, the results of this experiment may be easily explained; but, if this hypothesis be not adopted, I cannot explain them.

It might, perhaps, be suspected that the air attached by a certain attraction, but with unequal forces, to the surfaces of the two bottles, might have been the cause of this remarkable difference in the time of their cooling; but those who will take the trouble to reflect attentively on the results of the experiments I have described in a preceding Memoir, which were made with a view to clear up this point, with a metallic vessel first naked, and afterwards with one, two, four, and five coatings of varnish, will be persuaded that this cause is not sufficient to explain the facts.

By a course of experiments made at Munich, last year, of which the details are given in a Memoir sent to the Royal Society of London,[1] I have found that a given quantity of hot water, included in a metallic vessel of a given form and capacity, always cools with the same quickness in the air, whatever may be the metal employed to construct the vessel; provided always that the external surface of the vessel be very clean, and the temperature of the air the same.

In order that the cooling shall be effected in the same time, nothing more is required than that the external surface of the vessel be truly metallic, and not covered with oxide, or other foreign bodies.

On the inquiry, what quality all the metals might have in common, and possess in the same degree, to which this remarkable equality of their susceptibility of cooling might be attributed, I found it in their opacity.

The rays which cannot penetrate the surface of a body must necessarily be thrown back, or reflected ; and as the rays of light, which have much analogy with the invisible calorific or frigorific rays, easily penetrate glass, though they are reflected, at least for the greatest part, by metallic surfaces, I suspected beforehand the result of the experiment with the two bottles, — one of glass, and the other of tinned iron.

The state of a heated body, or a body which contains a certain quantity of caloric, has been compared to that of a sponge which contains a certain quantity of water. Supposing this comparison to be just, we might compare the loss of heat by the emission of the calorific rays to the loss of water by evaporation. Let us try if this comparison can supply us with the means of throwing some light on the interesting subject of our researches.

Instead of the sponge filled with water, let us substitute the earth, and suppose, for a moment, that the earth is everywhere equally heated, and its surface, in all parts, covered with a bed of the same kind of soil, equally moist.

As a square league in a mountainous country contains more surface, or more superficial acres than a square league situated in the plain, it is evident that more water would be evaporated from the whole surface of the earth in a given time, if the earth were covered with mountains than if its surface were an immense plain, and, consequently, that more caloric ought to be projected from the surface of any solid body broken with asperities, than from the surface of another body, of the same form and dimensions, which is smooth or well polished.

This reasoning appears to me to be just, and, if I am not deceived, the conclusions which may be drawn from the facts in question, well confirmed by experiment, ought to be considered as demonstrative. I have taken every possible care to establish these facts; and the results of all my experiments have constantly shown that more or less perfect polish, or the greater or less brightness of the surface of a metallic vessel, does not sensibly influence the time of its cooling.

I took two equal vessels of brass, and polished the external surface of one of them as highly as possible; and I destroyed the polish of the other by rubbing it in all directions with coarse emery. When these two vessels were filled with hot water, I did not find that the unpolished vessel employed more or less time in cooling than that which was polished.

I was careful to wash the surface of the unpolished vessel effectually with water, before the experiment; as I knew that if I did not take the precaution of removing all the dirt which might be lodged in the asperities of the surface, the presence of these small foreign bodies would influence the result of the experiment in a sensible manner.

We ought carefully to distinguish those surfaces which appear unpolished to our eyes, but which in fact are not so, from those which reflect little or no light.

It is more than probable that the surface of a metal is always polished, and even always equally so in all the cases wherein the metal is naked and clear and clean, notwithstanding all the mechanical means which may be used to scratch its surface and break the glare of its lustre.

Let us return to the comparison of the evaporation of water from the surface of the earth, with the emission of caloric radiating from the surface of a heated body, and let us suppose, for an instant, that the evaporation of the water from the surface of the earth does not depend on the heat of the earth itself, but that it is caused merely by the influences of surrounding bodies, — as, for example, by the rays of light received from the sun. It is evident that, in this case, the evaporation could not be sensibly greater in a mountainous country than in the plain; and by an easy analogy we see that if hot bodies be cooled, not in consequence of the emission of some material substance from their surfaces, but by the positive action of rays sent to them by colder surrounding bodies, the more or less perfect polish of their surfaces ought not sensibly to influence the rapidity of their cooling.

This is precisely what all my experiments concur to prove.

I have long sought, and with that patience which the love of the sciences inspires, to reconcile the results of my experiments with the opinions generally received concerning the nature of heat and its mode of action, but without being able to succeed.

It is in the hands of two of the most illustrious bodies of learned men that ever existed that I have thought it incumbent on me to deposit my labours, my discoveries, my doubts, and my conjectures.

I am earnestly desirous of engaging the philosophers of all countries to turn their attention towards an object of inquiry too long neglected.

The science of heat is not only of great curiosity, from the multitude of astonishing phenomena it offers to our contemplation, but it is likewise extremely interesting from its intimate connection with all the useful arts, and generally with all the mechanical occupations of human life.

Without a knowledge of heat, it is not possible either to excite it with economy or to direct its different operations with facility and precision.

Experiments tending to show that Heat is communicated through solid Bodies, by a Law which is the same as that which would ensue from Radiation between the Particles.

Having made a considerable number of experiments on the passage of heat through fluids, and through different substances in the state of powder, I was curious to ascertain the laws of its propagation through solid bodies, particularly metals.

I hoped this discovery would furnish some additional data to confirm or refute the opinions I had adopted concerning heat and its manner of acting; and it will be seen by the results that my expectations were not frustrated.

Having procured two cylindrical vessels of tin, each six inches in diameter and six inches high, I fastened them together, by means of a solid cylinder of copper, six inches long and an inch and a half in diameter, which was fixed horizontally between the two tin vessels. The extremities of the cylinder passed through two holes, an inch and a half in diameter, made for the purpose in the sides of the vessels, midway be-

9

tween the bottom and top, and were soldered fast in them.

Each of the vessels was made flat on the side where the copper cylinder was fastened, so that the extremity of the cylinder did not project into the vessel, but was level with the flattened part.

This instrument was supported at the height of eight inches and a half above the table on which it stood, by means of three feet, — two fixed to one of the vessels, and one to the other.

One of these vessels being filled with boiling water, the other with water at the freezing point, as the two extremities of the cylinder were placed in immediate contact with these two masses of fluid, a change of temperature must necessarily take place by degrees in all the interior parts of the cylinder. For the purpose of observing this change, three vertical holes were made in the cylinder, into which were introduced the bulbs of three small mercurial thermometers. One of the holes was in the middle of the cylinder, the others midway between the centre and either extremity.

Each of these holes is four lines in diameter, and eleven lines and a half deep; so that the bulbs of the thermometers, which are three lines in diameter, were all in the axis of the cylinder.

When the thermometers were put in their places, the holes were filled with mercury, in order to facilitate the communication of heat from the metal to the bulb of the thermometer.

To keep the hot water constantly boiling, a spirit-lamp was placed beneath the vessel containing it; and to keep the cold water constantly at the temperature of melting ice, fresh portions of ice were added to it, from time to time.

The thermometers are graduated to Fahrenheit's scale, the freezing point being marked 32°, and that of boiling water 212°.

As the first and most important object I had in view was to learn at what temperature the three thermometers would become stationary, I did not very carefully notice the progress of the thermometers toward this point; but as soon as they appeared nearly stationary, I observed them with the greatest attention for near half an hour.

To distinguish the three thermometers, I shall call that nearest the boiling water B, that in the centre C, and that nearest the cold water D.

The following are the progress and results of an experiment made the 28th of April, 1804, the temperature of the air being 78° of Fahrenheit.

Time.	Temperature of the hot water.	Temperature marked by the thermometer B.	Temperature marked by the thermometer C.	Temperature marked by the thermometer D.	Temperature of the cold water.
h. m. s.	Degrees.	Degrees.	Degrees.	Degrees.	Degrees.
1 52 15	212	160	130	105	32
— 53 30	—	$160\frac{1}{2}$	131	$105\frac{1}{2}$	—
— 55	—	161	$131\frac{3}{4}$	106	—
— 56 30	—	$161\frac{3}{4}$	132	$106\frac{1}{2}$	—
— 58	—	162	$132\frac{1}{2}$	107	—
2 0 0	—	162	$132\frac{3}{4}$	$107\frac{1}{2}$	—
— 1 30	—	162	133	$107\frac{1}{2}$	—
— 4	—	162	$132\frac{1}{4}$	$106\frac{1}{2}$	—
— 6	—	162	132	106	—
— 9	—	162	$132\frac{3}{4}$	$106\frac{1}{2}$	—
— 11	—	162	$132\frac{3}{4}$	$106\frac{1}{2}$	—
— 28	—	162	$132\frac{3}{4}$	$106\frac{1}{2}$	—

Before I proceed to examine more minutely the results of this experiment, I will endeavour to show

those results which it ought to have exhibited, on the
supposition that heat is propagated, even in the interior
of solid bodies, by *radiations* emanating from the surfaces of the particles composing these bodies.

On this supposition, we must necessarily consider
the particles that compose bodies as being *separate from
each other*, and even by pretty considerable distances,
compared with the diameters of these particles; but
there is nothing repugnant to the admission of this supposition; on the contrary, there are many phenomena
which apparently indicate that all the solid bodies with
which we are acquainted are thus formed.

To see now by what law heat would be propagated
in a solid cylinder, let us represent the axis of this cylinder by a right line A E, Plate IV. Fig. 1 ; and let
us begin with supposing that the cylinder consists of
three particles of matter only, A C E, placed at equal
distances in that line.

Let us farther suppose that the extremity, A, of the
cylinder is constantly at the temperature of boiling water,
while its other extremity, E, remains invariably at the
freezing point.

By an experiment, of which I have already given an
account to the Class,[2] I found that when two equal
bodies, A B, one hotter than the other, are isolated and
placed opposite each other, the intensities of their radiations are such, that a third body, C, placed in the
middle of the space that separates them, will acquire a
temperature, by the simultaneous action of these radiations, which will be an arithmetical mean between
those of the two bodies A and B.

From the result of this experiment we have ground

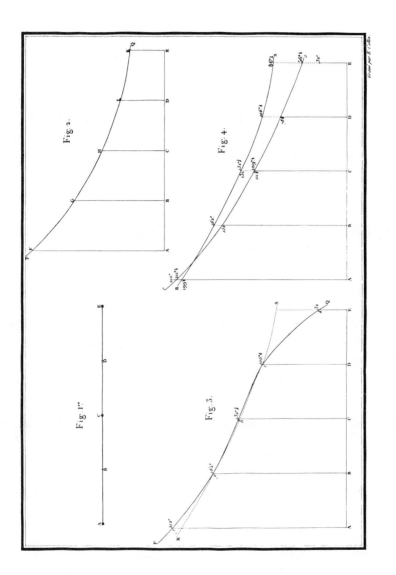

Fig. 2.

Fig. 1er.

Fig. 4.

Fig. 3.

to conclude that if the cylinder were composed of three particles of matter only, A, C, E, the particle C, which is in the middle of the cylinder, must necessarily have the arithmetical mean temperature between that of A and that of E, which are at the two extremities of the cylinder; that is to say, between 212° and 32° of Fahrenheit, which is 122°.

Now let us interpose between the particles A, C, and E, two other particles, B, D, and see whether the introduction of these two particles will make any change in the temperature of the particle C that occupies the middle of the cylinder.

If the particle B be placed in the middle of the space comprised between the extremity, A, of the cylinder and its middle, C, it ought to acquire a mean temperature between that of the extremity, A, of the cylinder, and that of the point C, namely that of 167°, the mean between 212° and 122°; and if the particle D be placed in the midst of the space comprised between the middle of the cylinder and its other extremity, E, this particle ought to acquire a mean temperature between that of the middle of the cylinder and that of its extremity, E ; it ought then to have the temperature of 77°.

From this new arrangement, the particle C, situate in the middle of the cylinder, will find for its neighbours, on one side the particle B, at the temperature of 167°, and on the other the particle D, at that of 77°. The point in question is, whether the presence of these two particles will make any change in the temperature of the particle C, or not.

In the first place, it is evident that if the calorific influences of the particle B on the particle C be as efficacious in heating it as the frigorific influences of the particle D be in cooling it, the temperature of the particle C ought not to be changed. But experience has shown that, at equal distances and equal intervals of temperature, the calorific influences of hot bodies and the frigorific influences of cold bodies are exactly equal; and as the distance from B to C is equal to the distance from D to C, while the interval of temperature between B and C, $= 45°$, is the same as that between D and C, $= 45°$, it is evident that the temperature of the particle C, which is in the middle of the cylinder, can be no way affected by the introduction of the intermediate particles B and D.

By the same way of reasoning may be proved, that the introduction of an indefinite number of intermediate particles would produce no change in the temperature of the middle of the axis of the cylinder, or in any part of it; and if the introduction of an indefinite number of intermediate particles make no change in the state of a thermometer placed in the middle of the axis of the cylinder, we may conclude that the thermometer would remain equally stationary if the number of intermediate particles were increased till they had that proximity to each other which is necessary to constitute a solid body. If, instead of a single row of particles in a right line, there were a bundle composed of an indefinite number of such rows placed side by side, forming a solid cylinder, the temperature in the different parts of the line A E would remain the same.

From this reasoning we may infer that the temperatures of the different parts of the cylinder should decrease in arithmetical progression from one extremity of the cylinder to the other.

But it is evident that this law of decrement of tem-

perature could take place only in the single case of the
surface of the cylinder being completely isolated, so as
to be no way affected by the action of surrounding
bodies, which is absolutely impossible.

The circumstances under which the experiments were
made are very different from those here taken for
granted. The bodies we subject to experiment are constantly surrounded on all sides by the air and other
bodies which act on our instruments continually, and
often in a very perceptible manner; and we can never
hope to isolate a cylinder so completely that the apparent progress of heat in its interior shall perceptibly
obey the law we have just discovered. In common
cases it deviates widely from this law.

As the causes of this deviation are well known, we
will see whether there be no means of appreciating their
effects.

The surface of the cylinder being surrounded by the
atmospheric air and other bodies, all which are of a
known and sensibly constant temperature, we may determine the comparative effects of these bodies on the
different parts of the surface of the cylinder.

In those parts of the cylinder which are hotter than
the air and other surrounding bodies, the surface of the
cylinder will be cooled by the action of these bodies;
but if one of the extremities of the cylinder be colder
than the atmospheric air, those parts of the cylinder
which are colder than the circumambient fluid will be
heated by its influence and that of the surrounding
bodies.

We will begin with examining the case where the
coldest extremity of the cylinder is at the same temperature as the surrounding air. Let us suppose,

then, that the experiment with boiling water at the one
end and freezing at the other be made when the temperature of the air is at the freezing point, or 32° of
Fahrenheit.

In this case it is evident that the surface of the cylinder must everywhere be *cooled* by the influence of the
surrounding atmosphere. The question, then, is to
determine the comparative effects, or the relative quantities of refrigeration or loss of heat, that must take
place *in the different parts of the cylinder;* and, in the first
place, it is clear that the hotter a given part of the cylinder is, the more heat it must lose in a given time, by
the influence of the surrounding cold bodies; whence
we may conclude that the refrigeration of the surface of
the cylinder by the influence of the air and other surrounding cold bodies must necessarily diminish from
the extremity of the cylinder, A, which is in contact with
the hot water, to its extremity, E, which is in contact
with the cold.

From reasoning which appears incontrovertible, and
which the results of a great number of experiments
appear to confirm, it has been concluded that the celerity with which a hot body placed in a cold medium is
cooled is always proportional to the difference between
the temperature of the hot body and that of the medium. Considering this conclusion as established, we
may determine *a priori* what ought to be the gradation
of temperatures in the interior of a given solid cylinder
surrounded by air, one extremity of which is in contact
with a considerable body of boiling water, while the
other is similarly in contact with cold.

We have seen that, if the surface of the cylinder were
perfectly isolated, the decrease of temperature from the

hottest extremity of the cylinder, A, to its other extremity, E, which is in contact with cold water, would be *in
arithmetical progression,* and it has just been shown that
the decrease must necessarily be accelerated by the action of the air and other surrounding cold bodies.

But the acceleration of the decrease of temperature in
those parts of the cylinder which are toward the cold
extremity, depending on the action of the air and surrounding bodies, must be continually diminishing in
proportion as the temperature of the surface of the cylinder approaches nearer and nearer that of the air; and
hence we may conclude that, if a given number of points,
at equal distances from each other, be taken in the axis
of the cylinder, the temperatures corresponding with
these points will be *in geometrical progression.*

We may represent the progress of the decrease of
temperature by Plate IV. Fig. 2.

In a right line A E, representing the axis of the cylinder, if we take the three points B, C, and D, so that
the distances A B, B C, C D, and D E shall be equal,
and, erecting the perpendiculars A F, B G, C H, D I,
EK, take A F = the temperature of the cylinder at
its extremity A, B G = its temperature at the point B,
and so of the rest; the ordinates A F, B G, &c. will
be in geometrical progression, while their corresponding
abscisses are in arithmetical progression; consequently
the curve, P Q, which touches the extremities of all
these ordinates, must necessarily be the *logarithmic* curve.

We will now see whether the results of experiment
agree with the theory here exhibited, or not.

To form our judgment with ease and, as it were, at a
single glance, of the agreement of our theory with the
results of the experiment of which I gave an account

at the beginning of this Memoir, we have only to represent these results by a figure in the following manner.

On the horizontal line A E, Fig. 3, representing the
axis of the cylinder employed in the experiment, we will
take three points, B, C, and D ; one, C, in the middle
of the axis, being the situation of the central thermometer, the other two, B and D, at the intermediate points
which the other two thermometers occupied between
the middle of the axis and its two extremities.

Erecting the perpendiculars A f, B g, C h, D i, and
E k, on the points A, B, C, D, and E ; and taking the
ordinate A $f = 212$, the temperature of boiling water,
B $g = 162$, the temperature indicated by the thermometer B, C $h = 132\frac{3}{4}$, the temperature indicated by
the thermometer C; D $i = 106\frac{1}{2}$, the temperature given
by the thermometer D, and lastly, E $k = 32$, the temperature of water mixed with pounded ice, — a curve,
P Q, passing through the points f, g, h, i, k, ought to
be the *logarithmic;* that is, supposing the temperature
of the surrounding air to be constantly at the temperature of melting ice during the experiment.

But the experiment in question was made when the
temperature of the air was at 78° F.; consequently,
reckoning from a certain point, taken in the length of
the cylinder, where the temperature was at 78°, to the
extremity, E, the influence of the surrounding air, instead of *cooling* the surface of the cylinder, *heated* it ; and
it is evident that the curve, P Q, must necessarily in this
case have a point of inflection.

In fact, it appears on a simple inspection of the figure,
that the curve, P Q, has a point of inflection ; but we see,
likewise, that this curve is not regular. That branch
which is concave toward the axis of the cylinder is not

similar to the adjoining portion of the curve, of equal
length, which is convex toward that axis, as it ought to
be according to our theory; and even the part of the
curve which is convex toward the axis, A E, differs sensibly from the logarithmic, particularly toward its extremity, P.

It ought necessarily to differ from this curve *as far
as the divisions of our thermometers are defective;* but the
deviation between the ordinates, A *f* and B *g*, indicated
by the results of the experiment in question, appears to
me much too considerable to be ascribed to the imperfection of our thermometers.

To see how far the curve, P Q, differs from the logarithmic, we have only to draw a logarithmic curve, R S,
through the points *g* and *i*, and we shall find, that the
ordinates corresponding to the points

	A,	B,	C,	D,	E,
	°	°	°	°	°
Instead of being	212.00	162	132¾	106½	32.00
Will be	199.55	162	131	106½	86.35
Difference	—12.45	0	—1¾	0	+54.35

The very great difference that exists between the temperature of cold water and that indicated by the results
of the experiment for the extremity of the cylinder
which was in contact with this water led me to suspect
that it was owing to the quality possessed by water in
common with other fluids, which renders it *a very bad
conductor of heat.*

If it be true, as I believe I have elsewhere proved,
that there is no sensible communication of heat between the adjacent particles of a fluid, from one to
another, and that heat is propagated through fluids
only in consequence of a motion of their particles, re-

sulting from a change in their specific gravity, occasioned by their being heated or cooled; as the specific
gravity of water is very little altered by an inconsiderable change of temperature when this fluid is near the
freezing point, it might have been foreseen that a solid
body a little heated, and plunged into cold water, would
be very slowly cooled.

The result of the following experiment, which I
made with a view to elucidate this point, will put the
fact out of all doubt.

The three thermometers being stationary, one, B, at
162°, the second, C, at 132¾°, and the third, D, at
106½°, the water in contact with one of the extremities
of the cylinder being still boiling, while the water mixed
with pounded ice, which was in contact with the other
extremity, was constantly at the temperature of melting
ice, I began to stir this mixture of ice and water pretty
briskly with a little stick, and I continued to stir it
uninterruptedly and with the same velocity for 22
minutes.

I had scarcely begun this operation when I had a
proof that my conjectures were well founded. The
mercury in the three thermometers immediately began
to descend, and did not stop till it had fallen very considerably.

The thermometer B fell from 162° to 152°; C from
132¾° to 111¾°; and D from 106½° to 78½°.

On comparing these numbers, we find that, in consequence of the agitation of the cold water for 22
minutes, the thermometer B fell 10° of Fahrenheit's
scale, the thermomete C 21°, and the thermometer D
28°.

As soon as I had ceased to stir the cold water the

three thermometers began to rise, and at the end of a quarter of an hour they had all reached the points from which they set out at the beginning of this operation.

To facilitate the comparison of the results of these two experiments, — one made with cold water at rest, the other with the same water in a state of constant agitation, — I have represented them in Fig. 4.

In the first place, we shall learn several very interesting facts by simple inspection of this figure; we shall see, —

1st. That the progress of refrigeration — or, to speak more properly, *the decrease of temperature* — was everywhere much more rapid when the cold water in contact with the extremity, E, of the cylinder was agitated than when it was at rest.

2dly. That the extremity of the cylinder in contact with this water was constantly near 30° colder in the first case than in the second.

3dly. We shall see that the progress of refrigeration was everywhere, and in both the experiments, such nearly as our theory points out.

The decrease of temperature toward the middle of the cylinder was so regular that it is more than probable the apparent irregularities toward the two extremities were occasioned solely by the difficulty which a body of water finds in communicating its mean temperature to a solid with which it is in contact.

The boiling water being in continual motion, owing to its ebullition, it had a great advantage over the cold water, which was at rest, in communicating its temperature to the extremity of the cylinder it touched; but I have found, notwithstanding this, that by agitating the boiling water strongly with a quill, and particularly when with the quill I made a rapid friction against the end of the cylinder immersed in the boiling water, I occasioned all the thermometers to rise several degrees.

It may perhaps be imagined, at first sight of the results of the experiment, that as the three thermometers, which occupied the parts about the middle of the axis of the cylinder, did not indicate a decrease perfectly agreeing with the theory, the theory itself cannot be true; but a moment's reflection will show that this inference would be too hasty, and that the difference between the theory and the results of our experiments, far from proving anything adverse to the theory, serve on the contrary to render it more probable.

The results of such experiments can never agree with the theory, except the divisions of our thermometers be perfectly accurate; but it is well known to every one who has any knowledge of natural philosophy that the divisions of our thermometers are defective.

One of the objects I had in view in the experiments of which I have just given an account to the Class, and in several others which I intend to make without delay, is to improve the division of the scale of the thermometer, in order to render this valuable instrument of greater utility in the delicate investigations of natural philosophy.

It appears certain that the increase of the elasticity of air by heat is much more nearly proportionate to the increase of temperature than the dilatation of mercury or any known fluid; consequently, it is the air thermometer we ought to endeavour to improve, and which must ultimately afford us the most accurate measure of heat that it is possible for us to procure.

DESCRIPTION

NEW INSTRUMENT OF PHYSICS.

FIRST MEMOIR ON HEAT

THE Count of Rumford is pleased to present the class with a description of a new instrument of physics of his own invention which was used last year in a series of experiments on heat. It will be accompanied by a brief exposition of the principal results of this research.

The instrument, which I have called a *thermoscope*, is designed to detect very slight changes in temperature.

The principal part of the instrument consists of a long glass tube curved at each end with very small glass globes at its extremities. The straight middle portion of this tube is placed horizontally while its two extremities, culminating in the two globes, are turned upward so as to form two "elbows" at right angles to the horizontal section of the tube. The latter is 15 or 16 inches in length while each of the vertical extremities is 6 or 7 inches long.

The diameter of the tube should be about a half-line; that of each of the glass globes should be an inch and a half or an inch and three quarters.

By means of a small glass reservoir one inch long which is soldered to the tube at one of the "elbows," a small quantity of colored alcohol is introduced into the interior of the instrument (just sufficiently to fill the reservoir) and, when this is done, the extremity of the reservoir is hermetically sealed; all communication is thus cut off between the air in the two globes of the thermoscope and in its tube, and the air of the atmosphere.

The instrument is put in order and prepared for experiment in the following manner.

A slight quantity of alcohol, enough to form a small cylinder or column within the tube, is drawn into the latter from the reservoir. One must take care to execute this delicate operation in such fashion that the small mobile column, which I shall call a bubble, comes to rest as nearly as possible at the center of the horizontal segment of the tube. When the bubble is so placed, the air in the two globes being equally warm, the instrument is in a state of rest.

When this operation (which requires great care and sometimes a great deal of time) is completed, the instrument is ready for use. It is made to function as follows. One of the two globes is covered by screens and thus insulated against the heat or cold of hot or cold bodies which are placed beside the other globe; within the latter, changes in temperature produce changes in air elasticity so that the little alcohol bubble within the horizontal part of the tube is put in motion and forced into a new position.

The *direction* of movement of the bubble indicates the nature of the change in air temperature within the globe, while the *quantity* of its movement is the measure of increases or decreases in the temperature of this air.

If the bubble moves away from the globe near which the chosen hot or cold body has been placed, it is evident that the air in the globe has been *heated* by the influence of this body; but, when the alcohol bubble moves toward the globe, it is clear that the air within the globe has cooled.

The speed with which the bubble serving as an index moves is proportional to the intensity of action of the body placed near the instrument.

To compare the heat- or cold-producing properties of two different bodies, the latter are simultaneously placed at either end of the instrument at such distance from their respective globes that the alcohol bubble remains immobile in its place. When this occurs, it is quite clear that the action of the two bodies upon their globes is precisely equal on each side. Thereupon one may calculate the intensity of the rays emanating from each body by the extent to which its surface is exposed to the globe and the square of its distance from it.

To compare the heat-producing capacities of a hot body with the cold-producing capacities of a cold one, one may mask one of the thermoscope globes by means of screens; the two bodies in question are then placed near the other globe, each on a different side. They are each placed at such a distance from the globe as to make their simultaneous action upon it precisely equal; in other words, one heats the globe in a given time as much as the other cools it.

This equality of action is announced by the immobility of the alcohol bubble, which serves as the index of the instrument. When such equality is established, one may calculate the intensity of radiation of the bodies in question by the extent to which their surfaces are exposed to the globe and the squares of their distances from it.

The sensitivity of this instrument is so great that, at 10–12 degrees Réaumur, the heat emanating from an open hand held at a distance of three feet from one of the globes causes the alcohol bubble to move; and a blackened metallic disk 4 inches in diameter having the temperature of melting ice and held at a distance of 18 inches causes it to move in the opposite direction with a speed quite visible to the eye.

With the help of this instrument, I have discovered that the surfaces of all bodies at all temperatures (cold bodies as well as hot bodies) are continually giving off rays, or rather (as I see it) waves comparable to those which sound-producing bodies emit into the air in all directions; and that these rays or waves gradually affect and change the temperature of all bodies that they strike without being reflected, in all cases where the bodies thus struck happen to be either hotter or cooler than the bodies whence the rays or waves emanate;

That the intensity of rays from different bodies of equal temperature differs greatly; and that it is less for polished, opaque bodies than for unpolished bodies;

That a copper surface, for example, gives off four times as many rays when it is covered with a coat of oxide, and five times as many when it is blackened as when the metal is *smooth* and well polished;

That the rays that any body at a given temperature gives off in all directions are either heat- or cold-producing for other bodies against which they may strike, depending upon whether the latter are more or less warm than the bodies whence the rays emanate; with the result that the same rays are at once heat-producing for certain bodies or, more accurately, for all bodies less warm than they, and cold-producing for all others.

According to these facts, it would appear that those bodies which, while hot, emit many heat-producing waves, would also, when they are colder than the bodies which surround them, transmit to the latter a large number of cold-producing waves; and my experiments have proved this to be so.

With bodies of the same kind and at equal temperature intervals, the cold-producing influences of cold bodies are as real and efficient as the heat-producing influences of hot bodies.

Two metal disks of equal diameter — one at a temperature of zero (that of melting ice) and the other at 40 degrees — were simultaneously placed at an equal distance from one of the thermoscope globes which was at a temperature of 20 degrees Réaumur; the fact that the index of the instrument remained at rest indicated that the globe was as cooled by the cold body as it was heated by the hot.

When one blackens the surface of either disk, the intensity of rays from this disk increases to such an extent that the other is no longer able to withstand its influence; however, by blackening the other disk as well, the balance is rapidly reestablished.

If emanations from (warm and cold) bodies are veritable waves in an extremely elastic fluid that has been designated under the name of ether, the communication of heat and cold must be analogous to the communication of sound, and all those means which have been devised to increase the effects of sound must be equally applicable to increasing the effects produced by emanations of hot and cold bodies; and, in fact, I have found that a megaphone which was well polished on its inner surface and which was placed between one of the globes of my thermoscope and a copper bowl three inches in diameter containing a mixture of water and crushed ice, more than doubled the cold-producing effects of this cold body upon the instrument.

The cold body *spoke* into the large opening of the megaphone, while the thermoscope globe *listened* behind its small aperture.

The details of my research on this interesting subject are set forth in a memoir sent to the Royal Society of London last autumn. An extract of this memoir will shortly appear in the *Bibliothèque Britannique;* but, since several members of the class to whom I had spoken of these experiments upon my arrival in Paris expressed a desire to see them performed, I have had my instruments sent from Munich, and I am willing to transport them to the laboratory of the Institute in order that they may be examined by interested members of the class, provided that I am assigned a room which can be securely locked.

Since these instruments are extremely fragile and so delicate that the mere presence of three or four persons in the same room disturbs them perceptibly, it is necessary to approach them with caution and put them away with the greatest care.

AN ACCOUNT

OF A

CURIOUS PHENOMENON OBSERVED ON THE GLACIERS OF CHAMOUNY;

TOGETHER WITH

SOME OCCASIONAL OBSERVATIONS CONCERNING THE PROPAGATION OF HEAT IN FLUIDS.

IN an excursion which I made the last summer, in the month of August, to the glaciers of Chamouny, in company with Madame Lavoisier and Professor Pictet of Geneva, I had an opportunity of observing, on what is called the Sea of Ice (*Mer de Glace*), a phenomenon very common, as I was told, in those high and cold regions, but which was perfectly new to me, and engaged all my attention. At the surface of a solid mass of ice, of vast thickness and extent, we discovered a pit perfectly cylindrical, about seven inches in diameter and more than four feet deep, quite full of water. On examining it on the inside with a pole, I found that its sides were polished, and that its bottom was hemispherical and well defined.

This pit was not quite perpendicular to the plane of the horizon, but inclined a little towards the south as it descended; and in consequence of this inclination, its mouth, or opening at the surface of the ice, was not circular, but elliptical.

From our guides I learned that these cylindrical holes are frequently found on the level parts of the ice; that they are formed during the summer, increasing gradually in depth, as long as the hot weather continues; but that they are frozen up and disappear on the return of winter.

I would ask those who maintain that water is a conductor of heat, how these pits are formed. On a supposition that there is no direct communication of heat between neighbouring particles of that fluid which happen to be at different degrees of temperature, the phenomenon may easily be explained; but it appears to me to be inexplicable on any other supposition.

The quiescent mass of water by which the pit remains constantly filled must necessarily be at the temperature of freezing, for it is surrounded on every side by ice; but the pit goes on to increase in depth during the whole summer. From whence comes the heat that melts the ice continually at the bottom of the pit? and how does it happen that this heat acts on the *bottom* of the pit only, and not on its sides?

These curious phenomena may, I think, be explained in the following manner. The warm winds which in summer blow over the surface of this column of ice-cold water must undoubtedly communicate some small degree of heat to those particles of the fluid with which this warm air comes into immediate contact; and the particles of the water at the surface so heated, being rendered specifically heavier than they were before by this small increase of temperature, sink slowly to the bottom of the pit, where they come into contact with the ice, and communicate to it the heat by which the depth of the pit is continually increased.

This operation is exactly similar to that which took place in one of my experiments (see my Essay on the Propagation of Heat in Fluids, *Experiment* 17),[3] the results of which no person to my knowledge has yet explained.

There is another very curious natural phenomenon which I could wish to see explained in a satisfactory manner by those who still refuse their assent to the opinions I have been led to adopt, respecting the manner in which heat is propagated in fluids. The water at the bottoms of all deep lakes is constantly at the same temperature (that of 41° Fahrenheit), summer and winter, without any sensible variation. This fact alone appears to me to be quite sufficient to prove that, if there be any immediate communication of heat between neighbouring particles or molecules of water, *de proche en proche,* or from one of them to the other, that communication must be so extremely slow that we may with safety consider it as having no existence ; and it is with this limitation that I beg to be understood when I speak of fluids as being non-conductors of heat.

In treating of the propagation of heat in fluids, I have hitherto confined myself to the investigation of the simple matter of fact, without venturing to offer any conjectures relative to the causes of the phenomena observed. But the results of recent experiments on the calorific and frigorific radiations of hot and of cold bodies (an account of which I shall have the honour of laying before the Royal Society in a short time)[1] have given me some new light respecting the nature of heat and the mode of its communication ; and I have hopes of being able to show *why* all changes of temperature in *transparent* liquids must necessarily take place at their surfaces.

I have seen, with real pleasure, that several ingenious gentlemen in London and in Edinburgh have undertaken the investigation of the phenomena of the propagation of heat in fluids, and that they have made a number of new and ingenious experiments, with a view to the further elucidation of that most interesting subject. If I have hitherto abstained from taking public notice of their observations on the opinion I have advanced on that subject in my different publications, it was not from any want of respect for those gentlemen that I remained silent, but because I still found it to be quite impossible to explain the results of my own experiments on any other principles than those which, on the most mature and dispassionate deliberation I had been induced to adopt; and because my own experiments appeared to me to be quite as conclusive (to say no more of them) as those which were opposed to them; and, lastly, because I considered the principal point in dispute, relative to the passage of heat in fluids, as being so clearly established by the circumstances attending several great operations of nature, that this evidence did not appear to me to be in danger of being invalidated by conclusions drawn from partial and imperfect experiments, and particularly from such as are allowed on all hands to be extremely delicate.

In all our attempts to cause heat to descend in liquids, the heat unavoidably communicated to the sides of the containing vessel must occasion great uncertainty with respect to the results of the experiment; and when that vessel is constructed of ice, the flowing down of the water resulting from the thawing of that ice will cause motions in the liquid, and consequently inaccuracies of still greater moment, as I have found from my own experience; and when thermometers immersed in a liquid at a small distance below its surface acquire heat in consequence of a hot body being applied to the surface of the liquid, that event is no decisive proof that the heat acquired by the thermometer is communicated by the fluid, from above, downwards, from molecule to molecule, *de proche en proche;* so far from being so, it is not even a proof that it is from the fluid that the thermometer receives the heat which it acquires; for it is possible, for aught we know to the contrary, that it may be occasioned by the radiation of the hot body placed at the surface of the fluid.

In the experiments of which I have given an account in my Essay on the Propagation of Heat in Fluids, great masses, many pounds in weight, of boiling-hot water, were made to repose for a long time (three hours) on a cake of ice, without melting but a very small portion of it; and on repeating the experiment with an equal quantity of very cold water (namely, at the temperature of 41° Fahrenheit), nearly twice as much ice was melted in the same time. In these experiments the causes of uncertainty above mentioned did not exist, and the results of them were certainly most striking.

The conclusions which naturally flow from those results have always appeared to me to be so perfectly evident and indisputable as to stand in no need either of elucidation or of further proof.

If water be a conductor of heat, how did it happen that the heat in the boiling water did not, in three hours, find its way downwards to the cake of ice on which it reposed, and from which it was separated only by a stratum of cold water half an inch in thickness?

I wish that gentlemen who refuse their assent to the opinions I have advanced respecting the causes of this curious phenomenon would give a better explanation of it than that which I have ventured to offer. I could likewise wish that they would inform us how it happens that the water at the bottoms of all deep lakes remains constantly at the same temperature; and above all, how the cylindrical pits above described are formed in the immense masses of solid and compact ice which compose the glaciers of Chamouny.

A remark, which surprised me not a little, has been made by a gentleman of Edinburgh (Dr. Thomson), on the experiments I contrived to render visible the currents into which liquids are thrown on a sudden application of heat or of cold. He conceives that the motions observed in my experiments, among the small pieces of amber which were suspended in a weak solution of potash in water, were no proof of currents existing in that liquid; as they might, in his opinion, have been occasioned by a change of specific gravity in the amber, or by air attached to it. I am sorry that so mean an opinion of my accuracy as an observer should have been entertained, as to imagine that I could have been so easily deceived. For nothing, surely, is easier than to distinguish the motion of a solid suspended in a liquid of the same specific gravity, which is carried along by a current in the liquid, from that of a body which descends, or ascends, in the liquid in consequence of its relative weight or levity. In the one case the motion is uniform; in the other, it is accelerated. In a current the body may be carried forward in all directions, and even in curved lines; but when it falls in a quiescent fluid by the action of gravity, or rises in consequence of its being specifically lighter than the fluid, it must necessarily move in a vertical direction.

The fact is, that I very often observed, in the course

of my numerous experiments, the motions of small particles of matter of different kinds in water, which Dr. Thomson describes ; but so far from inferring *from them* the existence of currents in that fluid, their cause was so perfectly evident that I did not even think it necessary to make any mention of them.

I cannot conclude this paper without requesting that the Royal Society would excuse the liberty I have taken in troubling them with these remarks. Very desirous of avoiding every species of altercation, I have hitherto cautiously abstained from engaging in literary disputes ; and I shall most certainly endeavour to avoid them in future.

I am responsible to the public for the accuracy of the accounts which I have published of my experiments; but it cannot reasonably be expected that I should answer all the objections that may be made to the conclusions which I have drawn from them. It will, however, at all times, afford me real satisfaction to see my opinions examined and my mistakes corrected ; for my first and most earnest wish is, to contribute to the advancement of useful knowledge.

AN ACCOUNT

SOME NEW EXPERIMENTS ON THE TEMPERATURE OF WATER AT ITS MAXIMUM DENSITY.

I N my seventh Essay on the Propagation of Heat in Fluids,[4] and in a paper published in the Philosophical Transactions for the year 1804,[5] in which I have given an account of a curious phenomenon frequently observed on the glaciers of Chamouny, I have ascribed the melting of the ice below the surface of the ice-cold water to currents of water slightly warmer, and consequently slightly heavier, which descend from the surface to the bottom of the ice-cold water; but the principal fact on which this supposition is founded having been called in question by various persons, I have endeavoured to establish it by new and decisive experiments.

If it is true that the temperature of water at its maximum density is considerably higher than the freezing-point of that liquid (as was announced many years ago by M. de Luc), and that the communication of heat in liquids is brought about by a movement of circulation caused by a change of density in the particles of the fluid resulting from a change of temperature, the explanation that I have given of the phenomenon of the melting of ice covered with a layer of ice-cold water by heat applied to the surface of the water, would seem

39

natural and admissible ; but if the density of water is greater at the temperature of melting ice than at any other more elevated temperature, as some philosophers assert, it is evident that the vertical descending currents of warm water which I have described cannot exist, and my explanation must be rejected.

This inquiry interested me all the more, because the fact in question had served as the foundation of the theory which I gave in my seventh Essay on the periodical winds of the polar regions, and as the basis of my conjectures on the existence of currents of cold water in the depths of the sea coming from the polar regions to the equator, and on the cause of the great difference which is found in the temperature of different countries situated in the same latitude and at the same height above the level of the sea.

After meditating on the means which I should employ to establish this important fact beyond doubt, I thought of the experiment which I am about to describe, and which is all the more interesting, since it not only demonstrates the existence in a mass of water which is warmed or cooled, of the currents assumed by my theory, but proves at the same time that the temperature at which the density of water is at a maximum is actually some degrees above that of melting ice.

Having provided a cylindrical vessel (A, Plate VI.), open above, made of thin sheet brass, $5\frac{1}{2}$ inches in diameter and 4 inches deep, supported on three strong legs $1\frac{1}{4}$ inches high, I placed in it a thin brass cup (B) 2 inches in diameter at its bottom (which is a little convex downwards), $2\frac{8}{10}$ inches wide at its brim, and $1\frac{3}{10}$ inches deep ; this cup stands on three spreading legs made of strong brass wire, and of such form

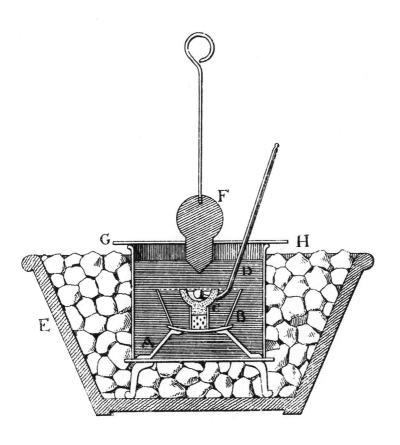

and length that when the cup is introduced into the cylindrical vessel, it remains firmly fixed in the axis of it, and in such a situation that the bottom of the cup is elevated just $1\frac{1}{4}$ inches above the bottom of the cylindrical vessel.

In the middle of this cup there stands a vertical tube of thin sheet brass $\frac{1}{2}$ of an inch in diameter and $\frac{6}{10}$ of an inch in length, open above, which serves as a support for another smaller cup (C), which is made of cork, the brim of which is on the same horizontal level with the brim of the larger brass cup in which it is placed.

This cork cup, which is spherical, being something less than half of an hollow sphere, is 1 inch in diameter at its brim, measured within, $\frac{4}{10}$ of an inch deep, and $\frac{1}{4}$ of an inch in thickness. It is firmly attached to the vertical tube on which it stands, by means of a cylindrical foot $\frac{1}{2}$ of an inch in diameter and $\frac{1}{4}$ of an inch high, which enters with friction into the opening of the vertical tube.

On one side of this cork cup there is a small opening, which receives and in which is confined the lower extremity of the tube of a small mercurial thermometer (D). The bulb of this thermometer, which is spherical, is $\frac{3}{10}$ of an inch in diameter, and it is so fixed in the middle of the cup, that its centre is $\frac{1}{4}$ of an inch above the bottom of the cup; consequently it does not touch the cup anywhere, nor does any part of it project above the level of its brim.

The tube of this thermometer, which is 6 inches in length, has an elbow near its lower end at the distance of 1 inch from its bulb, which elbow forms an angle of about 110 degrees, and the thermometer is so fixed in the cork cup, that the short branch of its tube, namely, that to the end of which the bulb is attached, lies in an horizontal position, while the longer branch (to which a scale made of ivory and graduated according to Fahrenheit is affixed) projects obliquely upwards and outwards in such a manner that the freezing-point of the scale lies just above the level of the top of the cylindrical vessel in which the cups are placed.

The cork cup, which was turned in the lathe, is neatly formed, and in order to close the pores of the cork, it was covered within and without with a thin coating of melted wax, which was polished after the wax was cold.

The thermometer was fixed to the cork cup by means of wax, and in doing this care was taken to preserve the regular form of the cup, both within and without.

The vertical brass tube which supports this cup in the axis of the brass cup is pierced with several small holes, in order to allow the water employed in the experiments to pass freely into and through it.

Having attached about 6 ounces of lead to each of the legs of the brass cup, in order to render it the more steady in its place, it was now introduced with its contents into the cylindrical vessel, and the vessel was placed in an earthen basin (E), and surrounded on all sides with pounded ice. ·This basin was 11 inches in diameter at its brim, 7 inches in diameter at the bottom, and 5 inches deep, and was placed on a firm table in a quiet room.

Several cakes of ice were then placed under the bottom of the brass cup, and the cup was surrounded on all sides by a circular row of other long pieces of ice fixed in a vertical position between the outer walls of the cup and the walls of the cylindrical vessel. These pieces were about 4 inches long, and extended from the bottom of the vessel to within a very short distance of the top.

All these pieces of ice having been fixed firmly in their places by means of some little wooden wedges, ice-cold water was poured into the cylindrical vessel until the surface of this liquid was an inch above the upper edge of the cork cup.

In this state of things it is evident that the two cups were filled with and surrounded on all sides by water at the temperature of melting ice, and that this temperature was maintained constant by the pieces of ice with which the water was in contact.

After having left the apparatus in this situation for about an hour, in order to satisfy myself that the temperature of the cold water was constant and uniform throughout its entire mass, I made the following experiment.

Experiment No. 1. — A solid ball of tin (F) having been provided, 2 inches in diameter, with a cylindrical projection on the lower side of it, 1 inch in diameter and $\frac{1}{2}$ of an inch long ending in a conical point which projected (downwards) $\frac{1}{2}$ of an inch farther, and having on the other side a strong iron wire 6 inches long, which served as a handle, — this ball, after having been immersed for half an hour in a considerable quantity of water at the temperature of 42° F., was withdrawn from the water, wiped dry with a handkerchief of the same temperature, placed without loss of time above the cylindrical vessel, and fixed in such a position that the entire conical point of the tin ball ($\frac{1}{2}$ of an inch in length) was submerged in the cold water contained in the vessel.

To fix and keep the metallic ball in its place, I used a strong slip of tin (G H), 6 inches long and $2\frac{1}{2}$ inches wide, with a circular hole in the middle of it 1 inch in diameter. This slip of tin being laid horizontally on the top or brim of the cylindrical vessel in such a manner that the centre of the circular hole coincided with the axis of the cylindrical vessel, when the short cylindrical projection belonging to the ball was introduced into that hole, the ball was firmly supported in its proper place.

The ball was placed in such a position that the end of the conical projection was immediately over the cork cup, at the distance of $1\frac{1}{2}$ inches above the level of its brim and consequently $\frac{1}{2}$ of an inch above the upper part of the bulb of the small thermometer which lay in this cup.

The quantity of cold water in the cylindrical vessel had been so regulated beforehand that when the conical point was entirely submerged, the surface of the water was on a level with the base of this inverted cone, so that the whole of the cylindrical part of the projection was out of the water.

I knew that the particles of ice-cold water which were thus brought into contact with the conical point could not fail to acquire some small degree of heat from that relatively warm metal, and I concluded that if the particles of water so warmed should in fact become *heavier* than they were before, in consequence of this small increase of temperature, they must necessarily *descend* in the surrounding lighter ice-cold liquid, and as the heated metallic point was placed directly over the cork cup, and fixed immovably in that situation, I foresaw that the descending current of warm water must necessarily fall into that cup and at length fill it, and that the presence of this warm water in the cup would be announced by the rising of the thermometer.

The result of this very interesting experiment was just what I expected: the conical metallic point had not been in contact with the ice-cold water more than 20 seconds when the mercury in the thermometer began to rise, and in 3 minutes it had risen three degrees and a half, namely, from 32° to $35\frac{1}{2}$°; when 5 minutes had elapsed it had risen to 36°.

Another small thermometer placed just below the surface of the ice-cold water, and only $\frac{2}{10}$ of an inch from the upper part of the conical point and on one side of it, did not appear to be sensibly affected by the vicinity of that warm body.

A third thermometer, the bulb of which was placed in the brass cup just on the outside of the cork cup and on a level with its brim, showed that the water which immediately surrounded the cork cup remained constantly at the temperature of freezing during the whole time that the experiment lasted.

As I well knew from the results of the experiments on the propagation of heat in a solid bar of metal,[6] that no one of the particles of cold water in contact with the surface of the conical projection, in the experiment which I have just described, could acquire by this momentary contact a temperature as high as that of the warm metal, I was by no means surprised to find that the thermometer belonging to the cork cup rose no higher than 36°.

In order to see if it could not be made to rise not only higher, but also more rapidly, by employing the metallic ball heated to such a temperature as it might be supposed would be sufficient to heat those particles of ice-cold water which should come into contact with its conical point, to the temperature at which the density of water is supposed to be a maximum, I made the following experiment.

Experiment No. 2. — Having removed the ball, I gently brushed away the warm water which in the last experiment had been lodged in the cavity of the cork cup, and which still remained there, as was evident from the indication of the thermometer belonging to the cup; I then placed several small cakes of ice in the cylindrical vessel, which ice, floating on the surface of the water in the vessel, prevented the water from receiving heat from the surrounding air, which at that time was at the temperature of 70° F. As the cork cup had been a little heated by the warm water in the foregoing experiment, time was now given it to cool.

As soon as the cup and the whole mass of the water in the cylindrical vessel appeared to have acquired the temperature of freezing, I carefully removed the cakes of ice which floated on the surface of the water, and introduced once more the projecting conical point belonging to the metallic ball into the ice-cold water in the vessel, placing it exactly in the same place which it had occupied in the foregoing experiment; but this ball, instead of being at the temperature of 42° F., as before, was now at the temperature of 60° F.

The results of this experiment were very striking, and, if I am not much mistaken, afford a direct, unexceptionable, and demonstrative proof, not only that the maximum of the density of water is in fact at a temperature which is several degrees above the point of freezing, but also that warm currents do actually set downwards in ice-cold water, whenever a certain small degree of heat is applied to the particles of that fluid which are at its surface, as I have already announced in my Essay on the Propagation of Heat in Fluids.[4]

The conical metallic point had been in its place no more than 10 seconds when I distinctly saw that the mercury in the thermometer belonging to the cork cup was in motion, and, when 50 seconds had elapsed, it had risen four degrees, viz. from 32° to 36°.

When 2 minutes and 30 seconds had elapsed, reckoning from the moment when the metallic point was introduced into the cold water, the thermometer had risen to 39°, and at the end of 6 minutes to $39\frac{1}{8}$°, when it began to fall; but very slowly, however, for at the end of 8 minutes and 30 seconds it was at $39\frac{3}{4}$°.

A small mercurial thermometer, the bulb of which was placed on one side of the cork cup at the distance of about $\frac{2}{10}$ of an inch from it, showed no signs of being in the least affected by the vertical current of warm water which descended from the conical point into the cup in this experiment.

This experiment was repeated four times the same day (the 13th of June, 1805), and always with nearly the same results. The mean results of these four experiments were as follows : —

Time elapsed, reckoned from the beginning of the experiment.			Temperature of the water in the cork cup, as shown by the thermometer.
m.	s.		Degrees.
o	o	. . . .	32
At o	10	began to rise	32+
At o	23	had risen to	33
o	28	" " "	34
o	35	" " "	35
o	48	" " "	36
1	3	" " "	37
1	35	" " "	38
2	32	" " "	39
3	41	" " "	$39\frac{1}{2}$
4	48	" " "	$39\frac{3}{4}$
6	5	" " "	$39\frac{7}{8}$

As I had found by some of my experiments made in the year 1797 (of which an account is given in my seventh Essay, Part I.) that water at the temperature of about 42° F., and consequently what we should call very cold, melted considerably more ice, when standing on it, than an equal quantity of boiling-hot water in the same situation, I was very curious to see whether the thermometer, the bulb of which lay in the cork cup, would not also be less heated by the ball when it should be applied *very hot* to the surface of the water, than when its temperature was much lower.

Seeing that this research ought to throw great light on the mysterious operations of the distribution of heat in liquids, I hastened to make the following experiment.

Experiment No. 3. — The cylindrical vessel with its contents having been once more reduced to the uniform temperature of freezing water, the metallic ball was heated in boiling water, and being as expeditiously as possible taken out of that hot liquid, its projecting conical point was suddenly submerged in the ice-cold water, as in the former experiments.

The result of this experiment was very interesting. It was not till 50 seconds had elapsed that the thermometer began to show any signs of rising, and at the end of 1 minute and 7 seconds it had risen only 2 degrees.

In the foregoing experiment, when the metallic ball was so much colder, the thermometer began to rise in 10 seconds, and at the end of 1 minute and 3 seconds it had risen 5 degrees.

This difference is very remarkable, and if it does not prove the existence and great efficacy of currents in conveying heat in fluids, I must confess that I do not see how the existence of any invisible mechanical operation, the progress of which does not immediately fall under the cognizance of our senses, can ever be demonstrated.

As the experiment made with the ball heated in boiling water appeared to me to be very interesting, I repeated it twice, and its results were always nearly the same. The mean results of these three experiments were as follows : —

Time elapsed, reckoned from the beginning of the experiment.				Temperature of the water in the cork cup, as shown by the thermometer.
m.	s.			Degrees.
o	o	. . . .		32
At o	50	the thermometer began to rise	. .	32+
At 1	2	had risen to		33
1	7	" " "		34
1	18	" " "		35
2	2	" " "		36
3	2	" " "		36½
4	17	" " "		37
6	12	" " "		38
7	17	" " "		38⅛
9	o	" " "		38¼
12	o	" " "		38¼
14	o	" " "		38¼

By comparing the mean results of these experiments with the mean results of those in which the ball was at the temperature of 60° or less, we may see how much more rapid the communication of heat in the cold water from above downwards was when the metallic ball was *relatively cold* than when it was much warmer; but we must not consider of too much importance the determination of the relative rapidity thus made, because it is more than probable that it was not till after the conical metallic point had been considerably cooled by contact with the cold water that the vertical descending currents could exist by which the thermometer was at length heated. At the beginning of the experiment made with the tin ball warmed in boiling water, the particles of water which were in immediate contact with the conical point while it was still very warm, were heated to a temperature higher than that at which the density of water is at a maximum, and the density of these particles being diminished by this high degree of heat, the vertical currents in the cold water were at the beginning ascending currents, as I satisfied myself by means of a small thermometer placed by the side of the conical point at a distance of $\frac{2}{10}$ of an inch from its base, and immediately below the surface of the cold water: this thermometer began to rise very rapidly as soon as the warm metallic point was plunged into the cold water.

Another small thermometer, the bulb of which was situated at about the same distance from the axis of the conical projection, but $\frac{1}{2}$ of an inch below the surface of the cold water, preserved throughout the entire experiment the appearance of perfect rest.

The results of this last experiment are all the more interesting because they afford a demonstrative proof that it was neither by a direct communication of heat in the water, which was at rest, from molecule to molecule, *de proche en proche*, nor by calorific radiations passing through the water, that heat was communicated from the metallic point to the bulb of the thermometer, but actually by a descending current of warm water ; for it is perfectly evident that if this heat had been communicated either by a direct transfer in the water from molecule to mole-cule or by calorific radiations passing from the surface of the metal through the water, which remained at rest, this communication would naturally have been the most rapid when the metallic point was the warmest. What did take place was exactly contrary to this, as we have just seen. Moreover, the small thermometer, which was placed close to the metallic body on one side, and which in this experiment was in no degree affected by the heat of this body, would not have failed to acquire as much heat at least as that placed in the cork cup, which was situated below the metallic body and at a greater distance from it.

The considerable amount of time which elapsed in the experiments performed with the tin ball heated in boiling water before the thermometer in the cork cup began to be so sensibly affected, and the rapidity with which it was then warmed through several degrees as soon as it began to rise, indicate a fact which it is important to notice. In order to throw light upon this fact, we must consider carefully the operation of the heating of cold water by the warm metallic surface with which it was in contact, and examine it in its progress and in all its details.

Let us begin by supposing that the conical point of the ball, at the temperature of boiling water, has just been submerged vertically up to the level of its base in a mass of undisturbed water at the temperature of melting ice. As the particles of water, which in this case are in contact with the warm metallic surface, cannot pass, all of a sudden, from the temperature of melting ice to that of boiling water without passing through all the intermediate degrees, and since these particles at the temperature of melting ice cannot become warmer without becoming more dense, it is evident that they must have a tendency to descend, and consequently to leave the surface of the metal, as soon as they begin to acquire heat; but experiment showed that, instead of descending, they were actually pushed upwards: this proves that they were heated so rapidly that, before they had time to leave the surface of the metal and to escape from its calorific influence, they had acquired a temperature so elevated that their density, after having passed rapidly the point of its maximum, became even less than it was at the temperature of melting ice. But after some moments, the metallic body having cooled somewhat, and the communication of heat to the particles of water taking place more slowly, these particles, having become more dense on account of a slight increase of temperature, had time to escape before becoming warmer, and at that time the descending current suddenly began.

This fact interests me the more, as it may serve in some sort to explain a phenomenon which I observed in an experiment made eight years ago, an account of which I gave in my Essay on the Propagation of Heat in Fluids.[4]

The phenomenon to which I have alluded was this: Having poured some mercury into a small cylindrical glass vessel 2 inches in diameter and $3\frac{1}{2}$ inches deep, until this fluid filled the vessel to the height of an inch, I poured on to the mercury twice as much water (that is, 2 inches), and, plunging the vessel up to the level of the upper surface of the mercury into a freezing mixture of pounded ice and sea-salt, the temperature of the air being 60° F., I allowed the whole to cool quietly, in order to see in what part of the water the ice would first

appear. It was at the bottom of the water, where this
liquid was in contact with the mercury, that the ice
formed.

The layer of water which rested immediately on the
surface of the mercury having been cooled to about the
temperature of 41° F., where the density of water is at
its maximum, the particles of this water, which were
then in immediate contact with the mercury, losing still
more of their heat, became of necessity less dense, and
had consequently a tendency to leave the bottom of the
water and to ascend upwards ; but the rapidity with
which they were cooled by the mercury was so great
that they were frozen before they could escape from the
cooling influence of this cold body.

After all that I have said about the warm and cold
currents which take place in a liquid which is warmed
or cooled, it might perhaps be thought that I regard
these currents as composed of single particles of the
liquid, which, having been in immediate contact with
the body which gives or which receives the heat, are
all of the same temperature. I am all the farther from
holding this opinion, since I know from the results
of several experiments made expressly for elucidating
this point, (and which I shall have the honor of presenting to the Class on another occasion,)[7] that a liquid
current cannot pass through another liquid mass which
is at rest, and which is of the same kind and of about
the same specific gravity, without producing a perceptible mixture of the two liquids ; much less, therefore,
can a small current of warm water pass without mixing
through a mass of cold water ; and the farther it advances
the more it will be mixed, and the more, in consequence,
will its temperature be found to be lowered.

For example, in the experiments of which I have just given an account, the cork cup, which received the current of warm water descending from the metallic point of the tin ball, was only $\frac{1}{2}$ of an inch below the extremity of this point; if this distance had been greater, the thermometer in the cup would certainly have risen to a less height: for this reason these experiments ought not to be regarded as suitable for determining with great exactness the temperature at which the density of water is at a maximum, but rather as proving that this temperature is really several degrees of the thermometric scale above that of melting ice; and this is all that I am particularly interested in showing at the present time.

Judging from the constant temperature which is found at all seasons at the bottom of deep lakes and from the results of several direct experiments, we may conclude that water is at its *maximum* density when it is at the temperature of about 41° of Fahrenheit's scale, which corresponds to 4° on that of Reaumur, and to 5° of the Centigrade scale.

INQUIRIES

THE MODE OF THE PROPAGATION OF HEAT IN LIQUIDS.

THE motions in fluids which result from a change in their temperature give rise to so great a number of phenomena, that philosophers cannot bestow too much pains in investigating that interesting branch of knowledge.

When heat is propagated in solid bodies, it passes from particle to particle, *de proche en proche*, and apparently with the same celerity in every direction; but it is certain that heat is not transmitted in the same manner in fluids.

When a solid body is heated and plunged in a cold liquid, the particles of the liquid in contact with the body, being rarefied by the heat that they receive from it, and being rendered specifically lighter than the surrounding particles, are forced to give place to these last and to rise to the surface of the liquid; and the cold particles that replace them at the surface of the hot body, being in their turn heated, rarefied, and forced up, — all the particles thus heated by a successive contact with the hot body form a continued ascending current, which carries the whole of the heat immediately towards the surface of the liquid, so that the strata of the liquid situated at a small distance under the hot body are not sensibly heated by it.

57

When a solid body is plunged in a liquid which is hotter than the body, the particles of the liquid in contact with the body, being condensed by the cooling they undergo, descend, in consequence of the increase of their specific gravity, and fall to the bottom of the liquid; and the strata situated above the level of the cold body are not cooled by it immediately.

It is true that the viscosity of liquids, even of those which possess the highest known degree of fluidity, is still much too great to allow one of their particles individually being moved out of its place by any change of specific gravity occasioned by heat or cold; yet this does not prevent currents from being formed, in the manner above described, by small masses of the liquid composed of a great number of such particles.

The existence of currents in the ordinary cases of the heating and cooling of liquids cannot any longer be called in question; but philosophers are not yet agreed with respect to the extent of the effects produced by those currents.

In treating of abstruse subjects, it is indispensably necessary to fix with precision the exact meaning of the words we employ. The distinction established between *conductors* and *non-conductors* of heat is too vague not to stand in need of explanation. An example will show the ambiguity of these expressions.

If two equal cubes of any solid matter, — copper, for instance, — of two inches in diameter, the one at the temperature of 60°, the other at 100°, be placed one above the other, the cold cube will be heated by the hot one, and this last will be cooled.

If the cold cube be placed upon a table and its upper surface covered by a large plate of metal, — of silver, for

instance,—a quarter of an inch thick, and if the hot cube
be placed upon this plate immediately above the cold
cube, the heat will descend through the metallic plate with
a certain degree of facility, and will heat the cold cube.

If a dry board of the same thickness with the metallic
plate be substituted in its place, the heat will descend
through the wood, but with much less celerity than
through the plate of silver.

But if a stratum of water or of any other liquid be substituted in place of the metallic plate or of the board,
the result will be very different. If, for instance, the cold
cube being placed in a large tub resting on the middle
of its bottom, the hot cube be suspended over it by
cords, or in any other manner so that the lower surface
of the hot cube be immediately above the upper surface
of the cold cube, at the distance of a quarter of an inch,
and the tub be then filled with water at the same temperature as that of the cold cube, the heat will not
descend from the hot cube to the cold one through the
stratum of water of a quarter of an inch in thickness that
separates them.

We may with propriety call silver a *good conductor* of
heat, and dry wood a *bad conductor;* but what shall we
say of water? I have called it a *non-conductor* for want
of a more suitable term, but I always felt that that word
expresses but imperfectly the quality that was meant to
be designated.

In the experiment of the two cubes plunged in water,
if the hot cube be placed below and the cold cube above
it, the heat will not only be communicated from the hot
to the cold cube, but it will pass even more rapidly than
when the two cubes are separated by a plate of silver.
But in this case it is evident that the heat is *transported*

by the ascending currents which are formed in the liquid in consequence of the heat which it receives from the hot body.

The existence of these currents in certain cases has been known a long time, but philosophers have not been sufficiently attentive to the many curious phenomena that depend upon them. It has not even been suspected with what extreme slowness heat passes in fluids, from particle to particle, *de proche en proche*, in cases where the effects of such communication become sensible.

For some time after I had engaged in this interesting inquiry, I conceived that this kind of communication was absolutely impossible in all cases; but a more attentive examination of the phenomena has convinced me that this conclusion was too hasty. As early as the beginning of 1800, in a note published in the third edition of my Seventh Essay, I announced a conjecture that the non-conducting power of fluids might perhaps depend solely on the extreme mobility of their particles; and it is certain, if this conjecture is well founded, liquids must necessarily become conductors of heat (though very imperfect ones) in all cases where this mobility of their particles is destroyed, as well as in these rare but yet possible cases, where a change of temperature can take place in a liquid without giving its particles any tendency to move, or to be moved out of their places.

The unequivocal results of a great many experiments have shown, that in ordinary cases, and perhaps in all cases where heat is propagated in considerable masses of fluids, its distribution is accomplished precisely in the manner that the new theory supposes, that is to say, by currents. And it is certain that the knowledge of that fact has enabled us to explain in a satisfactory manner several interesting phenomena of nature, which before were enveloped in much obscurity.

When a hot solid body is plunged in a cold liquid, there can be no doubt concerning the existence of the vertical ascending currents which are formed in the liquid, and which convey to the surface the heat which its particles have received; but with respect to the strata of liquid situated under the hot body, *are they or are they not heated by this body by means of a direct communication of heat from above downwards, from particle to particle, these particles remaining in their places?* This is a question on which philosophers are not yet agreed. As it is a question of great importance, I have long meditated on the means of deciding it; and after several unsuccessful attempts, I have at last succeeded in making an experiment which I think is decisive.

As the apparatus which I used for this experiment, and which I have the honour of laying before the assembly, is somewhat complicated; and as it is indispensably necessary to be intimately acquainted with it, in order to form a judgment concerning the degree of confidence which the results of the experiment may deserve, — it is necessary to give a detailed description of this machinery. The annexed figure gives a distinct representation of its principal parts. It is drawn on a scale of a quarter of an inch to the inch, English measure.

A B (Plate VII.) is a board, of oak, seen in profile; it is $1\frac{1}{2}$ inches thick, 18 inches long, and 11 inches in breadth. It serves to support two square upright pillars, C C, $18\frac{1}{2}$ inches in height and $1\frac{1}{2}$ inches square. They are firmly fixed in the board at the distance of 11 inches asunder, and serve to support the two cross-pieces, D E, F G, at different heights.

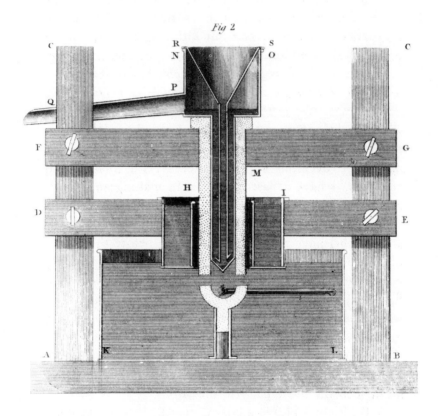

Fig 2

These cross-pieces are each pierced with two square holes, at the distance of 11 inches one from the other, into which the upright pillars C C enter, and the cross-pieces are supported at any height that is required, by means of a screw of compression. These screws are represented in the figure.

The cross-piece F G, which is represented in profile, is 17 inches in length, and $1\frac{1}{2}$ inches thick, and 3 inches in breadth. It is pierced in the middle by a cylindrical hole of 2 inches in diameter.

The cross-piece D E is 17 inches in length by $1\frac{1}{2}$ inches in thickness. It is 3 inches wide at each end and 6 inches in the middle, where it is pierced by a circular hole 5 inches in diameter.

The cross-piece D E serves to support the annular vessel H I, of which a vertical section passing through its axis is seen in the figure. This vessel, formed of thin brass plates, is 5 inches in diameter without, 3 inches in diameter within, and $27\frac{1}{8}$ inches in depth. This vessel is filled with water during the experiments to the height of $2\frac{1}{2}$ inches; and its form is such, that, if the water that it contains were frozen into a solid mass of ice, this piece of ice would have the form of a tube or perforated cylinder of 1 inch in thickness and $2\frac{1}{2}$ inches high by 5 inches in diameter without. Its cylindrical cavity would be precisely 3 inches in diameter.

K L is a vertical and central section of a cylindrical vessel of tin of 10 inches in diameter by $4\frac{1}{2}$ inches in depth. It is filled with water to the height of 4 inches, as it is seen in the figure.

The cross-piece D E is placed at such a height that the bottom of the annular vessel H I is plunged a quarter of an inch under the surface of the water contained in the great cylindrical vessel K L.

In the axis of this last vessel is placed a small hemi-spherical, cup of wood 2 inches in diameter without and $\frac{1}{2}$ of an inch thick. It is kept in its place by a short vertical tube of tin, soldered to the bottom of the cylindrical vessel K L, into which the stalk of the cup fits tightly.

The middle of the cavity of this cup is occupied by the bulb of a small mercurial thermometer of great sensibility. Its tube, which has an ivory scale, is laid down horizontally, and fixed in one side of the cup, through which the tube passes, in such a manner that the lowest part of the bulb is elevated $\frac{1}{10}$ of an inch above the bottom of the cup. The diameter of the bulb being $\frac{3}{10}$ of an inch, and the hemispherical cup having $\frac{1}{2}$ inch of radius within, it is evident that the upper part of the bulb is $\frac{1}{10}$ of an inch below the level of the brim of the cup that contains it. To avoid charging the figure with too many details, the scale of the thermometer is not drawn, but the tube is distinctly represented.

The horizontal cross-piece F G serves to support a very essential part of the apparatus, which remains to be described.

This cross-piece supports, in the first place, a vertical tube of wood, M, $6\frac{6}{10}$ inches in length and 2 inches in diameter without. Its interior diameter is $1\frac{1}{20}$ inch. This tube is supported by a projecting collar (represented in the figure), $2\frac{1}{2}$ inches in diameter, which rests on the cross-piece F G. It is a vertical and central section of this tube that is represented in the figure, and it is dotted in order to distinguish it from the surrounding parts of the apparatus.

The lower part of this tube is plunged $\frac{6}{10}$ of an inch under the surface of the water in the large cylindrical vessel K L; and it is placed precisely above the wooden cup in the prolongation of its axis, the lower extremity of the tube being at the distance of $\frac{3}{10}$ of an inch above the horizontal level of the brim of the cup.

On the top of the tube of wood is placed a cylindrical vessel N O, of sheet brass, 3 inches in diameter, $2\frac{3}{4}$ inches high, which has a lateral spout, P Q, placed a little above the level of its bottom.

From the middle of the bottom of this vessel, there descends a cylindrical tube of brass, 6 inches in length and 1 inch in diameter, which ends below in a hollow conical point, as represented in the figure.

R S is a vertical and central section of a funnel of brass, which ends below in a cylindrical tube of $\frac{3}{10}$ of an inch in diameter and $6\frac{6}{10}$ inches long. This funnel is kept in its place in the axis of the cylindrical vessel N O by the exact fitting of its upper edge upon that of the vessel into which it is adjusted.

The lower end of the tube of this funnel is surrounded by a projecting edge or flange in the form of a hollow inverted cone. The diameter of this conical projecting brim above, at its base, is $\frac{7}{10}$ of an inch, and it is soldered below to the end of the tube.

When hot water is poured into the funnel, this liquid, descending by the tube of the funnel, strikes against the inner surface of the hollow inverted cone which terminates the vertical tube that belongs to the vessel N O, and then, rising up through this last tube into that vessel, it runs off by its spout. It was with a view to force this water to come into more intimate contact with the hollow cone that the projecting edge, in form of an inverted cone, was added to the lower end of the tube of the funnel.

The object chiefly in view in the arrangement of this apparatus was to give to the conical point which terminates the vertical tube of the vessel N O, an elevated temperature, which should remain constant during some time, for the purpose of observing if the heat, which must necessarily be communicated by this metallic point to the small quantity of water with which it is in contact, and which is confined in the lower part of the wooden tube M, would descend, or not, to the thermometer which was placed in the wooden cup.

There was still one source of error and uncertainty against which it was necessary to guard. The heat communicated through the sides of the wooden tube to the water contained in the great cylindrical vessel K L might be transported to the sides of that vessel, and, being then communicated from above downwards through these sides, might heat successively the lower strata of the liquid, and at last that stratum in which the thermometer was.

It was to prevent this that the annular vessel H I was used, and it performed its office in the following manner: The particles of water contained in the great vessel K L, which, being in contact with the exterior surface of the wooden tube, were heated by that tube, could not fail to rise to the surface, and there they necessarily came into contact with the interior sides of the annular vessel, to which they communicated the excess of heat they had received from the wooden tube.

This heat, passing readily through the thin metallic sides of that vessel, was given off as fast as it was received to the particles of cold water contained in the vessel which were in contact with its sides, and these particles, rising to the surface of the water contained in the annular vessel in consequence of their acquired heat and levity, the progress of the heat from the wooden tube to the sides of the large vessel K L, was interrupted, and all the heat that passed through the sides of the wooden tube was by these means turned aside in such a manner that it could no longer disturb the progress of the experiment, nor affect the certainty of its results.

Before I proceed to give an account of the result of this inquiry, I shall take the liberty to recall the attention of the Assembly to the most important circumstances of the experiment.

On pouring boiling water in a small uninterrupted stream into the funnel, the hollow conical point which terminates the vertical tube belonging to the vessel N O was heated, and kept at a constant temperature little under that of boiling water.

This point was surrounded by a small quantity of water contained in the cavity of the lower part of the wooden tube, and as this water could not change its place nor be displaced by the surrounding cold water, being enclosed and protected by the sides of the wooden tube, it would necessarily become very hot in a short time.

But this small quantity of hot water lay immediately upon a stratum of cold water, which separated it from the bulb of the thermometer, placed directly under it at the distance of only half an inch.

If heat could pass in the water from above downwards, it would no doubt pass from the lower stratum of hot water contained in the open end of the wooden tube to the bulb of the thermometer, which lay immediately below it and at so small a distance.

Three experiments were made with this apparatus, and always with exactly the same results. In the first, a stream of boiling water was poured into the funnel during 10 minutes ; in the second, during 12 minutes ; and in the third, during 15 minutes.

The thermometer, whose bulb was in the wooden cup, remained *at perfect rest* from the beginning of the experiment to the end of it without showing the slightest sign of being in any way affected by the hot water which was so near it.

These experiments were made at Munich in the month of July, 1805 ; the temperature of the air and of the water contained in the vessel K L being 70° Fahrenheit.

A small thermometer placed in the water contained in the annular vessel H I, in such a manner that its bulb was scarcely submerged, marked that this water had received a little heat in each of the three experiments.

Another similar thermometer placed in the water contained in the large vessel K L, immediately under its surface and near one side of the vessel, showed that this water had not acquired any sensible increase of temperature during the experiments.

From the results of these experiments we are authorized to conclude, that heat does not descend in water to a sensible distance, in cases where the particles of the liquid which receive heat are exposed to be displaced and forced upwards by the surrounding colder and denser particles, that is to say, in all the cases (and they are the most common) where heat is applied to the strata of the liquid situated under its surface.

But the results of the experiments in question do not prove that heat cannot in any case descend in water ; and still less can it be inferred from them, that all direct communication of heat in this liquid, from particle to particle, *de proche en proche,* is impossible. They do not even prove that heat did not descend, *to a small distance,* below the level of the end of the wooden tube in these experiments; for it is certain that that event could take place without the thermometer, which was situated a little lower, being in any way affected by that heat.

The particles of water situated at a very small distance below the level of the lower end of the wooden tube, being heated by the stratum of hot water which rested immediately on them, might have been displaced by the surrounding colder and denser particles, and forced to rise to the surface; and these last being in their turn heated, forced upwards and replaced by other cold particles, it is evident that the heat could not make its way downwards so far as to arrive at the thermometer through a stratum of liquid, which, though apparently at rest, was nevertheless in part composed of particles which were continually changing.

I have long suspected that the apparent impossibility of a direct communication of heat between neighbouring particles of fluids depends solely on the great mobility of those particles (see note, p. 202, Vol. II. of my Essays, 3d edition, London, 1800);[4] and if this suspicion be well founded, it is certain that when this mobility ceases, the effect which depends on it must cease likewise.

When I speak of the mobility of the particles of a liquid amongst each other, I am very far, as I have already observed, from supposing that individually they can enjoy a free motion. I was formerly of that opinion, but a more attentive investigation of the phenomena has convinced me that I was mistaken. But although one

individual particle of a liquid can never be put in motion in consequence of a change of its specific gravity
occasioned by a change of temperature, yet what cannot
happen to a single particle may easily and must necessarily happen to small masses of the liquid consisting
of a great number of these particles united, as is abundantly proved by the currents which are so easily excited
by the contact of a hot or cold body plunged in a
liquid.

The force by which the particles of liquids adhere together is very great, and it is more than probable that it
is the cause of many very interesting phenomena, and
amongst others of the suspension of the heavy bodies
which much lighter liquids so frequently hold in solution.

From the result of an experiment which I made some
years ago[8] in order to determine the measure of the viscosity or the want of perfect fluidity in water at the temperature of 64° F., I found reason to conclude that a
solid body, having a surface equal to 368 square inches,
which should weigh only one grain Troy more than an
equal volume of water, would remain suspended in that
liquid; and from this datum it is easy to find by calculation what ought to be the diameter of a small solid
spherule of the heaviest matter, — of gold, for instance,
— in order to its remaining suspended in water in consequence of the viscosity of that liquid.

Having made this calculation in order to satisfy my
curiosity, I found that a solid spherule of pure gold, of
the diameter of $\frac{1}{300000}$ (or exactly $\frac{1}{298719}$) of an inch,
ought to remain suspended in water in consequence of
the adhesion of the particles of that liquid to each other.
But I shall return to this subject on a future occasion.

[THE preceding paper is printed from Nicholson's Journal, XIV. (1806), pp. 355 – 363. The paper, in a somewhat modified form, appears in the Bibliothèque Britannique (Science et Arts), XXXII. (1806), pp. 123 – 141, to which periodical it was contributed by Count Rumford in manuscript, which was translated by the French editor. In this version the beginning of the paper is much abridged; but in the latter part Rumford elaborated his speculations in regard to the effects produced by viscosity on the propagation of heat in liquids more at length than in Nicholson's Journal. In order to give fully his views on the subject, this portion of the French paper is here appended.

" If there is no direct communication of heat between contiguous particles of water at different temperatures, then the *apparent* mean temperature, which results so quickly when cold water is poured into a mass of warm water, must be produced by currents caused by differences in the specific gravity of the masses of the liquid at different temperatures. And if it is asked why the hot and cold particles, thus mixed together, do not separate again, on account of the difference in their specific gravity, we must seek for the reason in the imperfect fluidity of the water. This cause may keep the particles of water suspended, out of their natural position, just as it keeps in suspension, as well in other liquids as in water, particles of foreign substances, which, although specifically heavier or specifically lighter than the medium, are so small that the amount by which they are heavier or lighter than the surrounding liquid is not sufficient to overcome its viscosity.

" This want of perfect fluidity, a condition common to all liquids in different degrees, gives rise to a great number of very interesting phenomena, and it is a subject worthy of the close attention of philosophers.

" From the result of an experiment which I made a long time ago in order to determine the measure of the viscosity of pure water at the temperature of 64° F., that is the force necessary to separate contiguous particles of that liquid, I found reason to conclude that a solid body, having a surface equal to 368 square inches, which should weigh only one grain Troy more than an equal volume of water, would remain suspended in that liquid; and from this datum it is easy to find, by calculation, what ought to be the diameter of a small solid spherule of the heaviest matter, — of gold, for instance, — in order to its remaining suspended in water in consequence of the viscosity of that liquid: and it is also easy to prove, from the inflection which light experiences in passing over the surface of an opaque solid body, that a considerable quantity of opaque solid matter could be held suspended in water without sensibly diminishing its transparency, and without changing its colour; that is to say, without giving any indication of its presence.*8

"I have long suspected that the suspension of solid substances which are held in solution by liquids is due solely to the imperfect fluidity of the solvents, and the results of a great number of experiments which I have made to elucidate this important subject have always confirmed this opinion. Since, then, bodies specifically heavier than water can nevertheless remain suspended in that liquid, there can be no difficulty in admitting that isolated particles of cold water can equally remain motionless in the warm water with which they find themselves accidentally mixed. But although this may be true of the individual particles, the same principle cannot apply to masses of sensible size made up of a great number of these particles. These masses must yield to the natural effect of the differences in their specific gravity, and form currents which will be ascending or descending according as the masses in question are warmer or colder than the surrounding liquid; and these currents must contribute very largely to the intimate mixture of the particles at different temperatures, and must soon bring about a certain uniform temperature throughout the entire mass of the liquid.

" I said this uniformity of temperature may be only *apparent ;* because, if water is really a perfect non-conductor of heat, the particles of cold water, at least those which have not been warmed by contact with the walls of the vessel in which they are contained, ought to remain cold in spite of their more or less intimate mixture with the warm particles ; but, notwithstanding this fact, the mean temperature of the liquid, as shown by the thermometer, will be precisely the same as if there had been an actual communication of heat among the particles.

" Long after I had had reason to persuade myself that all the heat acquired by liquids, when they are warmed, is communicated by the vessel containing them to the individual particles, which are successively

---

* In order to satisfy my curiosity, I found, by calculation, the diameter which a small solid spherule of gold ought to have in order to its remaining suspended in water in consequence of the viscosity of that liquid. I found this diameter to be $\frac{1}{298719}$, or, in round numbers, $\frac{1}{300000}$ of an inch; that is to say, about two hundred times smaller than the diameter of a single fibre of raw silk, as spun by the worm, — an object which is so fine as scarcely to be visible.

brought into contact with its sides by the currents formed in· the mass of the liquid, — a long time, I say, after I had adopted this opinion, I continued to doubt whether single particles of warm water, when completely surrounded by particles of cold water, and remaining undisturbed in the midst of them, might not be able to communicate to these neighbouring particles that excess of heat which the shortness of the time of contact when the particles are in motion does not allow them to impart to each other. Indeed, if the property of water by which it is an apparent non-conductor of heat depends solely upon the extreme mobility of its particles, it is evident that the communication of heat, under the circumstances just supposed, must necessarily take place; and it must be remembered that in this case the fluidity of the liquid, as far as the particles in question are concerned, is as truly destroyed as if the entire mass were converted into ice.

"The inquiry as to the non-conducting power of fluids—an inquiry to which my experiments and observations have given rise — is, no doubt, of great interest to science; and, whatever may be the final result of its investigation, I shall regard myself fortunate in having drawn the attention of a great number of enlightened philosophers towards an object which was long neglected, and which was so worthy of being studied."]

OF THE SLOW PROGRESS

OF THE

SPONTANEOUS MIXTURE OF LIQUIDS

DISPOSED TO UNITE CHEMICALLY WITH EACH OTHER.

I N order to obtain the most exact knowledge of the nature of the forces which act in the chemical combination of various bodies, one must study the phenomena of these operations, not only in their results, but more especially in their progress.

When we mix together two liquids which we wish to have unite, we take care to shake them violently, in order to facilitate their union; it might, however, be very interesting to know what would happen, if, instead of mixing them, they were simply brought into contact by placing one upon the other in the same vessel, taking care to cause the lighter to rest upon the heavier.

Will the mixture take place under such circumstances? and with what degree of rapidity? These are questions interesting alike to the chemist and to the natural-philosopher.

The result would depend, without doubt, on several circumstances which we might be able to anticipate, and the effects of which we might perhaps estimate *à priori*. But since the results of experiments, when they are well made, are incomparably more satisfactory than conclusions drawn from any course of reasoning, especially in the case of the mysterious operations of Nature, I

74

propose to speak before this illustrious Assembly simply of experiments that I have performed.

Having procured a cylindrical vessel of clear white glass 1 inch 8 lines in diameter, and 8 inches high, provided with a scale divided from the bottom upwards into inches and lines, I put it on a firm table in the middle of a cellar, where the temperature, which seemed to be tolerably constant, was 64 degrees of Fahrenheit's scale.

I then poured into this vessel, with due precautions, a layer of a saturated aqueous solution of muriate of soda, 3 inches in thickness, and on to this a layer of the same thickness of distilled water. This operation was performed in such a way that the two liquids lay one upon the other without being mixed, and when everything was at rest I let a large drop of the essential oil of cloves fall into the vessel. This oil being specifically heavier than water, and lighter than the solution of muriate of soda on which the water rested, the drop descended through the layer of water; when, however, it reached the neighbourhood of the surface of the saline solution it remained there, forming a little spherical ball, which maintained its position at rest, as though it were suspended, near the axis of the vessel.

I then poured, with proper precautions, a layer of olive oil four lines in thickness on to the surface of the water, to prevent the contact of the air with the liquid, and having observed, by means of the scale attached to the vessel, and noted down in a register, the height at which the little ball was suspended, I withdrew, and, locking the door, I left the apparatus to itself for twenty-four hours.

In a preliminary experiment, made to determine in

what proportions the saturated solution should be mixed with distilled water, that the mixture might have the same specific gravity as the oil of cloves, I found that a mixture composed of 1 measure of the solution and 9 measures of distilled water had a slightly higher specific gravity than the oil; but with 10 measures of distilled water the oil sank in the mixture.

As the little ball of oil, designed to serve me as an index, was suspended a very little above the upper surface of the layer of the saturated solution, this showed me that the precautions which I had taken were sufficient to prevent the mixing of the distilled water and the saline solution when I put one upon the other, and I knew that this mixture could not take place subsequently without causing at the same time my little *sentinel,* which was there to warn me of this event, to ascend.

There was, however, a single source of error which I was obliged to guard against. I had observed, in other experiments of this kind, that the air which was disseminated through or dissolved in water containing in solution a small quantity of muriate of soda left the liquid, and attached itself to the little ball of oil of cloves which I had introduced into it, and, having formed on top of it a little bubble scarcely visible, caused it to ascend in the liquid, even when the density of the liquid had not changed at all.

To prevent this accident, I boiled for some time both the saturated solution and the distilled water employed in the experiment, in order to free them from air, and, for the same reason, I subsequently covered the water with a layer of olive oil to prevent the contact of this water with the atmospheric air.

After the little apparatus mentioned above had been left to itself for twenty-four hours, I entered the cellar, taking a light in order to note the progress of the experiment, and I found that the little ball had risen 3 lines.

The next day, at the same hour, I observed the ball again, and I found that it had risen about 3 lines more ; and thus it continued to ascend about 3 lines a day for six days, when I put an end to the experiment.

I afterwards made nearly similar experiments with saturated aqueous solutions of nitrate of potash, carbonate of potash, and carbonate of soda. In each of these experiments the surface of the saturated solution was covered with a layer of distilled water 3 inches in thickness, but the surface of this layer of water was not covered by a layer of olive oil; it was exposed to the air, and this circumstance was, without doubt, the reason that the daily results of a single experiment were not always the same two days in succession.

The little ball of oil of cloves, which served as an index to mark the progress of the mixture of the saturated solution with the distilled water resting upon it, ascended usually 2 or 3 lines in twenty-four hours, but sometimes I found that it had left its position and had risen to the very surface of the water.

In such cases it was, without doubt, borne upwards by the air which it had attracted from the liquid; for when I allowed a fresh drop of the same oil to fall into the water, I found that it never failed to descend immediately.in the liquid, and to take up its position 2 or 3 lines above the level at which, the day before, I had found the ball which had now left its place.

In the experiments made with solutions of carbonate of soda and carbonate of potash, the balls of oil changed in appearance by the end of two or three days; from being transparent, they became semi-opaque and of a whitish color; they changed at the same time with regard to their specific gravity as well, and became a little lighter. These changes were evidently due to the beginning of saponification.

This accidental circumstance made it necessary for me to renew each day the drop of oil which served as the index, allowing the others to pursue their way to the surface of the liquid without paying any further attention to them.

By using as indices little glass balloons of proper size and thickness, instead of the drops of oil, the inconveniences arising from the saponification of the oil might be avoided.

But without spending more time on the details of these experiments, I hasten to return to their results. They showed that the mixture went on continually, but very slowly, between the various aqueous solutions employed and the distilled water resting upon them.

There is nothing in this result to excite the surprise of any one, especially of chemists, unless it is the extreme slowness of the progress of the mixture in question. The fact, however, gives occasion for an inquiry of the greatest importance, which is far from being easy to solve.

Does this mixture depend upon a peculiar force of attraction different from the attraction of universal gravitation, a force which has been designated by the name of chemical affinity? Or is it simply a result of motions in the liquids in contact, caused by changes in their temperatures? Or is it, perhaps, the result of a peculiar and continual motion common to all liquids, caused by the instability of the equilibrium existing among their molecules?

I am very far from assuming to be able to solve this great problem, but it has often been the subject of my thoughts, and I have made at different times a considerable number of experiments with a view of throwing light into the profound darkness with which the subject is shrouded on every side.

At a subsequent sitting of the Class, I shall have the honour of giving an account of the continuation of my researches on this interesting subject.

RESEARCHES

HEAT DEVELOPED IN COMBUSTION AND IN THE CONDENSATION OF VAPOURS.

SECTION I. — *Description of a new Calorimeter.*

ATTEMPTS have been long ago made to measure the heat that is developed in the combustion of inflammable substances ; but the results of the experiments have been so contradictory, and the methods employed so little calculated to inspire confidence, that the undertaking is justly considered as very little advanced.

I had attempted it at three different times within these twenty years, but without success. After having made a great number of experiments with the most scrupulous care, with apparatus on which I had long reflected, and afterward caused to be executed by skilful workmen, I had found nothing, however, that appeared to me sufficiently decisive to deserve to be made public. A large apparatus in copper more than twelve feet long, which I had made at Munich fifteen years ago, and another, scarcely less expensive, made at Paris four years ago, which I have still in my laboratory, attest the desire I have long entertained of finding the means of elucidating a question that has always appeared to me of great importance, both with regard to the sciences and to the arts.

At length, however, I have the satisfaction of announcing to the Class, that, after all my fruitless attempts, I have discovered a very simple method of measuring the heat manifested in combustion, and this even with such precision as leaves nothing to be desired.

That the Class may be the better able to judge of my method of operating, and the reliance that may be placed on the results of my experiments, I place my apparatus before it.

The principal part of this apparatus is a kind of prismatic receiver, eight inches long, four inches and a half broad, and four inches three quarters high,* formed of very thin sheets of copper. This receiver, which well deserves the name, already celebrated, of *calorimeter*, is furnished with a long neck, near one of its extremities, three quarters of an inch in diameter, and three inches high, intended to receive and support a mercurial thermometer of a particular shape. The receiver has also another neck, an inch in diameter and the same in height, situate in the centre of its upper part, and closed by a cork.

Within this receiver, two lines above its flat bottom, is a particular kind of worm, receiving all the products of the combustion of the inflammable substances burned in the experiments, and transmitting the heat manifested in this combustion to a considerable body of water, which is in the receiver.

This worm, which is made of thin copper, occupies and covers the whole bottom of the receiver, yet without touching either its bottom or its sides. It is a flat tube, an inch and a half broad at one end and an inch at the other, and half an inch thick throughout. It is

* French measure.

bent horizontally, so as to pass three times from one end of the receiver to the other, and is supported in its place, two lines above the bottom of the receiver, by several little feet.

The aperture that forms the mouth of the worm is a circular hole in its bottom, near its broadest end. Into this hole is soldered a perpendicular tube, an inch in length and an inch in diameter, reaching within the worm to the height of a quarter of an inch above its bottom.

This tube passes through a circular hole in the bottom of the receiver, to which also it is soldered. Its lower aperture is seven lines below the bottom of the receiver; and through this the products of the combustion enter into the worm.

The other extremity of the worm passes horizontally through the perpendicular end of the receiver, opposite to that near which the products of the combustion enter the worm.

The worm, before it passes through the end of the receiver, is fashioned into the shape of a round pipe, half an inch in diameter; and an inch in length of this pipe is seen without the receiver. This piece is made to fit tight into another similar tube, belonging to the worm of another receiver, which I call the *secondary receiver*; the purpose of which is to receive the heat that might still be found in the products of combustion, after they have passed through the worm of the principal receiver.

To support these two receivers in the air, so as not to touch the table that supports them, each of them is fixed in a frame of dry linden wood, made of rods an inch square. Round the bottom of each receiver is a

copper rim, three lines deep, which is fastened by a row
of very small nails to the wooden frame. The body of
the receiver itself enters about a line into the frame, to
which it is very accurately fitted.

The flat form of the worm is essential to the perfection of the apparatus, as is evident when its purpose is
considered.

All the products of the combustion being elastic
fluids, and consequently substances incapable of communicating their heat but by proceeding particle after
particle to deposit it on the surface of the cold and
fixed body intended to receive it, it was indispensable so
to construct the apparatus that the hot fluids should of
necessity be spread *beneath* and *against* a large flat surface, placed horizontally, and always cold.

Before I employed horizontal worms made of flat
tubes, I had more than once tried those of the common
form ; but they never answered my purpose otherwise
than so imperfectly that I could never make any account
of the experiments in which they were employed.
There is no doubt but the shape I have adopted for the
worm of my calorimeter would be very advantageous
for every kind of apparatus for distillation.

One thing very important in the construction of my
apparatus is the shape of the thermometer which I
employ to measure the temperature of the water in the
receiver. This thermometer — which I made myself, and
which, after having undergone every kind of trial, has
always appeared good — is a mercurial thermometer,
divided according to Fahrenheit's scale. It is one of
four, all similar, that I employed at Munich, in the
winter of 1802, in my experiments on the refrigeration
of liquids enclosed in vessels.

The reservoir of this thermometer is cylindrical, about two lines in diameter only, and four inches high; and as the water in my calorimeter is four inches deep, this thermometer always indicates the mean temperature of the fluid, whatever may be the temperature of its different strata.

In my various inquiries concerning heat, I have had frequent opportunities of seeing the importance of this precaution; and I cannot conceive how any one can expect to avoid great mistakes in measuring the temperature of liquids heated or cooled, if we do not attend to this. For my own part, I confess I pay little regard to the experiments of which I am told, when I know they are so negligently made; and assuredly I shall never waste my time in attempting to build theories on their results.

In using the apparatus I have described, several precautions are necessary. In the first place, it is obvious that when the object is to ascertain the quantity of heat developed in the combustion of any inflammable substance, it is indispensably necessary so to arrange matters that *the combustion shall be complete.* I have thought that it might be so considered, whenever the substance burned leaves no residuum, and burns with a clear flame, without smoke or smell.

The least smell, particularly that peculiar to the inflammable substance burned, is a certain indication that the combustion is imperfect.

I had long sought, before I was able to find, to my satisfaction, a mode of burning very volatile liquids, such as alcohol and ether; but I have at length discovered it, as will soon appear. I have frequently succeeded in burning highly rectified sulphuric ether, without the least smell of ether being diffused through the room ; and it was in these instances alone that I considered the experiments as accurate.

As to wood, I have found a very simple method of burning it completely, without the least appearance of smoke or smell. I got a joiner to plane me shavings about half an inch wide, a tenth of a line thick, and six inches long; and holding these in the hand or with pliers, elevated at an angle of 45° or thereabout, and with the edges perpendicular, they burned like a match, with a very clear flame.

The slip of wood that burns being very thin, and placed between two flat flames which press on it closely, it is exposed to the action of so strong a heat that it burns perfectly and entirely.

If the shavings employed be too thick, a portion of the charcoal of the wood remains, particularly if it be oak, or any other wood of slow and difficult combustion ; and in this case the experiments are defective. But if the shavings be sufficiently thin, and well dried, I have found that any kind of wood may be burned completely.

In burning candles, wax tapers, or fat oils in lamps, the only precautions necessary are so to arrange the wick as to yield no smoke ; to place the flame properly in the aperture of the worm ; and to surround the apparatus on all sides by screens, to prevent the flame from being deranged by the wind.

In these experiments there is one source of error, too obvious to escape the most superficial observer, and to which it was important to attend. While the calorimeter is warmed by the heat developed in the combustion of the inflammable substance which is burning at

the aperture of the worm, it is continually cooled by
the ambient air that surrounds it on all sides. It would
be possible, no doubt, by calculations founded on a
knowledge of the law of refrigeration of the receiver,
which might be found by separate experiments, to ascertain the quantity of the effect produced by the refrigeration in question; and this even with a certain degree of
precision : but it would have been impossible by this
method, or by any other known, to calculate the effects
of another cause of error, less obvious perhaps, but
certainly more weighty, than that of the refrigeration
of the external surface of the receiver.

The nitrogen which is mixed with the oxygen of the
atmospheric air is necessarily carried into the worm
with the proper products of the combustion ; and without a precaution, which it occurred to me to employ to
prevent the effects of this cause of error by making
a compensation for them, all the experiments would
have been of no value.

Fortunately the method I employed to obviate the
effects of this cause of error was sufficient to prevent at
the same time those that might have arisen from the
cooling of the outer surface of the receiver.

As the receiver is cooled, whether by the atmospheric
air in contact with its external surface or by the nitrogen and other gases traversing the worm with the
products of combustion, only so far as the worm is
hotter than the surrounding air, while, on the contrary,
it is heated by these elastic fluids whenever it is at a
lower temperature than they are, — by arranging matters
so that the temperature of the water in the receiver shall
be a certain number of degrees, 5° for instance, below
the temperature of the air at the beginning of the ex-

periment, and putting an end to the experiment as soon as the water in the receiver has acquired a temperature precisely the same number of degrees higher than the air, the receiver will be heated by the air during half the time of continuance of the experiment, and cooled by it during the other half; so that the calorific and frigorific effects of the air on the apparatus will counterbalance each other, and produce no perceptible effect on the results of the experiments; consequently they will require no correction.

When we are making experiments to elucidate natural phenomena, it is always more satisfactory to avoid errors, or to compensate them, than to trust to calculation for appreciating their effects.

As the law of the variation of the specific heat of water at different temperatures is not known, and as we have but an imperfect knowledge of the true measure of the intervals of temperature marked by the divisions of our thermometers, to prevent the effects that our uncertainty on these points would have on the subject of inquiry, I took care to make my experiments in a room where the temperature varied very little, and to confine them to a few degrees of elevation of the temperature of the water in the receiver.

It is true, I made some experiments in a room where the air was much colder, and in which I employed ice instead of water to fill the receiver; but these experiments were for a particular purpose, and are not classed with the others. Besides, they never afforded such uniform and satisfactory results as those made under other circumstances.

It has been fully proved, not only by the results of my experiments, but by the experiments of others also, that the vapour of water in contact with ice frequently freezes, while this same ice is melting by the heat, or that its thaw appears fully established.

To give an idea of the reliance that may be placed on the results of the experiments made with the new apparatus I have just described, I will introduce here the particulars of an experiment made purposely to discover its degree of perfection.

Having filled two receivers, properly connected with each other, with water at the temperature of the air of the room, $55°$ F., I burned a wax taper under the mouth of the principal receiver, so that all the products of the combustion passed through the worm of the secondary receiver, after having traversed that of the principal. Each of the receivers contained 2371 grammes [36621.5 grains] of water.

The following are the results of the experiment : —

TIME OF THE OBSERVATION.			TEMPERATURE OF THE WATER	
Hours.	Min.	Sec.	in the principal receiver.	in the secondary receiver.
9	37		$55°$	$55°$
	49	42	65	55
	56	15	70	55
10	2	52	75	$55\frac{1}{4}$
	9	32	80	$55\frac{3}{4}$
	16	34	85	$55\frac{3}{4}$
	23	54	90	$55\frac{3}{4}$
	27			56
	31	40	95	$56\frac{1}{4}$
	39	35	100	$56\frac{3}{4}$
	47	40	105	$56\frac{3}{4}$

From the results of this experiment it appears that the water in the secondary receiver did not begin to be heated perceptibly till that in the principal receiver had been heated $15°$ or $20°$; and, as I had intended from the beginning never to continue an experiment longer than was necessary to raise the temperature of the water in the principal receiver 10° or 12° F., it may be supposed that, as soon as I found by this experiment how little heat remained in the products of combustion after they had passed through the worm of the principal receiver, I relinquished my original design of operating with the two receivers joined together. As it was evident, from the above results, that the second receiver could never be sensibly affected, or indicate anything except the confidence I might place in the indications of the first, I resolved to dispense with the trouble of using it.

It may be seen by the description I have given of this apparatus, that it may be used very conveniently for ascertaining the specific heat of gases, as well as that made apparent in the condensation of vapours, and generally in all researches where the quantity of heat communicated by an elastic fluid in cooling is to be measured. And as it would be extremely easy, by very simple means, to separate completely the products of the vapours condensed in the worm from the gases that pass through it without being condensed, I cannot avoid hoping that this apparatus will become useful as an instrument to be employed in chemical analyses. This, however, would only be an extension of the method already employed with so much success by M. de Saussure, and by Messrs. Gay Lussac and Thénard.

As soon as my apparatus was finished, I was eager to see what quantity of heat I should find in the combustion of wax and in that of olive oil, that I might afterward compare the results of my experiments with those of M. Lavoisier's; and, as I have the most implicit reliance on everything published by that excellent man, I sincerely wished to find in this comparison a proof of the accuracy of my method, and at the same time a confirmation of the estimates of M. Lavoisier.

Section II. — *Experiments made with white Wax.*

The air of the room being at the temperature of 61° F., 2781 grammes of water, of the temperature of 56° F., were put into the receiver of the calorimeter (including the quantity of this liquor that represents the specific heat of the instrument), and, a lighted wax taper having been properly placed at the entrance of the worm, the calorimeter was heated for 13 minutes and 26 seconds, when, the thermometer announcing that the water had acquired the temperature of 66° F., the taper was extinguished.

As I took care to weigh the taper before it was lighted, I found, by weighing it at the end of the experiment, that 1.63 grammes of wax had been burned.

To express the results of this experiment so as to render them obvious, and at the same time easy to be compared with the results of other similar experiments, we will see how much water of the temperature of melting ice would have been made to boil, at the mean pressure of the atmosphere, by the heat made apparent in the combustion of the 1.63 grammes of wax burned.

The distance on Fahrenheit's scale between the temperature of melting ice and boiling water being 180°, if the burning of 1.63 grammes of wax were requisite to raise the temperature of the water in the calorimeter 10°, the burning of 29.34 grammes would have been necessary to raise it 180°; and, if 29.34 grammes of wax could furnish by combustion sufficient heat to raise the temperature of 2781 grammes 180°, a gramme of this inflammable substance must furnish enough to heat 94.785 grammes of water to the same point.

Consequently, one pound of white wax, or wax taper, should furnish, in burning, sufficient heat to raise 94.785 pounds of water from the temperature of melting ice to the boiling point.

To find how many pounds of ice this quantity of heat would melt, we have only to add to the number of pounds of water at the temperature of melting ice it would cause to boil the third part of this number, and the sum would express the weight of the ice in pounds.

This, then, for white wax is : —

$$94.785$$
$$+ \; 31.595$$

$= 126.380$ lbs. of ice melted for 1 lb. of the wax burned.

Before I compare the result of this experiment with that of an experiment made with the same substance by M. Lavoisier, I will give an account of two other experiments I made with wax, as the reader will undoubtedly be struck with the uniformity of their results. This is so remarkable that I should scarcely venture to publish them had I not proofs that all my experiments were actually made and minuted down before I began my calculation of their results, and were I not assured that any person who will follow my method, using the same apparatus, will find the same results on repeating my experiments.

As the mode of operating in making these experiments must now be well known, I may suppress the particulars in what follows without inconvenience, and give only the results of the experiments.

I will begin with three experiments made with white

wax; and to render them more easy to compare, I will give them together in a tabular form.

Results of three Experiments on the Burning of white Wax, showing the Quantity of Water that would be heated 180°, or of Ice that would be melted, by one pound Weight of it.

No. of the Exp.	Quantity of wax burned.	Time employed in burning.	Quantity of water heated.	Elevation of its temperature.	Temp. of the water at the beginning of the exp.	Temp. of the water at the end of the exp.	Temperature of the air.	Results. Pounds of water heated 180°	Results. Pounds of ice melted.
	Grms.	m. s.	Grms.	Degrees.	Degrees.	Degrees.	Degrees.	lbs.	lbs.
1	1.63	13 24	2781	10° F.	56° F.	66° F.	61° F.	94.785	126.38
2	2.36	19 30		14½	51	65½	58	94.926	126.608
3	2.17	18 15		13¼	51¾	6	58	94.337	125.783

If we take the mean term between the results of these experiments, we shall find that the quantity of heat developed in the combustion of wax is such that one pound of this substance is sufficient to raise 94.682 pounds of water from the temperature of melting ice to the boiling point, and consequently that it should melt 126.242 pounds of ice.

According to the experiments of M. Lavoisier, the heat developed in the combustion of one pound of white wax was sufficient to melt 133.166 pounds of ice.

The difference between the results of our experiments with this substance is not very great; and if those of M. Lavoisier were made at a time when the temperature of the air was only a few degrees higher than that of melting ice (which I have no means of ascertaining), the quantity of nitrogen that must have entered into the calorimeter, with the oxygen employed to support the combustion, would have been so great as to account sufficiently for the difference. But the very great difference between the results of our experiments made with olive oil proves that one or other of our processes must have been defective.

The mean result of several experiments made with olive oil gave me for the measure of the quantity of heat developed in the combustion of one pound of this substance 90.439 pounds of water heated 180° F., or 120 pounds of ice melted, neglecting the fraction.

In the experiments of M. Lavoisier, more than 148 pounds of ice were melted by the heat that appeared to result from the combustion of one pound of this oil.

It is true that this result was considered by that eminent philosopher himself as too great to be capable of explanation; and he added, with that modesty which rendered him so engaging and so respectable: "We shall probably find ourselves under the necessity of making corrections, perhaps pretty considerable ones, in most of the results I have given; but I did not think this a sufficient reason to delay affording their assistance to those who might intend to pursue the same object."

As it appears very probable that all the fat oils, when perfectly pure, are composed of the same principles, I was curious to see whether rape oil, purified by sulphuric acid, would not afford more heat in its combustion than olive oil, when burned in its natural state. The result of three experiments showed me that rape oil, thus purified, does, in fact, yield more heat than olive oil. The difference is, indeed, pretty considerable, and more than I could have suspected.

The combustion of 1 lb. of purified
    rape oil gave   .   .   .   93.073 lbs. of water heated 180°.
1 lb. of olive oil gave   . . 90.439 "   "   "   "

Chemists may tell us whether the quantity of incombustible matter separated from rape oil in purifying it be sufficient, or not, to account for this difference.

On comparing the results of the experiments made with white wax and those with the purified oil, it appears that equal weights of these substances afford nearly equal quantities of heat in their combustion; and as, in fact, this ought to be the case, from the quantities of combustible matter they contain, the result tends to strengthen our confidence in this method of measuring the heat developed in combustion.

The combustion of
   1 lb. of white wax gave .   94.682 lbs. of water heated 180°.
   1 lb. of purified oil .   . 93.073 "   "   "   "

As the object I had chiefly in view in this series of experiments was to ascertain the quantities of heat developed in the combustion of pure hydrogen and carbon, in order to render this method useful in some chemical analyses, I examined particularly those inflammable substances that had been analyzed with most care.

Several attempts have been made to ascertain these interesting questions by direct experiments, in burning pure hydrogen, or pure hydrogen and carbon; but the results of these researches have varied so much that they cannot be relied on.

According to Crawford, the heat developed in the combustion of one pound of hydrogen gas is sufficient to raise the temperature of 410 pounds of water 180° F. But the estimation of M. Lavoisier is much lower. According to him, this heat would raise only 221.69 pounds of water the same number of degrees.

On the other hand, M. Lavoisier estimates the quantity of heat developed in the combustion of charcoal much higher than Dr. Crawford. I have many reasons to believe that they both estimate it too high; and, if this opinion be confirmed, we must estimate the heat developed in the combustion of hydrogen a little higher even than Crawford has done, to be able to account for that manifested in my experiments.

From several experiments, which I made five years ago, it appeared to me that one pound of charcoal, dried as much as possible before it was weighed by heating it red-hot in a crucible, was not capable of raising more than from 52 to 54 pounds of water from the temperature of melting ice to a boiling heat.

According to Crawford, this heat should suffice to boil 57.606 pounds, and according to Lavoisier, 72.375 pounds.

We shall see how these estimates agree with the results of my experiments.

As the experiments made with wax yielded very uniform results, and as the analysis of this substance has been made with great care, I shall examine how the quantities of hydrogen and carbon in this substance agree with the quantity of heat that it afforded me in combustion.

According to the analysis of Messrs. Gay-Lussac and Thénard, a pound of this substance contains

Carbon . . . . 0.8179 lb.
Free hydrogen 0.1191

If we adopt the calculations of Dr. Crawford, both for the heat furnished by the hydrogen and that furnished by the carbon, we shall have for the heat that should be furnished by the combustion

Of 0.1191 lb. of hydrogen, after tne ratio of 410
lbs. of water raised from 32° to 212° by burning
1 lb. of hydrogen . . . . . 48.831 lbs.
Of 0.8179 lb. of carbon, after the ratio of 57.606
lbs. of water raised from 32° to 212° by burning
1 lb. of carbon . . . . . . 47.116 "

Total of the heat that ought to be furnished by the
quantity of combustible matter (hydrogen and carbon) in 1 lb. of white wax . . . . 95.947 "
Quantity of heat furnished by 1 lb. of white wax,
during its combustion, according to my experiments . . . . 94.662 '

If we adopt the calculations of M. Lavoisier for the
heat furnished by carbon and hydrogen in their combustion, we shall have for the heat that ought to be furnished
by the burning

Of 0.8179 lb. of carbon, after the ratio of 72.375 lbs.
of water heated 180° by 1 lb. . . . 59.059 lbs.
Of 0.1191 lb. of hydrogen, after the ratio of 221.69
lbs. of water heated 180° by 1 lb. . . . 26.403 "

Total of the heat that ought to be furnished by the
combustible matter in 1 lb. of white wax . 85.462 "

From the results of these calculations it appears that
the estimations of Dr. Crawford agree much better with
the experiments than those of M. Lavoisier.

Let us see how the results of the experiments made
with fat oils agree with the estimations of these gentlemen.

According to the analysis of Messrs. Gay-Lussac and
Thénard, a pound of olive oil contains

Carbon . . . . 0.7721 lb.
Free hydrogen 0.1208

According to the calculations of M. Lavoisier, we have,

For 0.7721 lb. of carbon .   55.881 lbs. of water heated 180°.
"   0.1208 lb. of hydrogen .   26.780 "   "   "   "

Total   .   .   .   .   82.661 "   "   "   "

According to the calculations of Dr. Crawford, it is,

For 0.7721 lb. of carbon .   44.478 lbs. of water heated 180°.
"   0.1208 lb. of hydrogen .   49.528 "   "   "   "

Total   .   .   .   .   94.006 "   "   "   "

According to the experiments, 1 pound of purified rape oil furnished heat sufficient to raise 93.073 pounds of water 180°; and 1 pound of olive oil enough to heat 90.439 pounds.

From all these comparisons it follows that the estimations of Dr. Crawford agree much better than those of M. Lavoisier with the results of my experiments.

SECTION III. — *Experiments made with Spirit of Wine, Alcohol, and Sulphuric Ether.*

As the component parts of these inflammable liquids may be considered as well ascertained by the results of the excellent investigation of M. de Saussure, I undertook to examine them for the second time, in order to discover what quantities of heat are developed in their combustion. I had begun this undertaking five years ago; but, after having made a considerable number of experiments, I desisted from it on account of the great difficulties that occurred. As soon, however, as I had found means of rendering my apparatus more perfect, I formed the project of recommencing it.

Before I enter into the particulars of my experiments, I must say a few words respecting the difficulties that occurred to me, even after I had my new apparatus, and of the means I employed to surmount them. I even found myself exposed to dangers, which it is necessary for me to mention as a caution to those who may undertake the same inquiry.

When I made the experiments with highly rectified alcohol, and more particularly with ether, I found it very difficult to prevent a portion of these volatile liquids from escaping in the state of vapour from the bulk of them remaining in the lamp. I procured a small lamp, resembling in shape a small round snuff-box, with a nozzle rising from the centre of the circular plate, which closed it atop; and on this plate was fixed a small pan, to hold cold water, for keeping the nozzle cool and preventing the heat from being communicated to the body of the lamp. But this precaution was not sufficient, when I burned ether, as I found to my cost; for, though the pan was twice the diameter of the lamp, and filled with very cold water, the water was so heated in a few minutes that an explosion took place from vapour of ether kindling in the air with a flame that rose to the ceiling. Indeed it was near setting the house on fire.

Warned by this accident, I procured a new lamp, much smaller than the former, being only an inch in diameter and three quarters of an inch deep; and its nozzle, which was only two lines in diameter, was three quarters of an inch high. To keep this small lamp cool while burning, it was placed in a small pan, and kept constantly immersed in a mixture of water and pounded ice to within a quarter of an inch of the extremity of the nozzle. These precautions were sufficient to prevent any explosion, though not the evaporation either of the ether or of the alcohol. This fact I learned from observing that, as often as I made two consecutive experiments without filling the lamp afresh, the alcohol constantly appeared weaker in the second experiment than in the first.

The cause of this phenomenon was not difficult to discover. The most volatile and consequently the most combustible parts of this liquid, being diffused in vapour in the interior of the lamp, found means of escaping through the nozzle with the part of the liquid that traversed the match, leaving the alcohol that remained in the lamp perceptibly weakened.

To remedy this imperfection, I constructed a third lamp, which I now submit to the inspection of the Class. It is made of copper, and has the shape of a small cylindrical vase, an inch and a half in diameter, and three inches high, swelling out a little atop, and closed hermetically by a copper stopple, which, being ground with emery, fits tight into the neck of the vase. Through the centre of this stopple passes a small perpendicular hole, which can be shut completely or left a little open, as may be required, by means of a small screw carrying a copper collar.

A small tube, about an eighth of an inch in diameter and two inches and a half long, proceeds horizontally from the side of the vase very near the bottom. At the distance of an inch and four lines from the vase this tube is bent at a right angle, rising upwards perpendicularly to form the nozzle of the lamp.

This little tube is everywhere very thin, except at its upper extremity, where it is made thicker, to admit of being shaped so as to fit tight into a very small cylindrical extinguisher, five lines high by three and a half in diameter, intended to close the nozzle hermetically without touching or deranging the wick, the moment the lamp ceases to burn, and to keep it constantly closed when the lamp is not lighted.

Without this precaution, in experiments made with ether, so large a quantity of this volatile liquid would evaporate through the nozzle of the lamp while weighing, that it would be impossible to ascertain the quantity burned.

The nozzle of the lamp is steadied by two pieces of wire, proceeding from it horizontally, and soldered to the body of the lamp.

To keep this lamp constantly cold, as well as the liquid it contains, it is placed in a small pan, and covered completely, except the extremity of its nozzle and that of its neck, with a mixture of pounded ice and water.

When the lamp is weighed, it is taken out of the pan, and well wiped with a dry cloth before it is put into the scale.

When the lamp is kindled, the operator must not forget, after it has burned two or three minutes, to open the screw that closes its stopple a little, though but *very little*, otherwise it might go out.

As the little horizontal tube, by which the liquid that is burned passes from the reservoir of the lamp to its nozzle, is always filled with liquid, so that it can have no communication with the vapour diffused in the upper part of the reservoir, this vapour cannot escape by the nozzle of the lamp, as it did before I thought of this method of preventing it.

If I have been very minute in my description of this lamp, it is because I thought it necessary to spare those who might be disposed to repeat my experiments, or make similar ones, all the difficulties I had to surmount before I found the means of having under command the combustion of very volatile inflammable liquids.

As the apparatus I have employed has now been described, it will be easy to follow the steps of my experiments, and to appreciate their results. I will endeavour to describe them clearly, but also as briefly as possible.

Having procured a stock of spirit of wine of the shops, and of alcohol of different degrees of purity, I ascertained with the greatest care their specific gravities at the temperature of 60° F., taking that of water at the same temperature as 1000000. I chose this temperature that I might afterward the more easily ascertain the quantities of water that each ought to contain, according to the tables constructed from the experiments of M. Lowitz.

The following table will show the specific gravity of each, and the quantity of pure alcohol of Lowitz and of water contained in it.

Liquid.	Specific gravity at 60° F.	Composition.	
		Pure alcohol of Lowitz	Water.
Alcohol of 42°	817624	0.9179	0.0821
Alcohol of the shops	847140	0.8057	0.1943
Spirit of wine of 33°	853240	0.7788	0.2212

The following are the results of the experiments made to ascertain the quantities of heat which these liquids furnished in burning.

In three experiments made with the spirit of wine the quantities of heat manifested were, —

In the 1st, 53.260 lbs. of water raised from the temperature of
" 2d, 51.727 " melting ice to that of ebullition.
" 3d, 52.855 "
The mean result is . . . . . . 52.614 lbs.

As a pound of this liquid contained but 0.7788 of the alcohol considered by Lowitz as pure, the other part

($=$0.2212) being only water, which does not burn, to find
how much water would be raised from the temperature
of melting ice to that of ebullition by a pound of the pure
alcohol of Lowitz, we have only to divide the quantity,
that is, the measure, of the mean heat developed in the
experiments with the spirit of wine by the fraction that
expresses the quantity of alcohol in a pound of this
liquid.

Thus, we have $\frac{52.614}{0.7788} = 67.558$ pounds, the measure of
the heat developed in the combustion of one pound of
pure alcohol of Lowitz, according to the mean result
of the experiments made with spirit of wine.

In two experiments made with the alcohol of the shops,
the mean result was 54.218 pounds ; and, as this contained 0.8057 pound of pure alcohol, we have for the
measure of the heat developed in the combustion of
1 pound of pure alcohol $\frac{54.218}{0.8057} = 67.293$ pounds of water
heated 180° F.

Of three experiments made with the alcohol at 42°, the
mean result was 61.952 pounds of water heated 180° F.
by the heat developed in the combustion of one pound
of this liquid.

Hence, 1 pound of pure alcohol should furnish heat
enough in burning to raise 67.57 pounds of water 180°
F.; for $\frac{61.952}{0.9179} = 67.101$.

Taking the mean between the results of these eight
experiments with three alcoholic liquors, we shall have
for the measure of the heat developed in the combustion
of one pound of pure alcohol of Lowitz 67.317 pounds
of water raised from the temperature of melting ice to
that of ebullition.

It will be extremely interesting, no doubt, to know
whether this quantity of heat agree with the quantities

of combustible matter (carbon and hydrogen) in alcohol.
We will see.

According to the analysis of M. de Saussure, 1 pound
of the alcohol of Lowitz contains

Carbon	0.4282 lb.
Free hydrogen	0.1018
Water	0.4700
	1.0000

Now, according to the calculations of Dr. Crawford,
we shall have for the measure of the heat developed in
the combustion of

0.4282 lb. of carbon .	24.667 lbs. of water heated 180° F.
0.1018 lb. of hydrogen .	41.738 " " " "
Total	66.405 " " " "
The experiments gave us .	67.317 " " " "

It is rare in a research of such delicacy to find the results of experiment agree so perfectly with those of calculation.

SECTION IV. — *Heat developed in the Combustion of Sulphuric Ether.*

I have already mentioned the difficulties which I
overcame before being able to regulate the combustion of this substance in such a way as to render the
results of my experiments regular and satisfactory ; but
I met with still further difficulties in the course of this
delicate inquiry.

As alcohol is necessarily employed in making sulphuric
ether, and as these two liquids may be united in any
proportions, it is extremely difficult, if not impossible,

to separate them entirely ; and as both are colourless and limpid, either when mixed or separate, we can scarcely judge of the degree of purity of the ether, except by its specific gravity, and even in this way but very imperfectly.

The most highly rectified sulphuric ether which I could procure, and which I employed in my experiments, was prepared in M. Vauquelin's laboratory. Its specific gravity is 728.34 at the temperature of 16° Reaumur. As that which was employed by M. de Saussure in his analysis was only of the specific gravity of 717 at the same temperature, by regarding the ether which I employed as being a mixture of the same degree of purity with that of M. de Saussure, and the pure alcohol of Lowitz having a specific gravity of 792, we shall find, upon making a calculation, that the ether which I employed was a mixture of 85 parts of ether of the specific gravity of 717, and 15 parts of pure alcohol of Lowitz of the specific gravity of 792.

On burning this mixture under my calorimeter, after having brought my apparatus to the highest degree of perfection, I obtained the following results : —

Duration of the experiment.		Ether burned.	Quantity of water heated.	Temperature of water in the calorimeter		Elevation of the temperature of the calorimeter.	Temperature of the air.	Result.
				At the commencement.	At the end of the experiment.			Quantity of water heated 180° of Fahrenheit with the heat developed in the combustion of one pound of the combustible.
m.	s.	Grms.	Grms.	Degrees.	Degrees.	Degrees.	Degrees	
11		1.96	2781	55½° F.	65⅝° F.	10¼° F.	60° F.	79.996
11	15	2.01		54¼	64¾	10½	60	80.710
9		2.		58¾	69⅜	10⅝	60	80.146
20		3.29		56¾	73¾	17	61	79.884
22		3.06		56½	72½	16	64½	80.784
Mean result of the five experiments								80.304

Continuing to make use of the estimates of Crawford, for the quantities of heat developed in the combustion of hydrogen and carbon, we shall see if these estimates are sufficient to account for the heat manifested in these five experiments.

As the ether employed was a mixture of 15 parts of pure alcohol of Lowitz, and 85 parts of ether of the specific gravity of 717 at the temperature of 16° Reaumur, and consequently similar to the ether analyzed by M. de Saussure, we shall begin by determining the quantity of heat which ought to be developed in the combustion of these fifteen parts of alcohol.

As M. de Saussure has shown that in one pound of Lowitz's alcohol (of the specific gravity of 792) there are 0.4282 pound of carbon and 0.1018 pound of free hydrogen, we ought to find in 0.15 pound of this same liquid 0.06423 pound of carbon, and 0.01527 of free hydrogen.

According to the estimate of Crawford, 0.06423 pound of carbon ought to furnish a sufficiency of heat in its combustion to raise the temperature of 3.7002 pounds of water 180° F.; and 0.01527 pound of hydrogen ought to furnish enough to raise 6.2607 pounds of water the same number of degrees; and these two quantities of water, making together 9.9609 pounds, afford a measure of the quantity of heat which must be developed in the combustion of the 15 parts of alcohol, which are found mixed with 85 parts of ether, in order to form the combustible liquid employed under the name of sulphuric ether in my experiments.

Now, as one pound of this mixed liquid has furnished in its combustion enough of heat to raise 80.304 pounds of water 180° F., if we deduct from this mass the quantity of water which the 15 per cent of alcohol must heat (= 9.909), that which remains (= 70.3431 pounds of water) will be the measure of the quantity of heat developed in the combustion of 85 per cent of ether of the gravity of 717, which exists in this combustible liquid.

According to the analysis of sulphuric ether made by M. de Saussure, we ought to find in one pound of this liquid (of the specific gravity of 717)

Carbon	0.590 lb.
Free and combustible hydrogen	0.194
Oxygen and hydrogen in the proportions necessary to form water	0.216
	1.000

Consequently, we ought to find in 0.85 pound of the same kind of ether the following quantities of combustible substances, viz. : —

Carbon	0.5015 lb.
Free and combustible hydrogen . . .	0.1651

We shall now see if these quantities of combustible substances are sufficient to account for the heat which is manifested in our experiments.

The 0.5015 pound of carbon ought to furnish sufficient heat to raise 28.89 pounds of water 180° F.; and the 0.1651 pound of hydrogen sufficient to heat 67.64 pounds the same number of degrees.

These two masses of water form together 96.53 pounds ; but we shall see that the quantity of heat furnished by the 85 parts of ether in the experiments cannot be greater than that which is necessary to heat 70.3431 pounds of water 180° F.

As the experiments have been made with the greatest

care, and frequently repeated, and always with very uniform results; and as the estimates which we have adopted, with respect to the quantities of heat which are developed in the combustion of hydrogen and in that of carbon, have been confirmed so as to leave little doubt upon this subject, — upon investigating the cause of the great difference between the quantity of heat actually developed in the combustion of the 85 parts of sulphuric ether burned in the experiments which we have examined, and the quantity given by calculation, we are compelled, in my opinion, to admit that there is an error in the analysis of this liquid, and that it does not contain so much free and inflammable combustible matter as M. de Saussure ascribes to it.

As it seems to me to be much more probable that an error has been committed in determining the quantity of free hydrogen in this substance than in determining the quantity of carbon. I shall suppose with M. de Saussure that there is really in one pound of sulphuric ether (of the specific gravity of 717) 0.59 of carbon ; but instead of estimating the quantity of free hydrogen in this liquid according to the results of M. de Saussure, I shall adopt the estimate of Mr. Cruickshanks.

This excellent chemist concluded, from his experiments, that in the vapour of sulphuric ether the carbon is to the hydrogen as 5 to 1.

In the 0.85 pound of sulphuric ether (specific gravity 717) which were mixed with the 0.15 pound of alcohol, in order to form one pound of the mixed liquid employed in my experiments, there were 0.5015 pound of carbon ; and dividing this number by 5, we shall see that this carbon ought to be united with 0.1003 pound of free hydrogen, instead of being united with 0.1651 pound, as we shall suppose according to M. de Saussure.

Let us now see if, by adopting the analysis of Mr. Cruickshanks with respect to the hydrogen, instead of that of M. de Saussure, the calculation will agree better with the experiment.

We have seen that the quantity of water heated 180° Fahrenheit, which represents the quantity of heat which must be developed in the combustion of the 0.15 lb. of alcohol, was   .   .   .   9.9609 lbs.

And that the quantity answering to 0.5015 lb. of carbon, which exists in the 0.85 of ether, was .   28.89

We shall for the present add that which answers to the combustion of 0.1003 lb. of free combustible hydrogen, which, according to Mr. Cruickshanks, ought to be found united to this quantity of carbon in order to form the ether   .   .   41.123

These three quantities of water together are the measure of the heat which must be developed in the combustion of one pound of sulphuric ether of the kind employed in my experiments   .   .   79.9739

The mean result of five experiments was .   .   80.304.

This coincidence between the calculation and the experiment is, doubtless, too remarkable to be owing to chance; but I am ready to prove that it occurred without being foreseen or expected.

From all these results we may conclude, that one pound of sulphuric ether, of the specific gravity 717 at the temperature of 16° Reaumur, or of the same species with that employed by M. de Saussure, should have furnished in combustion enough of heat to raise 82.369 pounds of water 180° F., viz. : —

That furnished by 0.59 lb. of carbon   .   .   .   33.989 lbs.

And that furnished by 0.118 lb. of hydrogen   .   48.380

82.369

If the proportion of free hydrogen in the ether analyzed by M. de Saussure was really such as he has determined it to be, one pound of this liquid ought to furnish a sufficiency of heat in its combustion to raise 113.566 pounds of water 180° F., viz. : —

That furnished by 0.59 lb. of carbon . . .	33.989 lbs.
And that which was furnished by 0.194091 of hydrogen	79.577
	113.566

But I can the less persuade myself that this liquid can furnish in its combustion so much heat, because one pound of white wax furnished no more than what was sufficient to heat 94.682 pounds of water the same number of degrees.

According to the analysis of M. de Saussure, 100 parts of sulphuric ether, of the specific gravity of 717, at 16° Reaumur, are composed of

Carbon	59 parts
Hydrogen	22
Oxygen	19
	100

Supposing that the 19 parts of oxygen are combined with 3.6 parts hydrogen, so as to form with them 21.6 parts water, 100 parts of this kind of ether ought to be composed of

Carbon	59
Free and combustible hydrogen	19.4
Consequently, inflammable substances . . .	78.4
Water	21.6
	100

From the result of my experiments, 100 parts of this kind of ether ought to be composed of

Carbon 59
Free or combustible hydrogen 11.8

Consequently, combustible substances . . . 70.8
Water 29.2
————
100

Or, reducing the water to its elements, —

Carbon 59
Hydrogen, free or combustible . . . 11.8
Ditto, non-combustible 3.5
———— 15.3
Oxygen 25.7
————
100

According to M. de Saussure's analysis, as well as from the results of my experiments, 100 parts of pure alcohol of Lowitz, of the specific gravity of 792, at the temperature of 16° Reaumur, are composed of

Carbon 42.82
Free or combustible hydrogen 10.18

Consequently, combustible substances . . . 53
Water 47
——
100

Or, reducing the water to its elements, 100 parts of this alcohol are composed of

Carbon 42.82
Hydrogen, combined and non-combustible . 5.64
Hydrogen, combustible 10.18
———— 15.82
Oxygen 41.36
————
100

By supposing that water exists completely formed both in alcohol and ether, the constituent parts of these two liquids would be, according to the results of our inquiries,

	Alcohol.	Ether.
Carbon	42.82	59
Combustible hydrogen	10.18	11.8
Water	47	29.2
	100	100

The elements of water exist most assuredly both in alcohol and ether; but there is good reason to believe that water does not exist in its natural state of condensation in these two substances, neither when they are in a state of liquidity, nor when, being sufficiently heated, they are transformed into elastic fluids.

When we mix water with alcohol, there is a considerable change both in temperature and volume, which indicates a new arrangement of elements, or a chemical action; and what proves in a still more certain manner that this action has taken place, the liquid which results from this mixture may be distilled, i. e. *vaporized* by heat, and afterwards *condensed*, without being decomposed: but it is, above all, in the little heat which is developed in the condensation of the vapour of alcohol and ether that we discover certain proofs that the oxygen and hydrogen which exist as elements in these liquids do not exist in the state of water. I shall recur to this subject again.

SECTION V. — *On the Quantity of Heat developed in the Combustion of Naphtha.*

The naphtha which I made use of in my experiments was supplied by M. Vauquelin: it had been purified by distillation, and its specific gravity at the temperature of 56° F. was 827.31.

The following are the details and results of two experiments made with this liquid on the 29th of January, 1812.

The capacity of the calorimeter for heat was equal to that of 2781 grammes of water.

	Duration of the experiment.	Quantity of naphtha burned.	Elevation of the temperature of the calorimeter in degrees of Fahrenheit.	Result. Pounds of water heated 180° with 1 lb. of this substance.
	m.	Grammes.	Degrees.	lbs.
1st Experiment	32	4.45	16°F.	73.881
2d Experiment	36	2.77	12¾	72.771
Mean result				73.376

The naphtha was burned in the same small lamp which I had employed in my experiments made with alcohol and sulphuric ether; but as I had not been able to succeed in burning the naphtha without smoke, I cannot rely implicitly upon the results of these experiments. Perhaps with pure oxygen gas we might succeed in burning it entirely.

I have met with the same difficulty in burning oil of turpentine and colophon; and for this reason I thought it would be useless to detail my experiments with these two substances.

SECTION VI. — *On the Quantity of Heat developed in the Combustion of Tallow.*

Having procured tallow candles of a good quality, those which are called *six in the pound*, I burned one under the calorimeter, taking care to keep it well snuffed, in order to avoid smoke.

The following are the details and results of two experiments made on the same day (16th of November, 1811) with one of these candles.

The capacity of the calorimeter for heat was equal to that of 2371 grammes of water.

	Time while the candle was burning under the calorimeter.		Quantity of tallow burned.	Elevation of the temperature of the water in the calorimeter.	Result. Quantity of water heated 180° with the heat developed in the combustion of 1 lb. of tallow.
	m.	s.	Grammes.	Degrees..	lbs.
1st Experiment	16	2	1.6	$10\frac{1}{4}°$F.	84.385
2d Experiment	16	50	1.7	$10\frac{1}{2}$	82.991
Mean result					83.688
We have seen that with white wax the result was .					94.682
With purified rape oil					93.073
And with olive oil					90.439

SECTION VII. — *Quantity of Heat developed in the Combustion of Charcoal.*

If we could burn under the calorimeter some pieces of wood made into charcoal with the same facility that we burn thin pieces of dry wood, the investigation in question would not be attended with difficulty; but the charcoal cannot be burned in this manner. We can light a piece of charcoal very well, and if it be very thin it continues to burn until it is entirely consumed; but the combustion is so slow, and furnishes so little heat, that it would require several hours to heat the calorimeter sufficiently to give an appreciable result; and for this single reason the result could not but be extremely uncertain.

I have long endeavoured, but without success, to find

a method, by steeping thin chips of wood in some inflammable liquid, to burn the charcoal more rapidly.

Some chips of wood of a known weight, perfectly dried and strongly heated, were plunged into white wax, melted and very hot, and the chips, when taken out and cooled, were again weighed.

Their augmentation in weight gave me the quantity of wax which they had imbibed; and as I knew accurately how much heat this quantity of wax should have given in its combustion, if the chips thus prepared had been burned properly under the calorimeter, I should certainly have discovered how much heat the charcoal would have furnished; but the experiment did not succeed.

The wax was entirely burned, and the chip of wood became very red; but it was not burned, at least not entirely, nor in such a way as to give me the least hope of being able to derive any advantage from my experiment; and I did not succeed any better by steeping my chips of charcoal in melted tallow, in oil, alcohol, sulphuric ether, naphtha, essential oil of turpentine, in a solution of gum-arabic, and in that of sugar. I have also tried colophon, but without more success.

I have made several experiments in order to determine directly the quantity of heat which is developed in the combustion of considerable masses of charcoal (80 grammes) burned in a small stove, under a calorimeter of a large size, which I procured at Paris four years ago, and which I have still in my laboratory; but the results of these experiments have been too variable to satisfy myself.

After all the care which I took, I found that the experiments of Crawford were better than mine; and as they furnished more heat than I could find, I have not hesitated to adopt their results instead of relying upon my own.

Section VIII. — *Quantities of Heat developed in the Combustion of Wood.*

In a memoir which I had the honour to present to the Class on the 9th of September, 1812, I gave an account of a considerable number of experiments (upwards of fifty) which I made in order to determine the quantities of heat which are developed in the combustion of different kinds of wood.

From the results of these experiments, it appears that, at equal weights, the light and soft woods give out a little more heat than the compact and heavy woods; but as the difference is very small, we may rather ascribe it to a greater degree of humidity in the latter.

It is certain that the compact retain humidity with more tenacity than the light woods, and a small difference in the dryness of a wood ought to produce a sensible effect on its apparent weight, and consequently upon the result of the calculations which we employ in order to determine the heat which it furnishes.

In physical and chemical researches, it is always satisfactory to be able to compare the results of new experiments with those of more ancient date, particularly when the latter have been made by persons remarkable for their accuracy.

M. Lavoisier has shown that equal quantities of heat are produced in the combustion of 1089 parts in weight of oak, and 600 parts of charcoal; consequently equal quantities of heat ought to be furnished in the combustion of one pound of oak and 0.55 of a pound of charcoal.

According to the experiments of Dr. Crawford, one pound of charcoal furnishes in its combustion enough of heat to raise the temperature of 57.608 pounds of water 180° of F.

Consequently the temperature of 31.684 pounds of water would be raised the same number of degrees by the heat furnished in the combustion of 0.55 pound of charcoal.

According to the result of the experiments of M. Lavoisier, this same quantity of heat ought to be furnished in the combustion of one pound of oak.

Having made four consecutive experiments with very good dry oak wood, and in very thin slips, burned so as to give out neither smoke nor smell, and which left but an inappreciable quantity of ashes and no charcoal, I obtained the following results : —

Number of experiments.	Quantity of wood burned.	Elevation of the temperature of the calorimeter.	Result. Pounds of water heated 180° with one pound of combustible.
	Grammes.	Degrees.	lbs.
1	5.10	$10\frac{1}{4}°$ F.	31.051
2	5.13	$10\frac{1}{2}$	31.623
3	5.12	$10\frac{3}{8}$	31.941
4	4.95	10	31.212
Mean result			31.457
Result according to Lavoisier and Crawford's experiments . . . . }			31.684

It is rare to find experiments made by different persons at distant periods, and with very different apparatus, which agree better together.

But experiments which are well made can never fail

in agreeing in their results, whatever be the difference of the methods employed : it is, nevertheless, necessary to remark, that the coincidence in question could not be so perfect as it appears, for everything depends upon the equality of the humidity which may exist in the wood and charcoal employed, — a circumstance which it is impossible to establish.

SECTION IX. — *On the greatest Intensity of Heat which it is possible to produce by the Combustion of inflammable Substances.*

It is well known that the heat of a small fire seems to be less intense than that of a large fire, even when the same species of combustible is employed ; but I do not know that it has been attempted to determine the limits of the intensity of a fire, or the greatest degree of heat which it is possible to produce by means of combustion.

In order to elucidate this subject, it is necessary to consider attentively what passes in the chemical operation which we call *combustion.*

In all known cases where two elementary substances unite together so as to form a new substance, there is a change of temperature, so that the new substance *at the moment of its formation* has a temperature differing strongly from that of the surrounding bodies. Consequently, the surrounding bodies are always either heated or cooled more or less by the new body which has been formed.

But in order that this effect may be sensible to our organs, or capable of acting in a sensible manner upon our apparatus, it is necessary that the quantity of the new substance formed should be considerable; for it is certain that the most intense heat, if it be developed in a very small particle of matter, may exist without producing any sensible effect which could give us any indications of its existence.

It is not less true that the chemical union of two atoms, two different elementary substances, ought always, under every circumstance, to be accompanied with one and the same change of temperature; for this union takes effect in a place so distant, relative to all the other bodies (if, in every case, all the interstices are not filled with particles of an ethereal fluid), that we cannot conceive how the change of temperature in question may be either augmented or diminished by the effect of the action of these surrounding bodies.

It is extremely probable, from what we have been able to remark in a great number of phenomena, that the approximation of the elementary particles of bodies is always accompanied by an elevation of their temperature; and as there cannot be new substances formed except in consequence of an approximation and the chemical union of elementary particles, we may conclude that there cannot be new chemical compositions without a development of heat.

We may form an idea of what passes in combustion, by considering the phenomena which take place when water freezes.

At a certain temperature, which is invariable, the molecules of the liquid are disposed to approximate in order to form a solid body, ice; and the first particle of ice which is formed is accompanied by a development of a certain quantity of heat, which quantity is invariable.

It is also very probable that it is at a temperature which is *invariable* that the oxygen and hydrogen are disposed to approximate and unite in order to form an atom of vapour, and that the intensity of the heat developed at the moment of this union is also invariable, and that it is always manifested in all its intensity in the atom of vapour which is formed.

But as the atom of vapour is extremely small, and surrounded by bodies relatively very cold, its heat is soon dissipated.

There is, however, a method, which appears certain, that we may employ in order to determine the temperature of an atom of vapour at the moment of its formation, and by this means we may know what is the highest temperature which it is possible to procure by means of combustion.

We have seen that, according to the results of the researches of Dr. Crawford, it seems that when 1 pound of hydrogen is burned, enough of heat is developed on this occasion to elevate the temperature of 410 pounds of water 180° F. (= 100 degrees centigrade).

Now as 1 pound of hydrogen perfectly dry is united by burning to 7.3333 pounds of oxygen, and forms with it 8.3333 pounds of steam, it is evident that the quantity of heat which exists in 8.3333 pounds of steam at the instant when this steam is formed, is equal to that which is necessary to raise the temperature of 410 pounds of water 180° F., or to elevate the temperature of 73,800 pounds of water one degree of the scale of Fahrenheit.

From this calculation we may conclude that the quantity of heat which exists in 1 pound of steam, at the instant when it is formed, is sufficient to raise the temperature of 1 pound of water 10,063 degrees.

If the capacity of the steam for heat was equal to that of liquid water, it is very certain that the temperature of the vapour *at the instant of its formation* would be that of 10,063° F.

In order to form an idea of this degree of intensity, we may compare it to an intensity of heat which is known.

A piece of iron heated until it becomes red even in daylight has then the temperature of 1000° F.; consequently the temperature of the steam at the instant of its formation would be ten times higher than that of red-hot iron : but as, according to Crawford, the capacity of the steam for heat is greater than that of water in the proportion of 1.55 to 1, the temperature in question will be less than that of 10,063° in the same proportion.　It will therefore be equal to 8750° F.

Here, therefore, is the limit of the intensity of the heat in the midst of the greatest fire, in which pure hydrogen would be employed as a combustible, and in which the fire would be fed by pure oxygen.　This is an intensity which we may approach more or less, but which we can never attain.

As Wedgwood's pyrometer indicates much higher temperatures, it seems demonstrated by the result of this calculation that the scale of this pyrometer is faulty. These doubts have been stated by other chemists.

But in order to decide definitively upon this interesting question, it would be indispensably necessary to know accurately the capacity of steam for heat *at different temperatures;* a thing unknown, and which is difficult to determine.

Upon examining the subject attentively, we shall find, however, reasons for thinking that the capacity of steam for heat ought necessarily to be diminished with the increase of its temperature. The following calculations may serve to elucidate this subject.

In order to determine the highest degree of temperature which can exist in the midst of the greatest fire when pure hydrogen is the only combustible employed and when the fire is fed by atmospheric air, it is necessary to remark that, as oxygen and nitrogen are intimately mixed in the atmosphere, the heat which results from the combustion of hydrogen ought to be immediately divided between the vapour which results from the union of the hydrogen with the oxygen, and the nitrogen which is found necessarily mixed with this vapour.

In order to simplify our inquiry, we shall commence by supposing that all the oxygen which exists in the atmospheric air is employed.

In this case, as it requires 7.3333 pounds of oxygen to be united to 1 pound of hydrogen in order to compose 8.3333 pounds of steam, and as the atmospheric air is composed of 21 pounds of oxygen gas mixed with 79 pounds of nitrogen, the 7.3333 pounds of oxygen which are united to 1 pound of hydrogen in order to form 8.3333 pounds of steam, ought to be found mixed with 27.587 pounds of nitrogen ; consequently the heat developed in the combustion of 1 pound of hydrogen ought to be also divided between 8.3333 pounds of steam and 27.587 pounds of nitrogen ; and this partition ought to take place in the direct ratio of the weights of these two fluids, and of their capacity for heat.

The capacity of the steam being to that of nitrogen

as 1.55 to 0.7036 (according to Crawford), all the heat in question will be divided so that the steam shall retain the part of it represented by the number 9.5832 ($= 8.3333 \times 1.55$); and the nitrogen will receive the other part of it, $= 19.41$ (being the product of 27.587 multiplied by 0.7036).

Now, as the two numbers 9.5832 and 19.41 are both in the proportion of 1 to 2.0254, it is evident that the temperature will be the same which we should have if all the heat in question was equally divided between the steam which would result from the combustion of 3.0254 pounds of hydrogen, i. e. between 25.2113 pounds of steam.

And as we have seen that the heat manifested in the combustion of 1 pound of hydrogen, which is in the 8.3333 pounds of steam which are the products of this combustion, is sufficient for raising the temperature of this steam to that of 8750° F., it is evident that if this same quantity of heat is divided among 25.2113 pounds of steam, the temperature of this steam could not be higher than 2891° F.

This is, therefore, the highest temperature which we ought to find in the midst of a strong fire fed by the atmospheric air in which the combustible burned is pure hydrogen.

As this temperature is much lower than that which we can excite by combustion, even without employing pure hydrogen or pure oxygen, the result of this calculation furnishes a demonstrative proof that the capacity for heat of steam, or rather that of nitrogen, is diminished when its temperature is increased. In all probability, the capacities of both, and generally of all elastic fluids, are diminished when their temperature is increased.

We shall now see what is the highest temperature which it would be possible to attain, by burning charcoal, and by blowing the fire with pure oxygen gas.

According to Crawford, 1 pound of charcoal gives heat sufficient in its combustion to raise the temperature of 57.608 pounds of water 180° F., or to raise the temperature of 9369.44 pounds of water 1 degree.

Now, as 1 pound of charcoal is united to 2.5714 pounds of oxygen in burning, and forms with it 3.5714 pounds of carbonic acid, the heat which is found in 3.5714 pounds of carbonic acid *at the instant of its formation* would be sufficient to raise the temperature of 9369.44 pounds of water 1 degree; consequently the heat which is in 1 pound of this acid at the moment of its formation would be sufficient to raise the temperature of 3643.6 pounds of water 1 degree.

Here we have the *quantity* of heat which exists in the carbonic acid at the instant of its formation. In order to know what is the *intensity* which it would indicate if we could measure it at this moment by means of a thermometer, it would be necessary to know precisely the *specific heat* of the carbonic acid. If, with Crawford, we take it at 1.0459 (that of water being taken $= 1$), we shall have 3811° F. for the measure of the intensity of the heat which exists in the carbonic acid at the moment of its formation, and consequently for the intensity of the greatest fire made with charcoal (without mixture of hydrogen), even in the case where the fire is fed by *pure oxygen.*

It remains to determine the temperature which we might hope to obtain by burning charcoal with *atmospheric air.*

As we have found that the temperature of the 3.5714

pounds of carbonic acid, which are the product of the combustion of 1 pound of charcoal, is that of 3811° F. at the moment of its formation, we have only to ascertain how much the temperature of this acid ought to be diminished by the mixture of the nitrogen which must necessarily be there when the oxygen employed in the combustion of the charcoal is furnished by the atmospheric air.

As, in the atmospheric air, every pound of oxygen is mixed with 3.7619 pounds of nitrogen, the 2.5714 pounds of oxygen employed in the combustion of 1 pound of charcoal ought to be mixed with 9.6735 pounds of nitrogen; consequently all the heat developed in the combustion of 1 pound of charcoal will be found divided between 3.5714 pounds of carbonic acid and 9.6735 pounds of nitrogen.

And as the specific heat of the carbonic acid is to that of nitrogen as 1.0459 to 0.7036, this heat will be divided between these two substances in the proportion of $(3.5714 \times 1.0459 =) 3.7354$ to $(96735 \times 0.7036 =)$ 6.8062, which is in the proportion of 1 to 6.8221 or of 3.5714 to 6.5075; and thence we may conclude that the temperature of the mixture of 3.5714 pounds of carbonic acid and of 9.6735 of nitrogen would be the same as if we had mixed with the 3.5714 pounds of carbonic acid 6.5075 pounds more of this same acid, making together 10.0789 pounds of carbonic acid.

Now, as the heat developed in the combustion of 1 pound of charcoal was sufficient to raise the temperature of the 3.5714 pounds of carbonic acid coming from this combustion to that of 3811° F., this same quantity of heat ought to be sufficient to raise the temperature of 10.0789 pounds of carbonic acid to the temperature of 1350° F.

This is, according to the results of this calculation, the highest temperature which we ought to expect to find amid the strongest charcoal fire fed by atmospheric air.

But we are very certain that the intensity of the heat of the strongest charcoal fire is far superior to the above calculation; consequently we are authorized to conclude that the capacity for heat of the carbonic acid, and that of nitrogen gas, are *much diminished* when these elastic fluids are exposed to a *very high temperature.*

If, in endeavouring to discover the limit of intensity of a charcoal fire, I have supposed the fire to be *very large,* it is not because I suppose that the heat developed in combustion is more intense *at the primitive source* in a large than in a small fire; but as a small fire is always surrounded by bodies relatively very cold, such as the bars of the grate, etc., the products of the combustion (which are always at the instant of their formation at the same temperature) are so rapidly cooled when the fire is small, that the temperature which we may find in such a fire is necessarily lower than that which we find in the midst of a larger fire, where a greater quantity of the same kind of combustible is employed.

When a large charcoal fire is well lighted up in a close stove, constructed with bricks or fire-stones, all the interior surfaces become excessively hot, and the heat accumulates and becomes very intense throughout the whole interior of the stove, so that iron and even stones are melted in it, and flow like liquids; but when the fireplace is small, it is with difficulty that it can be heated so much as to make the sides red-hot; and if the fireplace be very small, a charcoal fire cannot be kept up at all, even with continual blowing. We may truly say that such a fire *dies of cold*, an expression which with as much force as justice describes the event as it really happens.

But if it be the cold communicated by the surrounding bodies which hinders a very small charcoal fire from burning, could we not make it burn by guarding it in a proper manner against the cold?

This is an experiment which I tried six years ago with the greatest success, and which ended in my causing to be made small portable cooking-stoves now in general use in Paris, and elsewhere for aught I know.

By surrounding the body of the stove with two strata of enclosed air, the cooling of the fireplace and the charcoal it contains is hindered; and in this way the charcoal burns perfectly well, and the fire is so well kept up that it obeys a small register, which regulates the quantity of air admitted into the body of the stove.

Some judgment may be formed of the advantages which ought to result from the use of these small portable furnaces in cooking, etc., arising from the saving of time and combustibles, when we are informed that the combustion may be regulated without any difficulty, so as to consume the charge of charcoal in 20 minutes with a brisk heat, or so as to keep up a moderate fire for three hours.

With these portable cooking-stoves it is indispensably necessary to use kettles or saucepans of a particular construction. They ought to be suspended by their rims, in large circles of wrought-iron or copper, the better to keep in the heat. The circle of a saucepan ought to be half an inch more in breadth than the saucepan is in depth.

But to return to the main branch of my subject. If

the present state of our knowledge does not admit of our establishing with a rigorous precision the highest temperature which it is possible to excite by means of the combustion of inflammable bodies, the calculation which I have submitted to the Class may nevertheless serve to guide our conjectures on this interesting subject. They will at all events show what is wanting to enable us duly to appreciate the subject.

SECTION X. — *On the Quantity of Heat developed in the Condensation of the Vapour of Water.*

Having filled the calorimeter and placed it on its stand, a current of vapour was introduced into the worm through a cork placed in the lower aperture of the worm. This cork having been perforated with a hole two lines in diameter, in the direction of its axis, a small cork (two lines in diameter and two in height) was fitted into it, and four other holes about a line in diameter, pierced horizontally through the sides of the large cork at two lines below its upper extremity, and communicating with the hole two lines in diameter in the axis of this cork, afforded a passage to the vapour, to admit of its entering by four small channels horizontally into the worm.

As the apertures of these small channels were higher than the level of the flat bottom of the worm, the water which resulted from the condensation of this vapour did not prevent the vapour from continuing to flow through these passages.

This vapour came from a long-necked matrass containing distilled water, which was put on a portable stove placed in a chimney at some distance from the calorime-

ter; and in order to stop all direct communication of
heat between the stove and the calorimeter, the former
was masked by plates, and the tube which conducted
the vapour to the calorimeter was well covered with
flannel.

The cold water which filled the calorimeter was of a
lower temperature than that of the chamber by 6° F.,
and when the thermometer of the calorimeter announced
an augmentation of temperature by 12° F., an end was
put to the experiment.

The water produced by the condensation of the vapour in the worm was carefully weighed; and from its
quantity, as well as from the heat communicated to the
calorimeter, the heat developed by the vapour in its condensation was determined.

As a small part of the heat communicated to the
calorimeter proceeded from the cooling of the water
condensed in the worm, after the vapour had been
changed into water, an account was kept of this heat.
It was supposed that the water at the moment of condensation was at the temperature of 212° F., being that
of boiling water; and it was determined, by calculation,
what part of the heat communicated to the calorimeter
must have been owing to this boiling water.

In making this calculation, no account was taken of
the difference in the capacity of water for heat which
depends on its temperature; this is but imperfectly
known, and besides, the correction which would have
been the result could not but have been very small.

The following are the details and results of two experiments made on the 21st of January, 1812.

Number of exp.	Temperature of the room.	State of the calorimeter (equal in capacity for heat to 2,781 grammes of water).			Quantity of vapour condensed into water in the worm.	Result.
		Temperature at the beginning of the experiment.	Temperature at the end of the experiment.	Elevation of its temperature.		Quantity of water which may be heated 1° F. with the heat developed in the condensation of 1 lb. of vapour.
	Degr's.	Degrees.	Degrees.	Degrees.	Grammes.	lbs.
1	61	55	$67\frac{1}{2}$	$12\frac{1}{4}$	29.61	1029.3
2	$62\frac{1}{4}$	$57\frac{1}{4}$	$67\frac{1}{2}$	$10\frac{1}{2}$	24.4	1052.3
Mean result						1040.8

By expressing the mean result of these two experiments in the way employed by Mr. Watt and others, I shall say that 1040 degrees of heat (Fahrenheit) are liberated in the condensation of steam, and that consequently this very quantity of heat is employed and rendered latent when the water, already at the temperature of boiling water, is changed into steam.

The duration of each of these two experiments was from ten to eleven minutes, and I had boiled the water some time in the matrass (to drive out the air which it contained) before I directed the steam from it into the worm of the calorimeter.

As the results of these experiments have been very uniform, and as they agree very well with the later experiments made by Mr. Watt with a view to determine the same question, I have not thought it necessary to repeat them.

I have, besides, been very much occupied with the following branch of my inquiries.

SECTION XI. — *Of the Quantity of Heat developed in the Condensation of the Vapour of Alcohol.*

As chemists are not agreed as to the state of the elements of the water which exist in alcohol, I thought that, by determining with precision the quantity of heat which is developed, we should be better able to form conjectures as to the state of the water, if it be at all times found in this inflammable liquid.

The results of the experiments which I made with alcohol are less regular than those of the experiments made with water, as might have been expected ; but they have nevertheless been sufficiently uniform to establish a fact which will be regarded, without doubt, as very curious and important.

As the vapour which is extracted from spirit of wine when boiled, varies a little with the intensity of the fire used in boiling it, I took care to note the time which was taken in every experiment, in order to be able to judge, by comparing the quantity of vapour condensed with the time employed to form it, of the intensity of the heat employed to boil the liquid.

In the following table we shall see the details and results of five experiments made on the same day (January 21, 1812) with alcohol of different degrees of strength. The capacity of the calorimeter was always equal to that of 2781 grammes of water, and the thermometer employed was that of Fahrenheit.

Number of experiments.	Specific gravity of the alcohol employed.	Time employed in the experiment.	Temperature of the apartment.	State of the calorimeter.			Quantity of alcohol condensed in the calorimeter.	Result. Quantity of water which may be heated 1° of F. with the heat developed in the condensation of 1 lb. of vapour.
				Temperature at the beginning.	Temperature at the end.	Elevation of its temperature.		
		Min.					Gram.	
1	85342	7	61°	$54\frac{1}{4}°$	$68\frac{1}{2}°$	$14\frac{1}{4}°$	69.86	499.54
2	85342	5	61	56	$66\frac{1}{4}$	$10\frac{1}{4}$	52.21	476.83
3	84714	8	$60\frac{1}{2}$	$55\frac{1}{2}$	$65\frac{1}{2}$	10	48.82	500.03
4	81763	$4\frac{1}{4}$	61	56	$66\frac{1}{4}$	$10\frac{1}{2}$	56.61	479.92
5	85342	$6\frac{1}{2}$	64	57	$71\frac{1}{2}$	$14\frac{1}{2}$	71.31	499.65

On determining, by calculation, the quantity of water which may be heated *one degree*, by the heat developed in the combustion of one pound of this vapour, I took care to keep an account of the difference between the capacity of water for heat and that of alcohol, when I determined how much heat should have been communicated to the calorimeter by the alcohol, and produced by the condensation of the steam, by being cooled in the worm.

In order to prove the state of the elements of the water which exist in the steam of alcohol, it must be shown how much water these elements ought to form.

We shall select the experiment which was made with alcohol of the specific gravity of 81,763, and which contained the least water. The quantity of steam condensed in this experiment was 56.61 grammes.

In 100 parts of this alcohol there were

> 91.79 parts of pure alcohol of Lowitz, and
> 8.21 parts of water.

Consequently there were in the 56.61 grammes of alcohol condensed in the calorimeter,

> 51.962 grammes of alcohol of Lowitz, and
> 4.648 grammes of water.

Now, as M. de Saussure has shown that there are 47 parts of water in 100 parts of alcohol of Lowitz, there must have been 24.422 grammes of water in the 51.962 grammes of alcohol of Lowitz, which were condensed in the calorimeter.

If to this quantity of water (= 24.422 grammes) we add the 4.648 grammes which were found mixed with 51.962 grammes of alcohol of Lowitz, in order to compose the 56.61 grammes of alcohol employed in the experiment, we shall have 29.07 grammes of water which ought to have existed ready formed either in the common state of water or in some other state, in the 56.61 grammes of alcohol condensed in the calorimeter.

But the condensation of 29.07 grammes of steam into liquid water ought to have of themselves furnished more heat than we had, in the experiment in question, in the condensation of these 29.07 grammes of elements of water with 27.57 grammes of carbon and hydrogen, which concur with these elements in forming the steam of the alcohol which was condensed.

If we apply a similar calculation to the results of the experiments made with alcohol which contained more water, the result of the inquiry will be still more striking.

In the experiment No. 5 the alcohol employed was of the specific gravity of 85,342 ; consequently 100 parts of this alcohol were composed of

> 77.88 parts of alcohol of Lowitz, and
> 22.12 water.

And in the experiment 71.31 grammes of vapour of alcohol were condensed.

There were, therefore, in these 71.31 grammes of condensed alcohol,

> 55.688 grammes of alcohol of Lowitz, and
> 15.622 grammes of water.

In the 55.688 grammes of alcohol of Lowitz there were 26.102 grammes of water, according to the analysis of M. de Saussure ; and this last quantity of water ($= 26.012$ grammes), added to the quantity found above, viz. 15.622 grammes, makes 41.727 grammes of water which ought to have existed, either as steam or otherwise, in the 71.31 grammes of alcoholic vapour condensed in the calorimeter, in the experiment in question.

In order to simplify our calculation, and to render our comparisons more striking, we shall show how much pure water, in vapour, ought to have been sufficient to furnish, in its condensation, the same quantity of heat which was furnished by the condensation of 71.31 grammes of alcoholic vapour, in the experiment in question.

In this experiment the temperature of the calorimeter was raised to $14\frac{1}{2}°$ of Fahrenheit.

In the second experiment, made with the steam of pure water, the temperature of the same calorimeter was raised $10\frac{1}{2}°$ of Fahrenheit, with the heat developed in the condensation of 24.4 grammes of this vapour.

Consequently the temperature of the calorimeter must have been elevated to $14\frac{1}{2}°$ of Fahrenheit, with the heat which must have been developed in the condensation of 33.695 grammes of steam from pure water.

Now, as the hydrogen and the oxygen forming the elements of 41.727 grammes of water, which are found to form constituent parts of the 71.31 grammes of vapour of alcohol condensed in the experiment in question, only furnished in their condensation the same quantity of heat as 33.695 grammes of steam of pure water should have furnished, it is clearly proved, in my opinion, that these elements are not so united as to form water, so long as they concur in the formation of alcohol.

I have discovered that the vapour of sulphuric ether furnishes about one half less of heat in its condensation than that of alcohol, and consequently one fourth only of what is furnished by the steam of water of equal weight; but, having been interrupted by an accident in

the course of my experiments with ether, I am desirous
of finishing them before I publish the results.

EXPERIMENTS AND OBSERVATIONS

ON THE

COOLING OF LIQUIDS IN VESSELS OF PORCELAIN, GILDED AND NOT GILDED.

NOTHING affords more entertainment than to compare the processes of the common arts of life and the ordinary habits of the people in their household operations with the principles of the physical and mathematical sciences. This comparison often presents very curious points of resemblance, and leads sometimes to very important improvements.

In all countries where the daily use of tea has become common among the rich, teapots of silver are preferred to those of porcelain or earthenware, and the reason given for this preference is that the beverage when prepared in the former is of a better quality than when prepared in the latter. I was, for a long time, of the opinion that this idea was owing simply to prejudice, and without foundation; but, having discovered some years since that metallic vessels, when clean and bright on the outside, possess the property of causing warm liquids which are put into them to retain their heat for a very long time, I began to see that the preference in question might be the legitimate result of long experience, as is almost always the case with those preferences which in the end are universally adopted.

In order to throw light on this subject, which had several points of interest for me, I made the following experiment. I procured (from M. Nast, a celebrated porcelain manufacturer, of Paris) two vessels of porcelain, of the same shape and of the same dimensions, the one white, the other completely covered on the outside with gilding; into these vessels I put equal quantities (250 grammes, or a quarter of a litre) of warm water, and then allowed them to cool gradually in a large room free from currents of air, having placed them three feet apart on a table in the middle of the room.

Each of the vessels was closed with a cork stopper, and by means of a mercurial thermometer with a cylindrical bulb, fixed in the axis of the vessel in such a way that while the thermometer was inserted in the cork the scale remained on the outside of the vessel, I noted very conveniently the progress of the cooling without touching the vessel, and without even approaching it sufficiently near for the heat of my body to interfere sensibly with the operation of cooling.

The result of this experiment was as I had expected. The gilded vessel cooled much more slowly than the plain one. Starting at the same time with both vessels at the same temperature, if it took *half an hour* for the plain vessel to cool down through a certain number of degrees, *three quarters of an hour* were necessary for the gilded one to cool down to the same point.

This comparative experiment was repeated several times, and invariably with the same result; the gilded vessel always cooled more slowly than the plain one in about the proportion of 3 to 2.

The advantage that can be gained from this remarkable property, possessed by metallic surfaces, of resisting the cooling (or heating) action of surrounding bodies, is too evident to need much explanation. Since, in household economy, use is often made of porcelain vessels for holding warm liquids, which it is desired to keep warm for a long time, — as, for example, tea, coffee, etc., — in all such cases it would be of advantage to use vessels gilded on the outside; or, if gilding be found too expensive, it is possible to use, and with equal advantage as regards retaining the heat, vessels which are silvered or covered with a layer, no matter how thin, of any other metal not liable to be readily oxidized in the air.

As to gilding the vessels on the inside, it would be to no purpose, for it would add nothing to the effect in question, as I have learned from the results of several experiments. This, however, applies only to simple vessels; for in case a double vessel were employed in order to retain more effectually the heat of any substance, the outside vessel must be gilded on the inside as well as on the outside; in no case is it necessary for the inner vessel to be gilded on the inside.

If it is a question of preserving the low temperature of liquids or other cold substances, such as ice-creams, etc., in this case, also, vessels having externally a polished metallic surface should be used; for a surface of this description throws off by reflection a large portion of the calorific rays which reach it from surrounding objects, and consequently the vessel grows warm very slowly.

Everybody knows how much time it takes to bring water to boiling in a silver coffee-pot which is clean and bright on the outside, especially before an open fire, or on glowing coals which burn without smoke. It is, however, very easy to hasten materially the heating of the liquid in this case; all that is necessary is to begin by blackening the outside of the coffee-pot over the flame of a candle or of a lamp, or to destroy or conceal in some other way the metallic lustre.

All the facts which I have just detailed are easily explained, and, to my mind, satisfactorily, by the theory of heat developed in the various memoirs on this subject which I have had the honour of presenting to this Assembly at different times.

If heat is nothing but a vibratory motion of the particles of a body, — a motion which always exists in all bodies, but which has greater or less rapidity or intensity according to the temperature of those bodies, — and if a body which is warmer than those which surround it is cooled on being exposed to their influence, not because it has transferred to them something material, to which the name of *caloric* has been given, but because of the effect of the action upon it of those bodies by means of their frigorific rays, that is to say, by the undulations caused in the surrounding mass of the fluid ether, — under these circumstances it is evident that the nature of the exterior surface of the warm body, which renders it more or less capable of reflecting the rays or undulations which reach it from surrounding objects colder than itself, ought to influence to a considerable extent the rapidity of the cooling process.

Now, we know that, of all the substances with which we are acquainted, the metals are the most impervious to light, and, at the same time, and perhaps as a necessary consequence, have for it the greatest reflecting power; moreover, the results of a large number of experiments have shown that they also possess in an

eminent degree the power of reflecting the invisible rays
or undulations which all objects in nature send off continually and in all directions from their surfaces in consequence of that peculiar motion of their particles which
constitutes their temperature.

Hence it appears that vessels having a metallic surface on the outside must be well adapted for preserving
the temperature of the substances which they contain,
whether that temperature be high or low, warm or cold.

I am far from maintaining that the sort of material
of which the vessel is made, and the thickness of its
walls, are matters entirely indifferent, provided that the
outer surface be covered with a thin metallic layer which
is clean and bright. I am aware that neither heat nor cold
can be communicated or propagated through the walls
of a vessel, or of any other solid body, instantaneously,
and that this communication takes place more quickly
in some substances than in others, more quickly
through a thin wall than through a thick wall of the
same material; and it is evident that this difference
must necessarily exert an influence on the rapidity of
the change of temperature of the vessel and of the
liquid it contains, whatever be the nature of the external surface of the vessel.

For example, as porcelain is a worse conductor of
heat than gold or silver, a vessel of given form and
dimensions, made of porcelain and well gilded on the
outside, if filled with warm water, would cool rather
more slowly in the air, or even in a Torricellian vacuum,
than another vessel of the same dimensions made of
gold or silver, and filled with water of the same temperature; but if the vessels were exposed at the same
time to a strong and very cold current of air, or were

plunged into cold water, the difference in the rate of cooling would be much greater.

Hence we may conclude that teapots and coffee-pots made of porcelain or earthenware and well gilded on the outside would be not only as good, but even better for common use than teapots and coffee-pots made of silver.

If equal quantities of warm water are placed in two porcelain vessels of the same form and dimensions, and with walls of the same thickness, the one gilded on the outside, the other plain, and these vessels are allowed at the same time to cool in still air, the gilded vessel is found to cool more slowly than the plain one in the proportion of 3 to 2, as has already been remarked; but if, instead of allowing the vessels to cool in air which is undisturbed, they are exposed to the action of a strong and cold current of air, the difference in the rapidity of cooling will be much less, as 6 to 5, for example; and if the current of air is very strong, and at the same time very cold, this difference will be still smaller.

If, instead of exposing the vessels in the air, they are plunged into cold water, the difference in the rapidity with which they cool will be reduced to almost nothing.

In the cases last mentioned we can say that the exterior surfaces of both vessels, although of different natures, yet, on being exposed to so great a degree of cold, are cooled to such an extent as to be in a condition to transmit the heat coming from the interior of the vessel as fast as it can reach them after making its way through the thickness of the walls, which offer all the while a certain amount of resistance to its passage.

To use another form of expression which I regard as more exact, and consequently more suitable, especially before this illustrious Assembly, it might be said that in the case in question the exterior surfaces (the one of white porcelain, the other of gold) of the two vessels being intimately exposed to the violent action of a rapid succession of the very cold particles of the surrounding fluid, became, both of them, cooled to such an extent that they were reduced to about the same temperature in spite of the continual heating action of the walls of the vessels in contact with them on the opposite side; and that, as a consequence, since these surfaces exercised on the walls of the vessels which they covered cooling actions which were sensibly equal, the two vessels were of necessity cooled with the same rapidity.

I will conclude this memoir with some observations which may serve to throw light on a point in the theory of heat which is of very great importance.

The great rapidity with which heat is communicated from one body to another, when two bodies of different temperatures are in contact, compared with the slowness of communication which takes place when the bodies are separated, however little, one from the other, has had a considerable tendency to give authority to the opinion quite generally adopted by chemists, that there are two modes by which heat can be transmitted from one body to another; that is, at a distance, by radiant *caloric*, and, on contact, by an actual transfusion of the same substance. If, however, attention be paid to a fact which no one up to this time has called into question, the phenomenon under consideration can, as it seems to me, be explained in a perfectly clear and satisfactory manner, without having recourse to such an extraordinary supposition as that there are two different modes by which heat is communicated.

It is generally recognized (I might say that it is proved) that the intensity of the action of calorific or frigorific rays is inversely proportional to the squares of the distances from the body from which they proceed; now, if this relation is constant, since the effect produced by these rays in a given time must necessarily be in proportion to the intensity of their action, it is evident that at the point of contact (if, indeed, there can be an actual contact between two bodies) the rapidity of the calorific action between two particles of different temperatures, and which are in contact, must be infinite.

But the time necessary to establish an equality of temperature throughout the entire masses of two bodies in contact, which are of sensible size and of different temperatures, will depend not only on the size of the bodies and on the extent of the surfaces by which they are in contact, but also, and above all, on the greater or less rapidity with which is propagated among the particles of the bodies that peculiar motion of those particles which constitutes their temperature.

I will observe here, in passing, that if in the communication of heat between two bodies in contact, it were only a question of the transfer from one to the other of the excess of a fluid as rare and as elastic as *caloric* is supposed to be, one would expect, it seems to me, an action as instantaneous as the discharge of a Leyden jar.

It cannot be said, in objection, that the warm body does not offer avenues enough for the escape of the caloric, for it is proved that the pores of all bodies, even of the most solid, are so considerable in comparison with the space occupied by the particles of those bodies, that a fluid as rare as caloric is supposed to be would be able to move about therein with great freedom. Besides, it often happens that a very large surface of the warm body is in contact with the cold body ; but, even in this case, there is nothing in the action taking place in the communication of the heat which resembles in any way the sudden explosion which takes place on the restoration of the equilibrium among the particles of an elastic fluid ; on the contrary, the slow and measured progress of this communication, as well as all the other phenomena that it presents, denote rather a gradual operation, like that which takes place when the motion of a body is accelerated or retarded.

The following experiment may serve to explain and confirm this important truth. If a ball of iron, three or four inches in diameter, fastened to a long handle of the same metal, be heated strongly in a forge until it is of a whitish-red heat, and then taken from the fire and plunged suddenly into cold water, the communication of the heat to the water will be so far from being instantaneous that a considerable time will pass before the ball ceases to be red and luminous at its surface ; and even after the surface of the ball has cooled so far as no longer to give off visible light, the interior will still be incandescent. It is easy to establish this last fact ; for if at this point the ball be taken from the water and held in the air for a few moments, the surface of the ball will again become red and luminous.

I confess frankly that I have never been able to reconcile these phenomena with that hypothesis which supposes that the increase of temperature of a body is due to the accumulation within it of a very rare and extremely mobile substance, especially when I have considered the great ease with which such a fluid ought to pass through the pores of all known bodies.

But whatever be the explanation given to the phenomena which present themselves in the heating and cooling of bodies, it is certain that every new fact relating to these actions which is discovered must tend towards perfecting the science of heat as well as the arts which depend upon it.

I flatter myself that this Assembly will find the results of the experiments which I have detailed sufficiently curious and interesting to deserve its attention.

ON THE CAPACITY FOR HEAT

CALORIFIC POWER OF VARIOUS LIQUIDS.

THIS subject is of rather an obscure nature, and it has been so little examined, that it will be useful to begin by elucidating it as well as I can.

Let us suppose two cylindrical vessels, with very thick sides, made of lead or any other metal, and perfectly equal in size, each being capable of containing a pint.

These two vessels being at the freezing-point, we shall pour into the one a pound of water at the temperature of 96° F. (= 28½° R.), being that of the blood, and into the other a pound of olive oil at the same temperature.

Each of these liquids will heat the cold vessel in which it is placed, the vessel in its turn will cool the liquid, and both the liquid and the vessel will latterly be of the same temperature.

If water and oil of olives had the same calorific power, a pound of water at the temperature of 96° would heat its cold vessel precisely as much and not more than a pound of oil would heat its vessel, the two vessels being of the same weight and at the same temperature at the commencement of the experiment.

But experience shows that water heats its vessel much more than oil does; consequently the calorific power of water is greater than the calorific power of oil of olives, when the *quantities* of these two liquids are estimated by their weight; and, if we designate the calorific power of water by 1, the calorific power of oil of olives will be expressed by a fraction under 1.

The power with which any given body, solid or liquid, being at a given temperature, resists the calorific or frigorific action of bodies warmer or colder than itself, is in proportion to its calorific power; and the greater is this power, the longer it resists these actions of surrounding bodies.

If, under equal surfaces, a pound of water and a pound of oil of olives, both at the same temperature (96° F.), are placed at the same time in a place where the temperature is lower (that of freezing, for instance), the oil of olives will be cooled much more rapidly than the water.

If it be in a *warm* place that the two liquids are exposed, the oil of olives will still have its temperature most rapidly changed; it will be more heated than the water.

In two cylindrical glass vessels, of equal size and very thin, place equal quantities of water, and at the same temperature (96° F.).

A piece of lead weighing a pound, and a piece of copper of the same weight, having been cooled in a mixture of pounded ice and water, remove them from this cold mixture and plunge each of them suddenly into one of the vessels of water.

The two masses of water will be cooled, but that which contains the copper most, for the calorific power of copper is greater than the calorific power of lead.

We may also say that the *frigorific power* of copper is greater than the *frigorific power* of lead, and, in the case in question, the expression, perhaps, will be most suitable.

It is always the same power; it is that by means of which any body resists the action of surrounding bodies, and which tends to change its temperature either by increase or diminution.

Much obscurity has been introduced into the science by vague ideas being attached to the words *hot* and *cold;* but it will not suit my purpose to enlarge upon this subject at present. I have already delivered my opinion in a former paper.

The little heat which I discovered in the condensation of alcohol having induced me to think that the specific heat of this liquid had not been accurately determined, and wishing to know it precisely, in order to enable me to finish the calculations which were necessary for elucidating the results of some of my experiments, I constructed a small and very simple apparatus, by the help of which I could easily, and as I presume accurately, determine it.

This apparatus consists of a small bottle of a particular form, constructed of thin leaves of red copper, intended to contain the liquid which is to be the subject of the experiment; and a small cylindrical vase, also constructed of thin pieces of red copper, in which I place water at a certain temperature. Into this water I plunge the bottle of copper containing the liquid which is the subject of the experiment; this liquid being of a different temperature from that of the water in the outer vase.

As the capacity of the vase for heat, as well as that of the bottle, is known, I determine, by a very simple calculation, the capacity for heat of the liquid contained in the bottle. This calculation, which is well known, is

founded in the changes which take place in the temperature of liquids, in the vase and in the bottle, by taking
a uniform temperature, when the bottle is immersed in
the water contained in the vessel.

In order that this equality of temperature may be
speedily brought about, the form of the bottle is such
that it has a very great surface relative to its small
capacity, and in order to manage it without touching it,
its neck, which is small, is closed by a long cork, which
serves as a handle.

In order to diminish as much as possible the effect
of the atmosphere and of surrounding bodies upon the
apparatus, while the experiment is going on, the quantity of water in the vessel is regulated so as to keep the
bottle wholly submerged in the liquid, and even the
upper end of the neck covered, when the bottle is immersed. The vessel which contains this water is placed
and suspended by a ring of cork in another vessel larger
and higher, and the interval between the two is filled
with eider-down.

The form of the bottle is such that its horizontal
section presents the figure of a rectangular cross. Some
idea may be conceived of its form and dimensions, if
we suppose a square piece of stick, each facet of which
is four lines broad by four inches three lines in length,
upon the four faces of which we have fixed four sticks of
the same length (i. e. four inches three lines), but each
of them being four lines thick by eight broad.

The four sticks last described will exhibit the figure
of the bottle; for the square piece of stick will be concealed by them from our view.

The neck of the bottle is in the prolongation of its
axis; it is four lines diameter by four high; it ought to

be circular; the cork should be an inch long, and the bottle weigh 76.07 grammes without its cork.

The cylindrical vase which contains the water is two inches diameter, and four inches nine lines high, and it weighs 74.65 grammes.

The exterior vessel, in which the latter is suspended by the cork ring, is five inches three lines high, and three inches diameter, so that the sides and bottom are everywhere separated by an interval of six lines; this interval is filled with eider-down, as already mentioned. ·

To prevent the water from touching the eider-down, the cork ring is covered with a thin coating of mastic.

In order to ascertain the temperature of the bottle, and of the liquid which it contains, without being obliged to plunge a thermometer into the bottle, which would in this case be inconvenient, I employed a very simple method.

I placed a large bucket filled with water in a room with a northern aspect. I allowed it to assume the temperature of the room, taking care to shut the door and windows day and night. I placed the small bottle on a stand in this bucket, keeping the upper part of the cork only out of the water. As the bottle is small and has a large surface, it speedily acquires the temperature of the bucket of water; but, in order to be well convinced that the bottle and the liquid which it contains have acquired the temperature in question, I leave the bottle a considerable time in the bucket, frequently half an hour and sometimes more.

In giving a detailed account of an experiment made with this apparatus, I shall have an opportunity of giving clear and precise ideas of the different parts of my apparatus, and of the particular objects which they are intended to attain.

Having found by various preliminary experiments made with water that the capacity for heat of the cylindrical vessel with that of the thermometer employed to determine the temperature of the water which it contained, was equal to that of 24.3 grammes of water, and that the specific heat of the bottle of copper was equal to that of 8.36 grammes of water, I made the following experiment with purified linseed-oil.

I put into the cylindrical vessel 180 grammes of water; the temperature of the room was $59\frac{1}{2}°$ F. I filled the copper bottle with the above oil, and corked it. I cooled it in a bucket of water at the temperature of $44\frac{1}{4}°$ F. The oil in the bottle weighed 82.55 grammes.

The bottle, having had time to acquire the temperature of $44\frac{1}{4}°$ F., was withdrawn from the bucket, and placed in a cylindrical vessel of tinned iron, of about four inches diameter and six high, filled to the height of four inches and a half with water at the temperature of $44\frac{1}{4}°$ F.

The bottle, being submerged in this vessel of cold water, was carried into the room where I had placed the small vessel of copper belonging to the apparatus; it was then taken out of the cold water, and plunged into the water contained in the small cylindrical vessel of copper, which contained 180 grammes of water at the temperature of $59\frac{1}{2}°$ F.

A thermometer having a cylindrical reservoir four inches long, which was placed in this vessel beside the copper bottle, soon fell, and in three or four minutes it marked $56\frac{1}{2}°$ of F., where it remained a long time stationary, and afterwards began to ascend slowly.

The capacities for heat of the warm bodies which

were cooled in this experiment were equal to that of 204.3 grammes of water; viz., —

That of the water employed . . . . 180 grammes.
That of the vases and thermometer . . . 24.3

Total . . . 204.3

The capacity for heat of the bottle containing the oil was equal to that of . . . . . . 8.36 grammes of water.

And to this we must add the cold water adhering to the bottle, when it came out of the cold water, and was plunged into the water contained in the copper vessel. I found by a particular experiment that this quantity of water was 1.04

Total 9.40

Now, as the temperature of the warm water in the cylindrical vase of copper was that of $59\frac{1}{2}°$ before the mixture, and $56\frac{3}{4}°$ after the communication of the heat had been obtained, it is evident that this water was cooled $2\frac{3}{4}°$. But if we multiply the number of grammes of water which the specific heat of this water represents, and that of the vessel ($= 204.3$ grammes), by the number of degrees which it has been cooled ($2\frac{3}{4}$) we shall have a product which will express the number of grammes of water which would have been cooled $1°$ F. by a loss of heat equal to that which the vessel and its contents supported in this experiment. It is $204.3 \times 2.75 = 561.84$ grammes.

We shall now see what part of this heat was communicated to the bottle and to the small portion of cold water attached to it, and what part to the oil contained in the bottle.

As the temperature of the bottle and its contents was $44\frac{1}{4}°$ F. before the mixture, and $65\frac{1}{2}°$ afterwards, it is evident that the bottle had acquired $12\frac{1}{4}°$ of heat; consequently, if we multiply 9.4 (the number which expresses the sum of the capacities for heat of the bottle, and of the cold water adhering to it) by $12\frac{1}{4}$ we shall have a product which will express the number of grammes of water which would have been heated one degree by the heat communicated during the experiment to the bottle, and to the small portion of water which adhered to it.

It is $9.4 \times 12.25 = 111.15$ grammes.

If from the heat lost by the vessel and the warm water, which we have found equal to that which is necessary for raising the temperature of 561.84 grammes of water one degree of F., . . .   561.84 grammes
we take the quantities which the bottle and the water adhering to the bottle have received . .   115.15

we shall have . . . .   446.69 grammes

of water heated one degree, expressing the quantity of heat employed for raising to $12\frac{1}{4}°$ F. the temperature of the 82.55 grammes of linseed oil which were put into the bottle.

On dividing this number (446.69) by $12\frac{1}{4}$, we shall see how many grammes of water would have been heated one degree by the quantity of heat in question.

It is therefore $\dfrac{446.69}{12.25} = 36.464$ grammes of water.

By the results of this calculation we find that the same quantity of heat which is necessary to raise the temperature of 36.464 grammes of water $12\frac{1}{4}$ degrees of Fahrenheit's thermometer is sufficient to raise the temperature of 82.55 grammes of oil the same number of degrees.

Consequently the capacity of water for heat is greater than that of oil of linseed in the proportion of 82.55 to

36.464 ; and if we express the capacity of the water by *unity*, as is usually done, the capacity of the above oil ought to be expressed by the fraction 0.44172.

These details must no doubt appear superfluous to those who are versed in the higher branches of knowledge, and who are accustomed to express the most complete relations by algebraical signs ; but it must be recollected that the subject of which I treat is familiar to few, and that it is necessary to explain with rigorous accuracy the principles upon which the method employed is founded, as well as the manner of using the apparatus which I recommend.

On repeating twice the experiment made with pure linseed oil I had as a result in one of these experiments a capacity for heat equal
to 0 44411
and in the other equal to 0.47193
If to these two results we add that of the first experiment, equal to 0.44172
___
we shall have as a mean result 0.45192

The following are the results of some experiments made with other liquids.   Olive oil furnished : —

	Specific heat of olive oil.
1st experiment	0.45944
2d experiment	0.43422
3d experiment	0.42183
Mean result . . .	0.43849

Three experiments made with naphtha gave the following results : —

	Specific heat of naphtha.
1st experiment	0.43408
2d experiment	0.39234
3d experiment	0.41905
Mean result . . .	0.41519

Three experiments made with spirits of turpentine gave the following results : —

Specific heat of turpentine.

1st experiment	0.29322
2d experiment	0.37031
3d experiment	0.34216
Mean result . . .	0.33856

Two experiments with alcohol, of the specific gravity of 817,624, gave the following results : —

Specific heat of alcohol.

1st experiment	0.54924
2d experiment	0.55063
Mean result . . .	0.54993

Two experiments with spirit of wine, of the specific gravity of 85,324, gave the following results : —

Specific heat of spirit of wine.

1st experiment	0.57840
2d experiment	0.58317
Mean result .	0.58078

Two experiments with sulphuric ether, of the specific gravity of 72,880, gave as results : —

Specific heat of sulphuric ether.

1st experiment	0.53711
2d experiment	0.54768
Mean result . . .	0.54329

I was at first much surprised to find so great a capacity for heat in sulphuric ether, but my astonishment was diminished when I recollected that this liquid can unite with alcohol in all proportions without exhibiting any symptoms of a chemical action ; for this reason, therefore, we ought to expect to find the same capacity for heat in both liquids.

OBSERVATIONS

RELATIVE TO

THE MEANS OF INCREASING THE QUANTITIES OF HEAT OBTAINED IN THE COMBUSTION OF FUEL.

I T is a fact which has been long known, that clays, and several other incombustible substances, when mixed with sea coal in certain proportions, cause the coal to give out more heat in its combustion than it can be made to produce when it is burned pure or unmixed; but the cause of this increase of heat does not appear to have been yet investigated with that attention which so extraordinary and important a circumstance seems to demand.

Daily experience teaches us that all bodies — those which are incombustible, as well as those which are combustible and actually burning — throw off in all directions heat, or rather calorific (heat-making) rays, which generate heat wherever they are stopped or absorbed; but common observation was hardly sufficient to show any perceptible difference between the quantities of calorific rays thrown off by different bodies, when heated to the same temperature or exposed in the same fire, although the quantities so thrown off might be, and probably are, very different.

It has lately been ascertained, that, when the sides and back of an open chimney fireplace in which coals are burned are composed of firebricks, and heated red-hot, they throw off into the room incomparably more heat than all the coals that could possibly be put into the grate, even supposing them to burn with the greatest possible degree of brightness. Hence it appears that a red-hot burning coal does not send off near so many calorific rays as a piece of red-hot brick or stone of the same form and dimensions; and this interesting discovery will enable us to make very important improvements in the construction of our fireplaces, and also in the management of our fires.

The fuel, instead of being employed to heat the room *directly* or by the direct rays from the fire, should be so disposed or placed as *to heat the back and sides of the grate*, which must always be constructed of firebrick or firestone, and *never of iron or of any other metal.* Few coals, therefore, when properly placed, make a much better fire than a larger quantity, and shallow grates, when they are constructed of proper materials, throw more heat into a room, and with a much less consumption of fuel, than deep grates; for a large mass of coals in the grate arrests the rays which proceed from the back and sides of the grate, and prevents their coming into the room; or, as fires are generally managed, it prevents the back and sides of the grate from ever being sufficiently heated to assist much in heating the room, even though they be constructed of good materials and large quantities of coals be consumed in them.

It is possible, however, by a simple contrivance, to make a good and an economical fire in almost any grate, though it would always be advisable to construct fireplaces on good principles, or to improve them by judicious alterations, rather than to depend on the use of additional inventions for correcting their defects.

To make a good fire in a bad grate, the bottom of the grate must be first covered with a single layer of balls, made of good firebricks or artificial firestone, well burned, each ball being perfectly globular, and about $2\frac{1}{2}$ or $2\frac{3}{4}$ inches in diameter. On this layer of balls the fire is to be kindled, and, in filling the grate, more balls are to be added with the coals that are laid on; care must, however, be taken in this operation to mix the coals and the balls well together, otherwise, if a number of the balls should get together in a heap, they will cool, not being kept red-hot by the combustion of the surrounding fuel, and the fire will appear dull in that part; but if no more than a due proportion of the balls are used, and if they are properly mixed with the coals, they will all, except it be those perhaps at the bottom of the grate, become red-hot, and the fire will not only be very beautiful, but it will send off a vast quantity of radiant heat into the room, and will continue to give out heat for a great length of time. It is the opinion of several persons who have for a considerable time practised this method of making their fires, that more than one third of the fuel usually consumed may be saved by this simple contrivance. It is very probable that, with careful and judicious management, the saving would amount to one half, or fifty per cent.

As these balls, made in moulds and burnt in a kiln, would cost very little, and as a set of them would last a long time, — probably several years, — the saving of expense in heating rooms by chimney fires with bad grates, in this way, is obvious; but still, it should be remembered that a saving quite as great may be made by altering the grate, and making it a good fireplace.

In using these balls, care must be taken to prevent

their accumulating at the bottom of the grate. As the
coals go on to consume, the balls mixed with them will
naturally settle down towards the bottom of the grate,
and the tongs must be used occasionally to lift them up;
and as the fire grows low, it will be proper to remove
a part of them, and not to replace them in the grate
till more coals are introduced. A little experience will
show how a fire made in this manner can be managed
to the greatest advantage and with the least trouble.

Balls made of pieces of any kind of well-burned hard
brick, though not equally durable with firebrick, will
answer very well, provided they be made perfectly
round; but if they are not quite globular, their flat sides
will get together, and by obstructing the free passage of
the air amongst them and amongst the coals will prevent the fire from burning clear and bright.

The best composition for making these balls, when
they are formed in moulds and afterwards dried and
burned in a kiln, is pounded crucibles mixed up with
moistened Sturbridge clay; but good balls may be made
with any very hard burned common bricks, reduced to a
coarse powder, and mixed with Sturbridge clay, or even
with common clay. The balls should always be made
so large as not to pass through between the front bars
of a grate.

These balls have one advantage, which is peculiar to
them, and which might perhaps recommend the use of
them to the curious, even in fireplaces constructed on
the best principles: they cause the cinders to be consumed almost entirely; and even the very ashes may
be burned, or made to disappear, if care be taken to
throw them repeatedly upon the fire when it burns with
an intense heat. It is not difficult to account for this

effect in a satisfactory manner, and in accounting for it we shall explain a circumstance on which it is probable that the great increase of the heat of an open fire where these balls are used may in some measure depend. The small particles of coal and of cinder which in a common fire fall through the bottom of the grate and escape combustion, when these balls are used can hardly fail to fall and lodge on some of them; and as they are intensely hot, these small bodies which alight upon them in their fall are soon heated red-hot, and disposed to take fire and burn; and as fresh air from below the grate is continually making its way upwards amongst the balls, every circumstance is highly favourable to the rapid and complete combustion of these small inflammable bodies. But if these small pieces of coal and cinder should, in their fall, happen to alight upon the metallic bars which form the bottom of the grate, as these bars are conductors of heat, and, on account of that circumstance, as well as of their situation, — *below* the fire, — never can be made very hot, any small particle of fuel that happens to come into contact with them not only cannot take fire, but would cease to burn, should it arrive in a state of actual combustion.

These facts are very important, and well deserving of the attention of those who may derive advantage from the improvement of fireplaces and the economy of fuel.

There are some circumstances which strongly indicate that an admixture of incombustible bodies with fuel, and especially with coal, causes an increase of the heat, even when the fuel is burned in a closed fireplace. No fireplace can well be contrived more completely closed than those of the iron stoves in common use in the Netherlands; but in these stoves, which are heated by coal fires, a large proportion of wet clay is always coarsely mixed with the coals before they are introduced into the fireplace. If this practice had not been found to be useful, it would certainly never have obtained generally, nor would it have been continued, as it has been, for more than two hundred years.

The combination of different substances, combustible and incombustible, to form, artificially, various kinds of cheap and pleasant fuel, particularly adapted for the different processes in which the fuel is employed, is a subject well worthy of the attention of enterprising and ingenious men. How much excellent fuel, for instance, might be made with proper additions and proper management, of the mountains of refuse coal-dust that lie useless at the mouths of coal-pits; and how much would it contribute to cleanliness and elegance if the use of improved coke, or of hard and light fire-balls, could be generally introduced in our houses and kitchens, instead of crude, black, powdery, dirty sea coal! Of the great economy that would result from such a change there cannot be the smallest doubt.

It is a melancholy truth, but at the same time a most indisputable fact, that, while the industry and ingenuity of millions are employed, with unceasing activity, in inventing, improving, and varying those superfluities which wealth and luxury introduce into society, no attention whatever is paid to the improvement of those common necessaries of life on which the subsistence of all, and the comforts and enjoyments of the great majority of mankind, absolutely depend.

Much will be done for the benefit of society, if means can be devised to call the attention of the active and benevolent to this long-neglected, but most interesting subject.

The Royal Institution seems to be well calculated to facilitate and expedite the accomplishment of this important object. Indeed, it is more than probable that this, precisely, is the object which was principally had in view in the foundation and arrangement of that establishment.

ACCOUNT

NEW EXPERIMENTS ON WOOD AND CHARCOAL.

HAVING had occasion to dry several kinds of wood, to ascertain how much water was contained in them, I procured a piece of each kind six inches long and half an inch thick, and planed off some pretty thin shavings, which I kept to dry for eight days in a room, the temperature of which was constantly about 60° F. The wood had been previously drying two or three years in a joiner's workshop.

Of each kind of shavings I took 10 grammes (154.5 grains), which I placed on a china plate in a kind of stove made of sheet-iron, and heated them moderately by a small fire under the stove for twelve hours, after which they were suffered to cool gradually during twelve hours more. The stove, being surrounded with brickwork, was still hot twelve hours after the fire had been extinguished.

On taking out the china plates in succession and weighing the shavings anew, their weight was found to be diminished about one tenth, some a little more, others a little less. When the shavings were put into the stove, their weight was 10 grammes; when taken out, it was about 9. Their colour was not perceptibly altered, and they had no appearance of having been exposed to a strong heat.

162

Desirous of knowing how far the drying of wood might be carried, I replaced them all in the stove, which I heated as before, neither more nor less, for twelve hours, and afterward left to cool slowly for twelve hours.

On taking out the shavings the next day, they had all changed colour more or less; from a yellowish-white they had become light brown, dark brown, more or less yellow, and some of a fine purple.

Their weight, which was at first 10 grammes, was now found to be

Oak	7.16	Cherry . . .	8.60
Elm . . .	8.18	Linden . .	7.86
Beech . . .	8.59	——— (after having	
Maple . . .	8.41	been in the open air	
Ash	8.40	twenty-four hours) .	8.06
Birch . . .	7.40	Male fir . . .	8.46
Service . . .	8.46	Female fir . . .	8.66

Wishing to know whether the wood might not be reduced to charcoal by continuing the moderate heat of the stove a long time, I took half the linden shavings, which weighed 4.03 grammes, placed them in a china saucer supported by a cylindrical earthen vessel 3 inches in diameter and 4 inches high, put this on an earthen plate, and covered it by a glass jar 6 inches in diameter and 8 inches high. On the earthen plate was a layer of ashes about an inch deep, serving to close the mouth of the jar slightly.

This little apparatus being placed in the stove, it was heated a third time for twelve hours, and then left twelve hours without fire, to cool gradually.

On taking out the apparatus, I found that the wood

was become perfectly black, and that the glass jar was yellowish, and its transparency diminished.

On weighing the shavings, which retained their original figure completely, I was surprised to find that they weighed only 2.21 grammes. As they were the remains of 5 grammes of wood, and as, from the experiments of Messrs. Gay-Lussac and Thénard, I had expected to find in this wood at least fifty per cent of charcoal, I did not think it possible to reduce the weight of the shavings to less than 2.5 grammes, particularly with the moderate heat I employed.

To clear up my doubts, I replaced the apparatus in the stove, and heated it again as before for twelve hours, and afterwards left it in the stove twelve hours to cool.

On taking out the apparatus, I found that the shavings weighed only 1.5 grammes. The jar was less transparent, and of a blackish-yellow colour throughout, but particularly in its upper part, above the level of the brim of the saucer in which the shavings were. These shavings were still of a perfect black.

Having heated the apparatus again for twelve hours, and then left it to cool, I was surprised, on taking it out of the stove the next day, to find that the jar had again become clear and transparent. Not the least trace of the yellow coating with which its inner surface had been covered now remained.

On examining the wood, I found that this also had changed its colour. It had assumed a bluish hue, pretty deep, but very different from the decided black it had before. Its weight was 1.02 grammes.

I put it twice more into the stove, and each time its weight was diminished, so that the 5 grammes of wood were reduced at last to 0.27 of a gramme, or about a twentieth of the original weight.

I am persuaded that I should have diminished it still more, if I had continued the experiment longer ; but it has been tried long enough to establish this remarkable fact, *that charcoal can be dissipated by a heat much less than has been considered necessary to burn it.*

It may be supposed that I was very desirous of knowing whether the same thing would occur to charcoal already formed by the usual process. Accordingly I took a piece of charcoal from my kitchen, heated it to a strong red heat, and, while it was still red, put it into a marble mortar, and powdered it. Having passed it through a sieve, I took 4.03 grammes of the powder, placed it in the saucer, heated it in the stove twelve hours, and then left it twelve hours to cool. On taking it out, it weighed but 3.81 grammes.

As this powdered charcoal was nothing but a collection of small bits of charcoal, which were in contact with the air only by a very small surface compared with that of the shavings, I made another experiment, the result of which was more striking and more satisfactory.

Having enclosed in a cloth a quantity of powdered charcoal, that had been passed through a sieve, I beat it strongly in a place where the air was still ; and when the air appeared to be well loaded with the fine dust of the charcoal, I placed on the ground a white china saucer, quitted the place, and left the dust to settle.

The saucer was covered with it, so as to appear of a very dark gray.

Before all the dust had settled, I wrote some letters on the saucer with the point of my finger, and these letters were afterward covered with a still finer dust.

I imagined it possible that the part covered by a very fine dust might be found whitened, while that covered with a stratum of coarser charcoal powder would be found perhaps still black.

The result of the experiment showed that this precaution was not necessary. All the charcoal powder disappeared completely in the stove, and the saucer came out perfectly white.

Another saucer, which had been blackened a little by rubbing it with lampblack, and placed in the stove by the side of that blackened with charcoal dust, came out of the stove as black as it went in. As soon as I saw that the linden shavings converted into charcoal might be dissipated by the moderate heat of the stove, I suspected that they had been consumed slowly by a silent and invisible combustion, and that the product of this combustion could be nothing but carbonic acid gas.

To clear up this point I made the following experiment.

Having procured a stock of very dry birch shavings in ribands about a twentieth of a line thick, near half an inch broad, and six inches long, I dried them for eight days in a room heated by a stove, where the temperature was about 60° F., the shavings being laid on a table at a distance from the stove. Of these shavings thus dried, I took 10 grammes, which I placed on a china plate, and heated in the stove, in the manner already described, for twenty-four hours. When taken out of the stove, they weighed but 7.7 grammes, and had acquired a deep brown colour inclining to purple. They were still wood, however; for, though deeply browned, they burned with a very fine flame.

Of these brown shavings I made three parcels, each

weighing 2.3 grammes. The first was placed in the stove on a white china plate, supported by a tile, but not covered. The second was put into it in a similar manner, except that it was covered with a glass jar, 6 inches in diameter and 6 inches high.

The third parcel was put into a glass vessel, 6 inches high, but only an inch and a quarter in diameter. This narrow vessel was put into a glass jar 3 inches in diameter and 7 inches high, which, being slightly closed with its glass cover, was also placed in the stove on a tile.

As the door of the stove (which is double, the better to confine the heat) does not shut so close as to prevent the free passage of air, and as the china plates on which two of the parcels were placed were flat, every circumstance was favourable for the free transmission of the carbonic acid gas arising from the decomposition of these two parcels by slow combustion, and there was nothing to prevent the progress of this operation. But the third parcel being enclosed in a narrow vessel, as this gas is much heavier than atmospheric air, the first portion of this gas arising from a commencement of combustion of the wood could not fail to descend in the vessel toward its bottom, gradually expel the air, and at length fill the vessel completely ; and as this sort of inundation by carbonic acid gas could not fail to stop the combustion, I expected to find that this parcel of shavings would be preserved, at least in part, even though both the others should be entirely consumed.

The stove having been heated in the usual manner, I found the next day that the results of the experiment had been such as I anticipated. The two parcels of shavings placed on the china plates had disappeared entirely, nothing at all remaining except a very small quantity of ashes, of a white colour inclining a little to yellow.

The yellow ashes in the plate that was not covered with a glass jar were deranged and dispersed by the wind occasioned by opening the door of the stove too suddenly; but those in the other plate, being protected by the glass, were found all together. As they still retained their original figure of shavings, though reduced to a very small bulk, this appeared to me a demonstrative proof that the shavings, whence they arose, had not been burned by a common fire. For this reason, and also on account of their extraordinary colour, approaching very near that of the wood in its natural state, I preserved them, to show them to the Class. They weighed only 0.04 of a gramme; and as the shavings, of which they were the remains, weighed 2.987 grammes on coming out of the hands of the joiner, these ashes make only one and one third per cent of the weight of the wood.

The third parcel of shavings, which had been placed in a narrow glass vessel, had not disappeared, but the wood was converted into perfect charcoal. I have the honour to present it to the Class, in the same vessel in which it was charred.

As the three parcels of shavings were of the same wood, and equal in weight; as they were exposed together to the same degree of heat, and for the same time; and as the two portions that were placed so as to facilitate the escape of the carbonic acid gas arising from their decomposition, disappeared entirely, while the third, which was so circumstanced that the escape of this gas was impossible, did not disappear; — it seems to me that there can be no doubt of the cause of the phenomena that presented themselves ; and it is certainly a curious fact, that charcoal, which has hitherto been considered as one of the most fixed substances known, can unite itself to oxygen, and form with it carbonic acid gas, at a temperature much below that at which it burns visibly.

INQUIRIES

RELATIVE TO THE STRUCTURE OF WOOD,

The specific Gravity of its solid Parts, and the Quantity of Liquids and elastic Fluids contained in it under various Circumstances; the Quantity of Charcoal to be obtained from it; and the Quantity of Heat produced by its Combustion.

SINCE the days of Grew and Malpighi, there have been but few regular inquiries into the structure of wood. The science of botany has, indeed, taken an excursive range; and the indefatigable zeal of modern naturalists, who have travelled over all the known world, has made us acquainted with an astonishing number of plants, unknown before in Europe, and therefore called *new*, by which our gardens and apartments are embellished with a profusion of gay flowers; but still the knowledge of the vegetable economy is scarcely at all advanced. The circulation of the sap in plants is still a subject of dispute, and the causes of its ascension are very imperfectly known. The specific gravity of the solid parts which form the wood of plants is unascertained, and, by consequence, the proportions of solids, of liquids, and of elastic fluids; the component parts of a plant, with the variations to which they are subject in different seasons, are matters of which we are still ignorant.

It is, indeed, known, that the wood of a tree remains and preserves its primitive form after it has been converted into charcoal; but no one has explained this extraordinary phenomenon, very little attention having been paid to it.

An earthen vessel becomes hard and brittle in the potter's furnace; the vessel shrinks during the operation of baking, but it undergoes no alteration of shape. This phenomenon is easily accounted for; the water, which distended the particles of the clay, kept them at a distance from each other, and rendered the mass soft and flexible, having been expelled by the power of the heat, the several particles contract themselves together, and form a hard brittle body, though the clay remains the same before and after the operation.

Is it not possible that wood is converted into charcoal by a similar process? For either the charcoal is already formed in the wood, or, the wood being decomposed, the charcoal is formed of its elements or a part of them. But is it not evidently impossible that the elements of a solid body should be so totally deranged as to separate them entirely from each other without destroying the form or figure of the body?

In the sequel of this paper it will be shown that the specific gravity of the solid parts of any kind of wood is very nearly the same as that of the charcoal obtained from it, — a circumstance that gives a great degree of probability to the hypothesis that the two substances are identically the same.

But I do not mean to amuse the Class with a detail of my own conjectures; it is to my experiments and their results that I now claim the honour of calling its attention.

I was by accident first induced to enter upon this

examination and inquiry into the structure of wood. In the course of a long series of researches upon heat, I wished to determine the quantities of that element produced by the combustion of different kinds of wood; but I had scarcely begun the inquiry when I found that, in order to procure satisfactory results to my experiments, it was indispensably necessary to obtain a better knowledge of wood itself; and therefore I immediately devoted myself to the study.

My first aim was to determine the specific gravity of the solid parts which compose the fabric of the wood, in order afterwards to determine the quantities of sap or water contained in wood under various circumstances.

Having found that very thin shavings filled with sap, or even with water, could be thoroughly dried in less than an hour, without injury to the wood, in a stove kept at a higher temperature than that of boiling water, or at about 500° of Fahrenheit's scale (= 260° French), I determined on using shavings of this description in my experiments.

SECTION I. — *Of the specific Gravity of the solid Parts of Wood.*

I began with the wood of the lime-tree, of which the texture is very fine and regular. From a small board, five inches long and half an inch thick, very dry, I took a quantity of thin shavings with a very sharp plane. These were exposed for eight days in the month of January upon a table in a large room not otherwise occupied, in order that they might attract from the atmosphere all that moisture which, as an hygrometric body, they were capable of imbibing. The temperature of the room was about 46° F.

Ten grammes (154.5 grains) of those shavings, laid on a china plate, were placed in a large stove made of sheet-iron, and there exposed to a regular heat of about 245° F. for two hours, in the course of which time they were frequently taken out and weighed in order to observe the progress of their desiccation. When they ceased to lose weight, the operation was stopped; when perfectly dried, their weight was 8.121 grammes.

By previous trials with my apparatus, I had learned that if the stove was too much heated the shavings became discoloured, which is always indicated by the emission of a particular odour, very readily to be perceived; but, by a careful regulation of the fire, this accident may be avoided and the shavings be thoroughly dried without injury, or even subjecting them to any sensible alteration.

I concluded that they had not undergone any change, because, upon again exposing them to the atmosphere, they regained the same weight which they had, under similar circumstances, prior to their being dried in the stove.

Being thus possessed of the weight of my shavings, as well under exposure to the air as in a dried state, which latter I could not but look upon as being perfect, it only remained to ascertain their weight in water when all their vessels and pores were completely filled with that liquid, to enable me to determine the specific gravity of the solid parts of this wood, which was accomplished without difficulty by the following process : —

A cylindrical copper vessel, 10 inches in diameter and as many deep, was filled with water from the Seine, previously well filtered, and, being set upon a common chafing-dish, was made to boil for some time, to expel the air contained in the water. The shavings were then thrown into the boiling water, and kept in that state for an hour. The water was not long in filling the vessels and pores of the shavings, from which it dislodged the air contained in them; so that the wood, specifically heavier than the water, was precipitated to the bottom of the vessel, and there remained.

When the vessel was removed from the chafing-dish, the water was suffered to cool to the temperature of 60° F., and then, plunging in both hands, I placed (under the water) all the shavings in a cylindrical glass vase, whose weight I had previously ascertained, which was suspended in the water by a silken cord, fastened at its other extremity to the arm of an accurate hydrostatic balance.

On weighing the shavings in the glass case thus immersed, I found their weight equal to 2.651 grammes.

As the shavings, while dry, weighed 8.121 grammes in the air, and 2.651 grammes in the water, they must have lost 5.47 grammes of their weight in the latter; consequently they must have displaced 5.47 grammes of water; and the specific gravity of the solid parts of this wood must be to that of the water at the temperature of 60° F. as 8.121 to 5.47, or as 14,846 to 10,000.

It may perhaps excite some surprise that the solid parts of so light a wood as that of the lime-tree should be heavier, by nearly one half, than water, taken in equal bulks. But this surprise will, without doubt, be increased when I declare that the specific gravities of the solid parts of all kinds of wood are so nearly alike as almost to induce a belief that there is the same identity in the ligneous substance of all sorts of wood as in the osseous substance of all species of animals.

I procured, from a joiner's workshop, dried wood of the eight following species; viz., poplar, lime, birch, fir, maple, beech, elm, and oak; and had them cut into small boards, 5 inches in length and 6 inches broad, from each of which I planed off some thin shavings, and exposed them to the air for eight days, in the month of January, in a large room, where the temperature, which varied but little, was about 40° to 45° F.

When these shavings had acquired their ordinary degree of dryness under existing circumstances, 10 grammes of each sort were weighed off, and, being laid separately in china plates, were thoroughly dried in the stove.

On being taken out of the stove, they were again weighed, and then thrown into boiling water, to expel the air from their pores and to moisten them thoroughly. When they had boiled for an hour, they were suffered to remain in the liquor till it was sufficiently cool; and after they had been weighed in the water, the specific gravity of their solids was calculated in the usual way.

The following table gives the details and results of this inquiry : —

Species of wood.	Weight.			Specific gravity of the solid parts of the wood.	Weight of a cubic inch of the solid parts of the wood.
	Exposed to the air in a room in winter.	Thoroughly dried in a stove.	In the water at 60° F.		
	Grammes.	Grammes.	Grammes.		Grammes.
Poplar	10	8.045	2.629	14854	29.45
Lime	10	8.121	2.651	14846	29.40
Birch	10	8.062	2.632	14848	29.44
Fir	10	8.247	2.601	14621	28.96
Maple	10	8.137	2.563	14599	28.93
Beech	10	8.144	2.832	15284	30.30
Elm	10	8.180	2.793	15186	30.11
Oak	10	8.336	2.905	15344	30.42
			Water	10000	19.83

The specific weight of the solid matter which composes the fabric of these woods is so nearly alike in them all, that the small variations to be observed in the different experiments may perhaps be accounted for otherwise than by supposing the ligneous substance to be essentially different in the several species.

The charcoal obtained from the various kinds of wood, if carefully prepared, has no sensible difference; and all the seerwoods give nearly the same chemical results when treated in the same manner. Hence, without doubt, we have good reason to suspect that the ligneous substance of all woods is identical. But without stopping to discuss this question at present, I shall endeavour to elucidate another, no less interesting, and which yields results more satisfactory.

Section II. — *Of the Quantities of Sap and Air discovered in Trees and in Seerwoods.*

Grew and Malpighi discovered in plants certain vessels which they suspected to be destined for the reception of air; and many physiologists have supposed that the air found shut up in the vessels of plants (if it be really confined there) would necessarily cause a reaction upon the neighbouring vessels, with an elastic force as variable as the temperature to which this elastic fluid is exposed, and might probably contribute to the circulation of the sap.

It would, doubtless, be an interesting question to determine precisely the quantity of air contained in plants in different seasons and under various circumstances. By examining the variations to which this quantity of air is subjected, and combining them with other simultaneous phenomena, we might hope to make some discovery which may assist us a little to elucidate the profound obscurity that at present conceals this part of the vegetable economy.

The specific gravity of the solid parts of a plant being known, it becomes very easy to determine in every case the quantity of air contained in its vessels and pores.

The following example will render this position perfectly clear.

An oak in complete health, in a growing state, was cut down on the 6th of September, 1812. A cylindrical piece, 6 inches long and rather more than an inch in diameter, taken from the middle of the trunk of this young tree, 3 feet above the earth, weighed, when full of sap, 181.57 grammes.

Upon plunging this piece of wood into a cylindrical vessel about $1\frac{1}{2}$ inch in diameter and $6\frac{1}{4}$ inches in height, filled with water at the temperature of 62° F., it displaced 188.57 grammes of the water; * whence I conclude with certainty, that this piece of oak, filled with sap, possessed a bulk equal to 9.5093 cubic inches, that its

---

* In order to determine and keep an account of the quantity of water remaining on the surface of this piece of wood at the instant of withdrawing it from the vessel, it was weighed when taken out, whilst still quite wet. As its weight had been taken previously to the operation, the augmentation it had acquired from the water was ascertained to a nicety.

The vessel when empty weighed 188.22 grammes, and when filled with water at the temperature of 60° F., 474.9 grammes; so that it contained 286.68 grammes of water. When the piece of wood was plunged into the water, a small glass plate, about two inches in diameter and two lines in thickness, ground with emery to fit it to the edges of the vessel, so as to close it hermetically, was laid upon its mouth, to shut up the piece of wood with the water still remaining in the vessel, whilst its outside was wiped with a dry cloth.

When the exterior of the vessel had been thoroughly dried, the glass cover was carefully removed, and the piece of wood withdrawn; the vessel was then weighed again with its remaining contents of water; and from its weight the quantity of water displaced by the wood was calculated.

specific gravity was 96,515, and, consequently, that a cubic inch of it weighed 19.134 grammes.

When the piece of wood had been reduced to the shape of a small board, about half an inch in thickness, I took from it forty very thin shavings weighing 19.9 grammes, but when thoroughly dried in the stove, at a temperature of 262° F., they weighed only 12.45 grammes.

From this experiment, it is evident that the wood in question, being full of sap, was composed of 12.45 ligneous parts, and 7.45 parts of water, or of sap, whose specific gravity is nearly the same as that of water.

Now, as one cubic inch of this wood weighed 19.134 grammes, it is very certain that it was composed of 11.971 grammes of ligneous parts, which were consequently solids, and of 7.163 grammes of sap.

But we have already seen, from the results of the experiments detailed in the former part of this memoir,[*] that a cubic inch of the solid parts of the wood of the oak weighs 30.42 grammes : —

Consequently the 11.971 grammes of solid parts found in one cubic inch of this wood, when the tree was alive, could have no greater bulk than . . . . 0.39353 cubic inch.

As one cubic inch of water weighs 19.83 grammes, the 7.163 grammes of sap found in the cubic inch of this wood must have occupied a bulk equal to 0.36122

Consequently a cubic inch of the wood in question contained a quantity of air whose bulk was equal to 0.24525

Making together 1.00000

[*] See the table, page 176.

We conclude from these results, that a young oak, in a growing state, at the beginning of September, when the wood appears to be diffused with sap, contains, nevertheless, about a fourth of its bulk of air, and that its solid ligneous parts do not make quite $\frac{4}{10}$ of its bulk. But we shall presently see that the lighter woods contain still less of ligneous parts, and more of air, than the oak.

A young Italian poplar, 3 inches in diameter, measured at 2 feet above the earth, was cut down on the 6th of September, while the tree appeared to be in a growing state. The specific gravity of a piece taken from the middle of the trunk was found to be 57,946; consequently a cubic inch of this wood weighed 11.49 grammes.

From a piece of this wood, apparently full of sap, forty thin shavings were taken, 6 inches in length and half an inch broad. The wood from which these shavings were planed weighed 12.37 grammes; and the shavings, when thoroughly dried in the stove, weighed 7.5 grammes.*

We hence conclude that a cubic inch of this wood, in its original state, while the tree was still alive, contained 7.1531 grammes of ligneous parts which formed the fabric of the wood, and 4.3369 grammes of sap, differing in its specific gravity little or nothing from common water.

---

* As the heat excited by the plane in taking off these shavings was sufficient to evaporate a very sensible quantity of sap belonging to the wood from which they were cut, the shavings became perceptibly dry during the operation; for I found that forty thin shavings sometimes lost more than one gramme (about $\frac{1}{12}$ of their weight) in less than a minute. In order to obtain their true weight, whilst they still remained part of the wood, I adopted the precaution of weighing the piece of wood both the moment before and the moment after the operation of planing. The difference in the weight of the wood, under these two circumstances, indicates the weight necessary to be given to the shavings, and which is here always attributed to them.

As one cubic inch of the solid parts of this kind
of wood weighs 29.45 grammes,* the 7.1531
grammes of ligneous parts found in a cubic
inch of the trunk of the living tree, in Sep-
tember, could only have occupied the space of   0.24289 cubic inch.
And the 4.3369 grammes of sap, contained in it,
only 0.21880
Consequently, in one cubic inch of this wood
there was a bulk of air equal to . . . 0.53831
                                     Total 1.00000

The difference between the structure of the oak and
of the poplar becomes very conspicuous on making a
comparison, according to the subjoined method, between
the constituent parts of these two kinds of wood, both
in a growing state.

Thus, a cubic inch of wood is composed of : —

	Ligneous parts.	Sap.	Air.
The oak . . .	0.39353	0.36122	0.24525
The poplar . .	0.24289	0.21880	0.53831

This striking difference in the proportions of the
ligneous substance, of sap, and of air, discovered in these
two species, sufficiently explain the difference observable
in their weight and hardness. This inquiry may probably lead to other discoveries of more general utility in
the study of the vegetable economy.

SECTION III. — *Of the relative Quantities of Sap and Air
found in the same Tree, in Winter and in Summer ; and
in different Portions of the same Tree, at the same Time.*

The following experiments were undertaken with a
view to discover the difference between the quantities of

---

* See the table, page 176.

sap and air found in the wood composing the trunk of a large tree, in winter and in summer.

On the 20th of January, 1812, I had a lime-tree felled of about twenty-five or thirty years' growth, which had stood among several others of the same age in my garden at Auteuil. On taking a piece of wood from the middle of the trunk, at about 3 feet above the ground, it appeared to be filled and even drowned in sap. Its specific gravity was 76,617; consequently, one cubic inch of the wood weighed 15.788 grammes.

Having planed off 10 grammes of thin shavings from this piece, and dried them thoroughly in the stove, I found their weight reduced 4.72 grammes.

Thus in possession of the specific gravity of the solid part of this wood, it was easy to determine, with the aid of these data, the constituent parts of a cubic inch, which were as follows : —

Ligneous parts	0.25353 cubic inch.
Sap	0.44549
Air	0.30098
	1.00000

On the 8th of September, in the same year (1812), I had a piece of wood (= 5.84 cubic inches) cut from the trunk of another lime, of equal age with the former (from 25 to 30 years), at the height of 3 feet above the earth. This tree was in a growing state, and the piece taken from it, after it had been trimmed by the joiner, weighed 87.8 grammes, and displaced 115.8 grammes of water, at the temperature of 62° F.; consequently, its specific gravity was 75,820. In the month of January the specific gravity of this same species of wood had been found to be 79,617.

From the piece of wood taken from the tree on the 8th of September, I had 14.19 grammes of thin shavings planed off, which, after they had been thoroughly dried in the stove, weighed only 7.35 grammes. Hence we have, as the constituent parts of a cubic inch of this wood: —

Ligneous parts	0.26489 cubic inch.
Sap	0.36546
Air	0.36965
	1.00000

From the results of these two experiments, we may conclude that the body of the tree contains more sap in the winter than in summer, and more air in summer than in winter. But the following experiments demonstrate the sap to be very disproportionately distributed in the several parts of the same tree, at the same season.

On the 8th of September, I had a branch, about 3 inches in diameter, cut from the lime just spoken of, and which issued from the trunk at the height of 10 feet above the surface of the earth. From the lower end of this branch I took a piece of wood, and subjected it to the investigation requisite to ascertain its constituent parts.

Its specific gravity was 70,201. The same day I found the specific gravity of a piece of the trunk of the same tree to be 75,820.

Surprising as this difference appeared, my astonishment was still more excited, on finding that a piece of wood of three years' growth, cut from the upper end of the same branch, where it was but one inch in diameter, had a specific gravity of 85,240.

There was, therefore, much more sap and less air in

the wood of the upper extremity of the branch than in the lower, which was nearer to the body of the tree.

I afterwards examined the young shoots of the current year, in the same tree, as well as in several other species of wood, and uniformly found that the specific gravity of the young wood, that is to say, of the current year, is always considerably greater than that of the same species of wood when grown older. Doubtlessly, because it contains more sap and less air than the old wood.

In the management of experiments for determining the specific gravity of wood of the current year, it is indispensably necessary to take an account of the space occupied by the pith, without which precaution we shall be led to false conclusions.

I found the specific gravity of the oak of the current year to be 116,530; that of the elm 110,540. Young shoots of these trees, deprived of their bark and pith, descend rapidly on being thrown into water; whilst species of the same tree, more advanced in age, swim on the surface, even when the wood is green, and more full of sap.

This fact is worthy the attention of persons occupied in the study of vegetable physiology.

I was next curious to examine the root of the lime from which I had already had one piece of wood from the trunk, and two pieces from one of its branches. With this view, on the 8th of September, 1812, I caused one of its roots, of about 2 inches diameter, to be taken up, and cut from it a piece weighing 93.25 grammes, which displaced 115.8 grammes of water. Its specific gravity was 80,527, and, consequently, greater than that of the wood extracted from the trunk of the same tree, but less than that cut from the upper end of one of its branches. 20.48 grammes of thin shavings, from this piece of the root of the lime, weighed only 10.85 grammes after being thoroughly dried in the stove.

From these data, I determined the constituent parts of a cubic inch of the root thus : —

Ligneous parts	0.28775 cubic inch.
Sap	0.37358
Air	0.33867
	1.00000

The constituent parts of a cubic inch of the body of the same tree were, as we have shown : —

Ligneous parts	0.26489 cubic inch.
Sap	0.36546
Air	0.36965
	1.00000

The constituent parts of a cubic inch of the wood of the same tree, taken the same day from the lower extremity of a branch, were: —

Ligneous parts	0.25713 cubic inch.
Sap	0.27513
Air	0.46774
	1.00000

Lastly, the constituent parts of a cubic inch of the wood, taken near the upper extremity of the same branch, were : —

Ligneous parts	0.25388 cubic inch.
Sap	0.47599
Air	0.27013
	1.00000

For the more easy comparison of the results of these four experiments upon the wood of the lime-tree, made on the same day, with different portions of the same tree, I have collected them together in the following table.

| | A cubic inch of wood was composed of | | |
	Ligneous parts.	Sap.	Air.
The root　.　.　.　.	0.28775	0.37358	0.33867
The trunk　.　.　.　.	0.26489	0.36546	0.36965
The lower end of a branch　.	0.25713	0.27513	0.46774
The upper end of　"　.　.	0.25388	0.47599	0.27013
Wood taken from the trunk of a lime-tree of the same age, on the 20th of Jan.　.　.	0.25353	0.44549	0.30098

Being desirous to ascertain whether a difference considerable enough to be valued existed between the wood of the heart, or core, and the sap-wood found between the rind and the body of the same tree, I took, on the 11th of September, an elm fagot, 5 inches in diameter, lopped from a large tree, which had been felled on the 20th of the preceding April, and had two cylindrical pieces, each 6 inches in length, cut out of it. The thickest of these taken from the core weighed 191.05 grammes, and displaced 194.45 grammes of water; the other, consisting of the sap-wood, weighed 93.61 grammes, and displaced 111.45 grammes of water.

The specific gravity of the core was, therefore, 98,251; that of the sap-wood, 81,764. But as the fagot had lain exposed to all the summer rains, the wood was far from being dry. I was, however, much surprised at discovering that the core of the wood was more charged with sap or water than that of the same kind of wood when in a growing state, — a fact which induces a suspicion that the sap in trees is not enclosed in vessels or tubes apparently impervious to that liquid.

To obtain a better knowledge of the wood in question, I planed off forty shavings, 6 inches in length and half an inch in breadth, from a small board cut from the core ; with an equal number of shavings, of similar dimensions, from another board cut from the sap-wood.

The forty shavings from the core, taken just as they were planed off, weighed 16.37 grammes, and 10.53 grammes after they had been thoroughly dried in the stove.

The forty shavings of sap-wood weighed 16.97 grammes before they were dried, and 11.99 grammes afterwards.

Thus possessed of the specific gravity of the solid parts of this kind of wood, it only remained to determine, from these data, the constituent parts of an inch of the wood, which was readily performed, as follows : —

	Ligneous parts.	Sap.	Air.
In the core of the elm	0.41622	0.35055	0.23323
In the sap-wood	0.38934	0.23994	0.37072

It appears, from the results of these experiments, that the sap-wood of the elm contains rather more ligneous parts in its timber than the core of the same tree ; and that it contains much less sap and more air. But as the tree had been felled nearly five months before it became the subject of investigation, it is very possible that the sap-wood had become much drier than the core of the tree.

I had purposed to repeat these experiments upon wood in a growing state and upon seerwood ; but the interference of other occupations has prevented a continuance of the inquiry. It cannot, however, but lead to results curious in themselves ; and I therefore recommend it to the notice of all students in vegetable economy, as well as to those who love that noble science, and feel a gratification in being able to remove the veil under which the mysterious operations of nature are concealed.

The particular object which I had in view in exploring the structure of wood has led me by a way by no means likely to be fertile in interesting discoveries; but I have begun the work, and feel myself bound to complete it, in preference to every other consideration. These fascinating researches, I am aware, have already carried me too far, and I must now resign them into the hands of others, in order to fulfil my engagements. This I do most cheerfully, and it will give me the greatest pleasure to behold a field, too long neglected, once more broken up.

SECTION IV. — *Of the Quantities of Water contained in Woods considered as dry, or Seerwoods.*

Wood is a hygrometric substance, and, when exposed to the atmospheric air, always imbibes a visible quantity of water, varying, however, with the temperature and humidity of the air.

If the moisture in the wood were confined in vessels so constructed as to be totally impervious to water, the fabric of the wood would be uniformly the same, with the exception only of the variations caused in its dimensions by change of temperature, in which case it would be very easy to determine the quantity of water contained in the wood, when the specific gravity of its solid parts was known. But, as the bulk of all woods is considerably diminished in drying, the experiment is rendered rather prolix, though by no means difficult, and its results are clear and satisfactory.

A few examples will suffice to point out the method to be pursued.

The composition of the oak, in a growing state, at the beginning of September, has been already given. In order to ascertain the change which this wood undergoes by the process of drying, I made the following experiment.

From a fagot of oak, $5\frac{1}{2}$ inches in diameter, which, covered with its bark, had been exposed to dry in the open air for eighteen months, I took a piece of rather more than an inch square and 6 inches in length; it was good firewood, and seemed very dry.

This piece, after being trimmed by the joiner, weighed 126.2 grammes, and displaced 157.05 grammes of water; its specific gravity was consequently 80,357, and a cubic inch weighed 15.939 grammes.

Forty-three shavings of this wood, 6 inches long and half an inch broad, weighed 17.9 grammes; but when thoroughly dried in the stove, they were reduced to 13.7 grammes. They were therefore, prior to being put into the stove, composed of 13.7 grammes of solid parts — that is to say, of dry or seerwood — and 4.2 grammes of water.

The results of this experiment indicate that 100 kilogrammes of this excellent firewood contained 76 kilogrammes of seerwood and 24 of water; which is, probably, the ordinary state of the best firewood sold in the timber-yards of Paris, and all other places.

Were the wood to be kept for several years in a dry place, secured from the rain, it is possible that it might become dry to such a degree as to contain only about 12 per cent of water, and 88 of seerwood. But it will appear in the sequel that wood of any kind, exposed to the atmosphere, could never become more dry, on account of its hygrometric quality, which it constantly preserves.

The following are the constituent parts of a cubic inch of firewood employed in this experiment : —

Ligneous parts, or seerwood . .	0.40166 cubic inch.
Sap, or water	0.18982
Air	0.40852
	1.00000

Thus we are enabled clearly to demonstrate the difference between the oak in a growing state, and the same kind of wood after it has been felled and dried in the air, secured from the rain, for eighteen months : —

	Dry wood.	Water.	Air.
In a cubic inch of oak, in a growing state	0.39353	0.36122	0.24525
In a cubic inch of the same kind of wood, after it had been felled and dried for 18 months	0.40166	0.18982	0.40852

By comparing the relative quantities of seerwood contained in a piece of timber while in a growing state, and in the same timber after it has been dried, we may ascertain how much its fabric has shrunk by desiccation.

It appears from these experiments, that the oak sold in the timber-yards of Paris for firewood contains rather more than one half of the sap which it formerly had in a growing state.

I have made several similar experiments upon other species of wood; but their results are better calculated for exhibition in a table than for circumstantial detail.

SECTION V. — *Of the Quantities of Water attracted from the Atmosphere by Woods of various Species, after being perfectly dried.*

It has been long known that charcoal imbibes the humidity of the atmosphere with considerable eagerness; but I have discovered that dry wood attracts it with still greater avidity. The following are the details and results of a series of experiments made last winter, with a view to elucidate this subject.

Having procured thin shavings, about 5 inches long and half an inch broad, of nine different species of the woods of our climate, in order more certainly to reduce them to an equal degree of dryness, I began my experiment by boiling them for two hours in water, that they might be thoroughly impregnated with that element.

I then dried them well in a stove, in which they were kept during 24 hours, exposed to a temperature higher than that of boiling water, at about 250° of Fahrenheit's scale.

On taking them out of the stove, they were carefully weighed, being still hot; they were then suffered to remain in the open air for 24 hours, in a large room, whose temperature was uniformly during the day and night at about 45° to 46° F. This was on the 1st of February, 1812.

The weight of the shavings, on being removed from the stove thoroughly dried, and after having been exposed to the air of the large room, was as follows : —

Species of wood.	Weight.	
	On being withdrawn from the stove.	After exposure for 24 hours, in a room at a temperature of 46° F
	Grammes.	Grammes.
Italian poplar	3.58	4.45
Lime-tree, seasoned, and fit for the joiner's use	5.28	6.40
" green wood	5.39	6.47
Beech	7.02	8.62
Birch	4.41	5.47
Fir	5.41	6.56
Elm	5.87	7.16
Oak	6.46	7.93
Maple	4.76	5.85

Hence it appears that 100 parts of the wood, after 24 hours' exposure in the large room, were composed of dry wood and water in the following proportions : —

100 parts of	Seerwood.	Water.
Poplar	80.55	19.55
Lime-tree, seasoned . . .	82.50	17.50
" green . . .	83.31	16.69
Beech	81.44	18.56
Birch	80.62	19.38
Fir	82.4	7.53
Elm	81.80	17.20
Oak	83.36	16.64
Maple	81.37	18.63

I suffered all these woods to remain in the large room during eight days, but their weight was very little augmented ; and as often as the temperature of the air of the room was raised above 46° F. they lost weight. So that the above may be considered as their habitual state of dryness during the winter, in our climate.

To ascertain the quantity of moisture habitually retained by these woods in the summer, I made the following experiments.

Thin shavings of the species of woods below, half an inch broad, were thoroughly dried in the stove, and then exposed for 24 hours in a room with a northern aspect, whose temperature was tolerably uniform at 62° F. The following are the results : —

Species of wood.	Weight.		In 100 parts of wood were found	
	When dry.	At the accustomed state of humidity in the air at 62° F.	Seerwood.	Water.
	Grammes.	Grammes.	Parts	Parts.
Elm, the core . .	10.53	11.55	91.185	8.815
" the sap-wood .	11.99	13.15	91.197	8.803
Oak, seasoned and fit for the joiner's use . .	13.70	15.05	91.030	8.970
Oak felled 6th Sept. .	12.45	13.70	90.667	9.333
Lime, seasoned . .	7.27	7.80	93.205	6.795
" when growing .	6.75	7.30	92.466	7.534
" the root . .	9.96	10.80	92.222	7.778
Elm, seasoned .	9.25	10.80	91.133	8.867
Italian poplar . .	7.50	8.00	93.750	6.250

With a view to ascertain the habitual state of the dryness of woods in autumn, I carefully preserved these same shavings till the 3d of November, in a northern chamber, not inhabited ; at which period its temperature had stood for several days at 52° F., with little variation. I then weighed the shavings, and from their weight calculated the quantity of water contained in them.

The following table, containing the results of all these experiments, displays, in a familiar and satisfactory manner, the customary state of the woods, in different seasons, in our climate.

Species of wood.	100 parts in weight of wood, cut into thin shavings and exposed to the air, contained water		
	In summer, at a temperature of 62° F.	In autumn, at a temperature of 52° F.	In winter, at a temperature of 45° F.
	Parts.	Parts.	Parts.
Poplar . . .	6.25	11.35	19.55
Lime . . .	7.78	11.74	17.50
Oak . . .	8.97	12.46	16.64
Elm . . .	8.86	11.12	17.20

From a comparison of these results it appears that these woods, when exposed to the air at a temperature of 45° F., contain twice the quantity of water that they do when the temperature of the air is at 60° F. But it is necessary that the wood be cut into very thin shavings, to enable it to become suddenly *in equilibrio* with the air, conformably to its quality of a hygrometric body; otherwise the state of the air may change, and that very frequently, before its humidity or dryness can have had sufficient opportunity to produce all its effect upon the wood.

To discover what is termed *the medium dryness* of any species of wood, in our climate, it is requisite that we be acquainted with the quantity of water contained in wood every day of the year, and even in every hour and every minute, which is obviously impossible; but there is another method to be pursued in this inquiry, much less laborious and which will lead to results as satisfactory as the nature of the subject will admit.

As a very large piece of wood, a large beam for instance, dries so very gradually in the air as not to attain a state of perfect dryness in less than 50 or 60 years, it is sufficient to examine the interior of such a beam, after having been sheltered for 80 or 100 years from the rain, to discover the state of such part of the wood as may still be considered sound.

In pulling down old houses, we meet with beams proper for the present inquiry.

An old castle in my neighbourhood being pulled down, I had an opportunity of examining the interior of a large oaken beam, which had, without doubt, been there more than 150 years, and as it formed part of the timbers of the edifice, had been secured from the rains.

A piece of this wood in a high state of preservation, after it had been planed by the workman, was accurately weighed, and then plunged into water, to ascertain its specific gravity. It weighed 75.05 grammes, and displaced 110 grammes of water, at the temperature of 61° F.; its specific gravity, therefore, was 68,227, and a cubic inch weighed 13.53 grammes.

Forty shavings of the wood weighed 11.4 grammes, which were reduced to 10.2 grammes when they had been thoroughly dried in the stove.

Hence we may conclude that a cubic inch of this old wood was composed of

Ligneous parts	0.39794 cubic inch.
Water	0.07186
Air	0.53020
	1.00000

We may also conclude from these results, that the wood of the centre of a large oaken post, though kept for ages out of the reach of the rain, can never contain, in our climate, less than 10 per cent of its weight in water; and that a cubic inch of such wood contains more than half a cubic inch of air.

The *yearly medium temperature* at Paris is about $54\frac{1}{2}°$ F.; now, as we have just seen that the habitual state of dryness in woods at the temperature of 52° F. is such as to give about 11 per cent of water for 100 parts of wood, we must not be surprised at finding 10 per cent of water in the interior of a large beam, after it had been sheltered from the rain during 150 years.

To ascertain whether the property of wood to attract moisture from the atmosphere was augmented or diminished by the beginning of carbonization, I made the following experiments.

Fourteen grammes of ash-shavings, after being highly dried on a marble slab over a chafing-dish, were exposed to the air, in the month of February, in a large room whose temperature was about 20° F., and in 15 hours they had gained 1.65 grammes in weight.

Fourteen grammes of the same sort of shavings, having been first scorched in the stove till they had assumed a brown color, were at the same time dried over the chafing-dish, and exposed with the others to the cold air for the same length of time; but they gained in weight only 1.01 grammes, while those which had not been scorched, as already stated, had gained 1.65 grammes.

Fourteen grammes of the shavings of lime-wood, in their natural state, and fourteen grammes of the same kind of shavings, after they had been violently scorched in the stove, were dried together over the chafing-dish, and then exposed in the open air, at the temperature of 40° F. for 15 hours. The shavings in their natural state gained 1.33 grammes in weight; while those that had been scorched gained 0.7 grammes.

A similar experiment, upon shavings of the cherry-tree, some in their natural state, and others scorched, was productive of the same result.

Whence we conclude that wood in its natural state attracts the moisture of the air more copiously than it does after having been subjected to the first degree of carbonization.

From similar experiments upon wood and charcoal, I find that dry wood attracts humidity more powerfully than dry charcoal.

It would be worth ascertaining, whether wood is not also more powerfully attractive of gas than charcoal; but as I have not time to enter upon this particular inquiry, I can only recommend it to those whose inclinations may lead that way. Leaving, therefore, this subject untouched, I must, without any further circumerration, pursue the original object I had in view in these disquisitions upon wood, namely, to endeavour to become acquainted with those inflammable substances which burn on setting fire to a piece of wood under a calorimeter.

Section VI. — *Of the Quantities of Charcoal to be obtained from different Kinds of Wood.*

Having discovered that pieces of wood, more or less thick, may be perfectly carbonized in glass vases with thin tops, closely covered, and exposed for two or three days to a moderate heat in a stove, I adopted this method in all my experiments on the carbonization of wood.

The glass vases which I make use of are what the chemists call *proofs*, with feet: they are small cylindrical vessels, about $1\frac{1}{2}$ inch in diameter and 6 inches in height; the covers consist of glass plates about 2 inches in diameter and from 2 to 3 lines in thickness, neatly ground with very fine emery, well diluted with water, on a large glass slab; and, the edges of the vases being ground with the same exactness, they become hermetically closed by the covers, so as to preclude every access of the air, especially if the edges of the vases and the whole surface of the covers be well rubbed with black-lead.

The elastic fluids, in escaping from the interior of the vases, occasionally raise the cover for a moment, on one side, even when surmounted by a considerable weight; but as it is only raised a very little, and falls again immediately, the vase is never open more than an instant at a time, and then not so as to admit the obtrusion of any extraneous matter.

When one of these vases is put into the stove, it is placed upon a square tile, or half-brick, of burned earth, and another of the same kind is also laid upon the cover to keep it steady.

During the carbonization of the wood, the interior of the vase is always clouded, assuming a very deep blackish-yellow colour; and during the operation a strong smell of soot or of pyroligneous acid issues from the stove; which is even insupportable at the commencement, if it be too nearly approached, as well as on withdrawing the vases from the stove, if the covers be removed without due precaution.

There is, therefore, a *decomposition* during the carbonization of wood, and a formation of pyroligneous acid. This fact has been long known; but, in some of my experiments, and particularly in those made upon fir, with a very moderate fire, I obtained a product, which, upon a very exact scrutiny, appeared to me to be *bitumen.*

This product had been condensed upon the glass cover, whence it had afterwards run in large drops upon the vertical surface of the side of the vase. It was hard and brittle, of a dark yellow colour; it was not affected by boiling water, nor by boiling alcohol, but was gradually dissolved by sulphuric ether.

It would be superfluous here to enter upon the details of all my experiments relative to the carbonization of wood. As the process I have employed cannot now but be well known, after what I have said in this memoir and in the one that I had the honour to present to the Class on the 30th of December in last year,[9] I shall here only give the result of those experiments.

The six following, made with different species of wood, were so uniformly alike in their results, that I was much surprised.

One hundred parts (10 grammes) of the six following kinds of wood, in thin shavings, and thoroughly dried, were carbonized at one time in the stove, in glass vases, well closed with flat glass covers. As the heat was managed with great care, in order to determine with precision, from the weight of the vases, the moment when the operation was finished, the experiment occupied four days and as many nights. When the vases with their contents ceased to lose weight, the process was stopped, and the charcoal was weighed while still hot.

The following were the results : —

	Poplar	43.57 parts.
100 parts in weight of dry wood gave in dry charcoal.	Lime	43.59 "
	Fir	44.18 "
	Maple	42.23 "
	Elm	43.27 "
	Oak	43.00 "

The medium term of the results of these six experiments gives 43.33 parts of charcoal in 100 parts of dry wood ; and as they were made with woods differing considerably in their apparent weight, their hardness, and other distinctive physical characters, we may conclude, from the great similarity of the results of these experiments, that none of the circumstances from which the woods derive their particular characters have any material influence upon the quantities of charcoal they are capable of yielding ; and hence we may deduce that the ligneous substance or seerwood, if not the same in all, is at least composed of identical substances.

There is still, however, a very interesting question remaining for discussion, namely, Is the seerwood charcoal?

To elucidate this question, I began by examining whether charcoal had the same specific gravity as seerwood.

I, therefore, reduced some common oak-charcoal, which appeared to be well manufactured, into pieces about the size of small peas, and then boiled them in a pretty good quantity of Seine water, previously well filtered; the pores of the charcoal were speedily so completely filled with this liquid, that, becoming heavier than the water, in equal bulk, it precipitated itself to the bottom of the vessel, and there remained.

On removing the vessel from the fire, the water was suffered to cool to the temperature of 60° F.; and then the charcoal, while still submerged, was put into the small glass vase of the hydrostatic balance, and weighed. Its weight in the water, at the temperature of 60° F., was 2.44 grammes.

When the charcoal was taken out of the water, it was put into a cylindrical glass vessel 1½ inch in diameter, and 6 inches in height, in which it was thoroughly dried in the stove at a temperature of about 265° F.

After it had been six hours in the stove, it was taken out and weighed while still hot, and found to be equal to 6.7 grammes; therefore its specific gravity was 157,273.

We have before shown that the specific gravity of the solid parts of oak, in the state of seerwood, is 153,440.

This is certainly very similar to that of charcoal made of the same kind of wood; but we have not yet proved seerwood to be charcoal; on the contrary, we have just seen that it requires 100 parts of seerwood to obtain 43.33 parts of dry charcoal.

Neither is seerwood simply a hydrure of dry wood, as we shall see in the sequel.

It should seem that the fabric of a plant, which may perhaps be nothing but pure charcoal, is always covered with a substance analogous to the flesh which conceals the skeleton of an animal. This vegetable flesh does not exist in considerable masses; for, as the plant is not under the necessity of moving from one place to another in search of nourishment, it has no need either of flexible joints in its skeleton, nor of muscles capable of exerting a great force; and it probably arises from the circumstances of the skeleton and the flesh being very intimately blended together, that they are not discriminated and distinguished from each other.

I consider seerwood as the skeleton of the plant, with the flesh, though quite dried, still adhering to it; and as we have seen that there are 43.33 parts of charcoal in 100 parts of seerwood, I should say that 100 parts of seerwood are composed of

Charcoal . . . . 43.33 parts.
Vegetable flesh, dried 56.67
_____
Making together . . . . 100.00 parts.

The beautiful analyses of Messrs. Gay-Lussac and Thénard have shown us that seerwood is composed of carbon, hydrogen, and oxygen; and that two different species of wood analyzed by them (the beech and the oak) were composed of these three elements in nearly equal proportions. They also discovered that the oxygen and hydrogen in these woods are in the requisite proportions for the formation of water; wherefore they concluded that carbon was the only combustible substance contained in wood.

It will appear in the sequel, how well the results of these ingenious inquiries accord with those of my experiments.

But first, I shall examine what quantity of charcoal it is possible to obtain from different species of woods, under various degrees of dryness, pursuing the method already adopted in my experiments.

From the mode in which charcoal is ordinarily made, a very considerable portion is lost and improvidently burned during the operation.

As it appears to be clearly proved, by the results of the six experiments above related, that the quantity of charcoal to be obtained from any given quantity of wood is invariably in proportion to the quantity of dry ligneous substance contained in the wood, the inquiry into the quantities of charcoal to be produced from different species of woods, at various degrees of dryness, becomes limited to that of the quantities of wood absolutely dry, contained in the woods in question.

It has been shown that 100 parts in weight of oak, thoroughly dried, give 43 parts of charcoal.

We have likewise seen, that 100 parts of oak as dry as it can be made in summer, at the temperature of 62° F., contain only 91 parts of seerwood, and, consequently, that 100 parts of such wood would furnish only 39.13 parts of charcoal.

From the results of an experiment of which I have given an account in this memoir, it appears that 100 parts of oak, in the state wherein it is found when exposed to the winter's air, at the temperature of 46° F., contain only 83.36 parts of seerwood; consequently, 100 parts of such wood would yield no more than 35.84 parts of charcoal.

From the examination we have made of the oak in that state in which it is deemed fit for burning, we have found that 100 parts of this kind of wood contain only 76 parts of absolutely dry wood; whence we conclude that 100 parts of such wood would produce 32.68 parts of charcoal.

It has been shown that 100 parts of an oak felled on the 6th of September, while in a growing state, contained only 62.56 parts of seerwood, and that, consequently, 100 parts of such wood would yield only 26.9 parts of charcoal.

In making these calculations, no account has been taken of the quantity of wood, or other combustible, burned in order to heat the closed vessel in which the wood was carbonized, pursuant to the process here adopted. But it may be remarked that such quantity will be increased or diminished according to the construction of the furnace and the arrangement of the other parts of the apparatus; and it will always be too considerable to be omitted in the list of expenses.

As M. Proust obtained only 19 or 20 parts of charcoal in 100 of oak, it is probable that some waste occurred in the process; but as it is certain that in the carbonization of wood some loss will happen, so in the ordinary method of making charcoal there is always a considerable reduction of the quantity that ought to be produced, arising from the quantity of wood consumed, either wholly or in part, to obtain heat sufficient to char the portion of wood that is reduced to a coal.

Messrs. Gay-Lussac and Thenard found from 52 to

53 parts of carbon in 100 of seerwood, but 100 parts of seerwood yielded me only 43 parts of charcoal; this difference, however, it is easy to explain, as will be seen in the sequel.

Section VII. — *Of the Quantities of Heat developed in the Combustion of different Species of Wood.*

Many persons have already endeavoured to determine the relative quantities of heat furnished by wood and charcoal in their combustion; but the results of their inquiries have not been satisfactory. Their apparatus has been too imperfect, not to leave vast incertitude in the conclusions drawn from their investigations. Indeed, the subject is so intricate in itself, that with the best instruments the utmost care is requisite, lest, after much labour, the inquirer should be forced to content himself with approximations instead of accurate results and valuations strictly determined.

All woods contain much moisture, even when apparently very dry; and as the persons alluded to have neglected to determine the quantities of absolutely dry wood burned by them, much uncertainty prevails in the results of all their experiments.

Another source of uncertainty lies in the great quantity of heat suffered to escape with the smoke and other products of the combustion.

As the calorimeter used in my experiments has been described in a memoir which I had the honour to present to the Class on the 24th of February, 1812,[10] it is unnecessary here to resume that subject; suffice it to explain, in a few words, the various precautions I adopted in burning wood under the calorimeter.

I picked out the woods intended for the experiment from a joiner's workshop, and they all appeared to be quite dry; I had them formed into small boards, 6 inches in length and $\frac{1}{2}$ an inch thick. From these boards I had some shavings planed off, about $\frac{1}{10}$ of a line thick, $\frac{1}{2}$ an inch broad, and 6 inches in length.

When these shavings were sufficiently dry, they were burned, one by one, under the mouth of the calorimeter; and I took care to hold them, by means of a small pair of nippers, so as to make them burn with a brisk flame, and without the least smoke or smell or calculable residuum in ashes.

The following is the method I pursued in making these experiments.

The calorimeter, filled with water at a temperature of about 5° of Fahrenheit's thermometer lower than that of the apartment in which the experiments were made, was placed upon its stand at the height of about 18 inches above the table on which the apparatus was laid.

The extremity of the calorimeter containing the opening, which I call its mouth, projects about 4 inches beyond the edge of the stand, so as easily to admit the point of the flame from the small piece of burning wood; and the height of the stand is so adjusted that the operator may rest both his elbows on the table, while his hands sustain the fragment of wood to be burned.

Near the calorimeter stands a small lamp, by which the pieces of wood, or rather shavings, may, without loss of time, be set on fire and burned in succession; and care is taken to have always in the hand a sufficient quantity of the shavings, of a known weight.

The very small portions of the shavings which remain

between the nippers are carefully preserved, and weighed at the close of the experiments, to determine precisely how much of the wood has been consumed.

An assistant keeps his eye constantly on the thermometer attached to the apparatus, and announces the moment when the water in the calorimeter has attained a temperature as much higher than that of the room as it was below it at the beginning of the operation ; and the flame from the piece of wood then burning is immediately blown out.

The remains of the shaving are laid aside, to be afterwards weighed with the other fragments.

The water in the calorimeter was then stirred, by shaking it, taking care to hold the instrument by its wooden frame, and the temperature of the water was minutely observed and set down in a register.

An experiment of this kind usually occupies about 10 or 12 minutes, according to the nature of the wood and the number of degrees to which the temperature of the calorimeter is raised.

I made choice of the birch for my first experiments, because the texture of its wood is very firm and even, and burns with a very regular flame.

To give the details and their results in few words, I have placed them together in the subjoined table.

The calorimeter, with the water it contained, was equal in capacity, as to heat, to 2781 grammes of water.

Heat developed in the Combustion of Birch Wood.

	No. of exp.	Quantity of wood consumed.	Heat communicated to the calorimeter.	Result. With the heat developed in combustion of 1 pound of combustible	
				Pounds of water heated 1° of Fahrenheit's thermometer.	Pounds of water at the temperature of melting ice, thrown into ebullition.
		Grms.	Degr's.		
Firewood, 2 years old	1	5.00	$10\frac{1}{2}$	} 5875 {	32.445
" " "	2	4.00	$8\frac{1}{2}$		32.841
Shavings dried in the air	3	4.55	$10\frac{1}{4}$	} 6261 {	34.805
" " "	4	4.54	$10\frac{1}{4}$		34.881
Shavings highly dried over a chafing-dish .	5	3.97	10		38.916
Shavings highly dried over a chafing-dish .	6	2.58	$6\frac{1}{2}$	} 7002 {	38.925
Shavings highly dried over a chafing-dish .	7	4.97	$12\frac{1}{2}$		38.858
Shavings highly dried and scorched in a stove	8	5.07	$10\frac{1}{4}$	} 5614 {	31.325
Shavings highly dried and scorched in a stove	9	5.10	$10\frac{1}{4}$		31.052
Shavings scorched, but not to so high a degree	10	4.89	$10\frac{1}{2}$	5971	33.174

On comparing the results of these six experiments, all made with the same kind of wood, in thin shavings, it will appear that the drier the wood, the greater was the quantity of heat produced from a given weight of shavings. But I found, in taking account of the quantities of moisture contained in the woods, the quantities of heat were always sensibly proportional to those of the dry wood burned, with the exception, however, of the three latter experiments, which were made with wood highly dried for 24 hours in a stove, and which gave several indications, by no means equivocal, of the beginning of a decomposition.

The shavings most scorched in the stove gave less

heat than those which had been less scorched ; the two sorts being taken in equal weights.

In all these experiments more or less water dripped from the worm, a certain proof that some hydrogen had been burned ; this fact I was very desirous to verify, on account of its great importance to science.

It is not, therefore, mere carbon which furnished all the heat developed in the combustion of woods ; of this important fact we shall shortly have an additional proof.

As the great quantity of nitrogen carried along with the products of the combustion, and which, after having passed through the worm, was lost in the atmosphere, also, without doubt, took with it a little more moisture than it had brought into the apparatus, a calculation of the quantity of water formed in the combustion of wood, grounded only on that found in the worm, would be erroneous, though there was always considerably more than necessary to demonstrate that water had been formed.

Before we close this paper, we shall point out a mode whereby the quantity of water thus formed may be estimated, even to such a degree of precision as to leave nothing more to be desired.   But it is first necessary to determine the quantity of heat developed in the combustion of the carbon found in this wood, and which was totally consumed.

Although our experiments on the carbonization of wood, in close vessels, by a moderate fire, leave no doubt as to the quantities of charcoal which the woods therein employed were capable of producing, still the knowledge of this fact is not alone sufficient to enable us to determine the quantity of carbon contained in the wood.

As 100 parts of wood are required for 43 of charcoal,

it is evident that the seerwood is at least partially decomposed when the charcoal is produced in the process of carbonization, that is to say, when the skeleton of the wood is deprived of its flesh, and left naked; and it is well known that a great quantity of pyroligneous acid is formed in the carbonization of wood, and this acid contains carbon.

From the process employed by Messrs. Gay-Lussac and Thénard, in their learned analysis, there can be no doubt that they discovered and kept an account of all the carbon found in the woods analyzed by them; and as there was no pyroligneous acid formed in my experiments when the wood was totally consumed without either smoke or smell, it is manifest that in this case all the carbon contained in the wood was burned.

According to the analyses of Messrs. Gay-Lussac and Thénard, 100 parts of oak, perfectly dry, contain 52.54 parts of carbon; and 100 parts of beech contain 51.45.

Now, as it seems to me extremely probable that the dry ligneous substance is palpably the same in all woods, I shall take the medium term of the results of these two analyses, and consider it as an indubitable fact, that 100 parts of perfectly dry wood contain 52 parts of carbon.

Therefore, as 100 parts of seerwood furnished me with only 43 of charcoal, we must conclude, if dry charcoal be considered as carbon, that of the 52 parts of carbon contained in 100 parts of seerwood, 9 are taken up in the composition of the pyroligneous acid formed in the carbonization of the wood, which 9 parts make more than 17 per cent of all the carbon contained in the wood.

Though charcoal should not be purely carbon, we

must, nevertheless, admit that there is still a much greater proportion of carbon employed in the formation of that acid, or of other substances which fly off into the atmosphere during the process of the carbonization of the wood.

In pursuing inquiries in natural philosophy, the first object that demands attention is to keep an accurate account of weights ; and so long as we proceed with the balance in hand, there is little hazard of being misled.

And here, before I proceed further in the inquiry into the sources of the heat developed during the combustion of wood, I shall exhibit a general table of the details and results of forty-three experiments made upon eleven different kinds of the woods of our climate. As I shall have occasion to refer to some of these experiments for the establishment of facts, it is requisite that they should first be known.

All these experiments having been made and registered long before I began the calculations ultimately adopted for the elucidation of their results, I have not hesitated to rely on them. And further, as they were made with all possible care, and with instruments to me apparently perfect, I can answer for their accuracy.

New experiments ever bear a certain value ; all the knowledge which constitutes the imperishable riches of mankind consists only of accurate statements of well-conducted experiments. Happy they who have the good fortune of contributing something to the general stock !

Heat developed in the Combustion of various Species of Wood.

Species of wood.	Quality.	Number of the experiment.	Quantity of wood burned.	Heat communicated to the calorimeter whose capacity was equal to 2781 grammes of water.	Quantity of water at the thawing temperature, boiled by the heat developed in the combustion of one pound of combustible.
			Grms.	Degr's.	Pounds.
Lime	Joiner's dry wood, 4 years old	11	4.52	10⅛	34.609
		12	4.55	10¼	34.805
	Same kind highly dried over a chafing-dish	13	4.06	10¼	39.605
		14	3.80	10	40.658
	Same kind, rather less dried .	15	5.57	14	38.833
Beech	Joiner's dry wood, 4 or 5 years old	16	4.74	10⅜	33.817
		17	4.72	10¼	33.752
	Same kind, highly dried over a chafing-dish.	18	5.07	12	36.334
		19	4.43	10⅜	36.184
Elm	Joiner's wood, rather moist .	20	6.34	11½	27.147
	Joiner's dry wood, 4 or 5 years old	21	5.28	10⅜	30.359
		22	5.45	10⅝	30.051
	Same kind, highly dried over a chafing-dish	23	4.70	10½	34.515
		24	5.28	11½	33.651
	Same kind, dried and scorched in the stove.	25	4.00	8	30.900
Oak	Common firewood, in moderate shavings	26	4.83	8	25.590
	Same kind in thicker shavings; leaving a residuum of charcoal	27	6.40	10¼	24.748
	Same kind in thin shavings .	28	6.14	10½	26.272
	Same kind, thin shavings, well dried in the air . . .	29	7.22	13	29.210
	Joiner's wood, very dry, in thin shavings	30	5.30	10¼	29.880
		31	5.33	10¼	19.796
	Thick shavings, leaving 0.92 gramme of charcoal . . .	32	6.48	11	26.227
Ash	Joiner's common dry wood .	33	5.29	10½	30.666
	Same kind, shavings dried in the air	34	3.78	8¼	33.720
	Same kind, highly dried over a chafing-dish	35	5.23	12	35.449

Heat developed in the Combustion of Wood.   (Continued.)

Species of wood.	Quality.	Number of the experiment.	Quantity of wood burned.	Heat communicated to the calorimeter whose capacity was equal to 2781 grammes of water.	Quantity of water at the thawing temperature, boiled by the heat developed in the combustion of one pound of combustible.
			Grms.	Degr's.	Pounds.
Maple	Seasoned wood, highly dried over a chafing-dish . . .	36	3.85	9	36.117
Service	Same kind	37	4.49	10½	36.130
	Same kind, scorched in a stove	38	4.30	9	32.337
Cherry	Joiner's dry wood	39	4.75	10¼	33.339
	Same kind, highly dried over a chafing-dish	40	4.36	10½	36.904
	Same kind, scorched in a stove	41	5.00	11¼	34.763
Fir	Joiner's common dry wood .	42	5.35	10½	30.322
	Shavings, well dried in the air	43	4.09	9	34.000
	Highly dried over a chafing-dish	44	3.72	9	37.379
	Dried and scorched in a stove	45	4.40	9½	33.358
	Thick shavings, leaving much charcoal	46	4.51	6½	28.695
Poplar	Joiner's common dry wood .	47	4.13	9¼	34.601
	Same kind, highly dried over a chafing-dish	48	3.95	9½	37.161
Hornbeam	Joiner's dry wood	49	4.98	10¼	31.800
		50	5.01	10¼	31.609
Oak	Dried to {81.4 wood / 19.6 water} imperfectly burned, leaving a residuum of charcoal, in the combustion, of 0.81 gramme	51	6.14	10½	26.421
	0.73 "	52	4.83	8	25.591
	0.94 "	53	6.71	11	25.917

These experiments might lead to a great number of observations ; but I shall endeavour to reduce them to the exposition of a few simple facts which they present.

One fact, certainly very curious and of the first importance to the knowledge of the vegetable economy, appears to be well established; namely, that the skeleton of trees is pure charcoal, and that it exists in a perfect state in wood.

If this charcoal did not exist perfectly formed in wood, it could not possibly preserve its form, while its envelope of vegetable flesh is destroyed by the fire in the process of carbonization.

As the vegetable flesh contains hydrogen as well as carbon, it is more inflammable than charcoal, and is consumed at a lower temperature; and, by proper management of the fire, it may be totally destroyed without the enclosed skeleton of charcoal being injured.

Some months ago I presented the Class with a small sprig of charcoal produced from a piece of oak partially burned under my calorimeter. It was nearly all the charcoal contained in the piece. All the coal or flesh of the wood burned with a brisk flame, and the skeleton of the wood had got red, but the heat was not sufficient to consume it.

The charcoal-maker seldom does more than burn the flesh of the wood, and leaves the skeleton of charcoal naked.

The dry vegetable flesh produces more heat in its combustion than an equal weight of dry charcoal.

Shavings scorched in the stove by a great heat yield less heat in their combustion than shavings of the same kind of wood, whose vegetable flesh has not been touched. See experiments Nos. 5, 6, 7, 8, 9, 10, 25, 38, 41, 45.

In tables of experiments similar to those registered in the preceding table, it is scarcely possible to have errors on the greater side; but they may easily enough happen on the lesser. We may, therefore, place the more confidence in those wherein the quantities of heat manifested have been the greatest.

In the experiments Nos. 13 and 14, the wood of the lime-tree, dried over a chafing-dish, was productive of more heat than any other wood that I examined.

The result, it will be seen, was for 1 pound of this wood burned in experiment,

No. 13 . . 39.605 pounds of water heated 180° F., and in
No. 14 . . 40.658 " " "
─────────
Medium . . 40.1315

In order accurately to ascertain how much water this wood contained, I dried thoroughly in the stove a parcel of shavings which had been previously dried over the chafing-dish, and found that it still retained 6.977 per cent of water.

Therefore we may conclude that 1 pound of this wood contains only 0.93023 pound of seerwood.

Now, if 0.93023 pound of seerwood will heat 40.-1315 pounds of water 180° F., 1 pound of the same wood ought to heat 43.141 pounds; and I therefore take this quantity of water heated 180° F. as the standard of the heat developed in the combustion of 1 pound of wood perfectly dried.

Many persons have endeavoured to account for the heat manifested in the combustion of wood, by attributing it altogether to the charcoal contained in the wood burned.

This hypothesis we have now to examine.

It has been seen, that 100 parts of the wood of the lime-tree, perfectly dried, yielded 43.59 parts of charcoal; consequently 1 pound of this wood, thoroughly dried, can contain only 0.4359 pound of charcoal.

According to the results of Crawford's experiments, which we have found to be very accurate, 1 pound of charcoal furnishes in its combustion only the necessary heat for raising 57.608 pounds of water 180° F. ; therefore the charcoal contained in 1 pound of dry lime-wood, equal to 0.4359 pound, can furnish in its combustion no more heat than is necessary to raise 25.111 pounds of water 180°; but as the experiment has given 43.141 pounds, there must certainly have been some other substance burned beside the charcoal, and which could have been none other than hydrogen.

Before we determine the quantity of hydrogen consumed, it is essential to ascertain how much heat has been furnished, not merely by the charcoal itself, but by the charcoal and the carbon contained in the wood; for it is very certain that all the carbon was burned, since no pyroligneous acid was formed.

According to the analyses of Messrs. Gay Lussac and Thénard, 1 pound of dry wood contains 0.52 pound of carbon.

If we adopt Crawford's estimate, we shall find that the combustion of 0.52 pound of carbon ought to furnish heat sufficient to raise 29.956 pounds of water 180° F.

Deducting this quantity of water from that given by the experiment, namely, 43.141 pounds, we shall have 13.185 pounds as the measure of the heat produced by the combustion of the hydrogen consumed in the experiment.

From the results of this inquiry we may conclude, that of the heat manifested in the combustion of wood, rather more than two thirds are produced by the combustion of the carbon, and a little less than one third by the hydrogen consumed.

These data supply us with an easy method of determining the quota of free and combustible hydrogen contained in seerwood.

According to Crawford's estimate, which we have followed all along, 1 pound of hydrogen yields in its combustion heat sufficient to raise 410 pounds of water 180° F.; therefore, the 13.185 pounds heated 180° in the experiment in question must have required 0.035158 pound of hydrogen, which is consequently the amount of free and combustible hydrogen contained in 1 pound of seerwood.

Assuming the medium term of the results of the two analyses of dry wood, made by Messrs. Gay Lussac and Thénard, 1 pound of seerwood would be composed of

Carbon	0.52 pound.
Hydrogen and oxygen, in the necessary proportions for forming water . .	0.48
	1.00

From the result of my experiments, 1 pound of seerwood is composed of two distinct substances; namely, —

A skeleton of charcoal weighing . . .	0.43 pound.
Vegetable flesh	0.57
	1.00

And these 0.57 pound of vegetable flesh are composed of

Carbon, free and combustible . . .	0.090 pound.
Hydrogen, free and combustible . .	0.035
Hydrogen and oxygen, in the necessary proportions for the formation of water . .	0.445
	0.570

In making these estimates, I have availed myself of the valuation of the total quantity of carbon contained in seerwood, given in the analysis of Messrs. Gay Lussac and Thénard ; and I have supposed the 43 per cent of charcoal, which I found to be contained in seerwood, to be pure carbon.

Should it ultimately appear that charcoal is not pure carbon, which is extremely probable, numerous alterations in all these estimates must follow, though the experiments made upon the woods will always retain their value. And I cannot but hope that they will be frequently repeated, with such variations as may conduce to important discoveries.

It will be a satisfaction to me to know that I have put into the hands of more skilful workmen than myself some instruments of which they may advantageously avail themselves : and to have pointed out, as well as a little smoothed, a new path, wherein they may walk without danger of being lost.

SECTION VIII. — *Of the Quantity of Heat lost in the Carbonization of Wood.*

In making charcoal, a considerable quantity of heat is dissipated and lost in the air; whence it is evident that the same amount of heat cannot be obtained from burning a given quantity of charcoal as would be furnished by the combustion of the wood of which it is formed.

We can now determine, with great precision, the loss of heat which is inevitable in making charcoal, even when all possible precautions have been taken; as well as that which happens every day in the process employed by the charcoal-maker.

As the combustion of 1 pound of charcoal, perfectly dry, yields heat sufficient to boil 57.608 pounds of water, at the thawing temperature; and as 1 pound of wood, thoroughly dry, furnishes 0.4333 pound of dry charcoal, it follows, that the charcoal produced from 1 pound of dry wood should furnish in its combustion heat sufficient to boil 24.958 pounds of water, at the thawing temperature.

But we have already seen that the combustion of 1 pound of wood, thoroughly dry, should furnish sufficient heat to boil 43.143 pounds of water at the freezing temperature; or, which is the same thing, to raise it 180° of Fahrenheit's thermometer.

These two numbers (43.143 and 24.958), which express the quantities of heat in question, being in the proportion of 100 to 57.849, it is evident that the loss of heat *inevitable* in the carbonization of wood is upwards of 42 per cent, or exactly 42.151 per cent of the total quantity that the wood will furnish.

In order to determine the loss of heat which occurs in the forests, by the ordinary process of the charcoal-burner, it is requisite to ascertain the precise product of charcoal from a given quantity of wood, though it is probable that this product is very variable. M. Proust estimates it at 20 per cent in weight at the highest.

Adopting, for a moment, this estimate, and supposing the carbonized wood in the same state of dryness as what is usually sold for firewood; as 100 pounds of such wood contains only 0.76 pound of perfectly dry wood, this quantity would furnish in its combustion only the degree of heat necessary to raise 32.043 pounds of water 180° F.

But the 0.20 pound of charcoal produced by the carbonization of 1 pound of this wood, according to the usual process, can only furnish by combustion a sufficient quantity of heat to raise 11.521 pounds of water 180° F. ; and as the numbers 32.043 and 11.521 are nearly in the proportion of 100 to 36, it should seem that the loss of heat in question is about 64 per cent.

One very important fact, which appears to be well ascertained by the results of this inquiry, is, that all the charcoal produced from the carbonization of 3 pounds of any kind of wood scarcely gives more heat in its combustion than would be furnished by 1 pound of the same sort of wood burned, and in its natural state.

CHIMNEY FIREPLACES,

PROPOSALS FOR IMPROVING THEM TO SAVE FUEL; TO RENDER DWELLING–HOUSES MORE COMFORTABLE AND SALUBRIOUS, AND EFFECTUALLY TO PREVENT CHIMNEYS FROM SMOKING.

Advertisement

The Author thinks it his duty to explain the reasons which have induced him to change the order in which the publications of his Essays has been announced to the Public.

Being suddenly called upon to send to Edinburgh a person acquainted with the method of altering chimney fire-places, which has lately been carried into execution in a number of houses in London, in order to introduce these improvements in Scotland, he did not think it prudent to send any person on so important an errand without more ample instructions than could well be given verbally; and being obliged to write on the subject, he thought it best to investigate the matter thoroughly, and to publish such particular directions respecting the improvements in question as may be sufficient to enable all those, who may be desirous of adopting them, to make, or direct the necessary alterations in their fire-places without any further assistance.

The following letter, which the Author received from Sir John Sinclair, Baronet, MP, and President of the Board of Agriculture, will explain this matter more fully:

"You will hear with pleasure that your mode of altering chimnies, so as to prevent their smoking, to save fuel, and to augment heat, has answered not only with me, but with many of my friends who have tried it; and that the Lord Provost and Magistrates of Edinburgh have voted a sum of money to defray the expences of a bricklayer, who is to be sent there for the purpose of establishing the same plan in that city. I hope that you will have the goodness to expedite your paper upon the management of Heat, that the knowledge of so useful an art may be as rapidly and as extensively diffused as possible.

With my best wishes for your success in the various important pursuits in which you are now engaged, believe me, with great truth and regard,

Whitehall, London 9th February, 1796

Your faithful and obedient servant, John Sinclair"

CHAPTER I.

Fireplaces for burning Coals, or Wood, in an open Chimney, are capable of great Improvement. — Smoking Chimneys may in all cases be completely cured. — The immoderate Size of the Throats of Chimneys the principal Cause of all their Imperfections. — Philosophical Investigation of the Subject. — Remedies proposed for all the Defects that have been discovered in Chimneys and their open Fireplaces. — These Remedies applicable to Chimneys destined for burning Wood, or Turf, as well as those constructed for burning Coals.

THE plague of a smoking chimney is proverbial; but there are many other very great defects in open fireplaces, as they are now commonly constructed in this country, and indeed throughout Europe, which, being less obvious, are seldom attended to; and there are some of them very fatal in their consequences to health; and, I am persuaded, cost the lives of thousands every year in this island.

Those cold and chilling draughts of air on one side of the body while the other side is scorched by a chim-

222

ney fire, which every one who reads this must often have felt, cannot but be highly detrimental to health, and in weak and delicate constitutions must often produce the most fatal effects. I have not a doubt in my own mind that thousands die in this country every year of consumptions occasioned solely by this cause, — by a cause which might be so easily removed ! — by a cause whose removal would tend to promote comfort and convenience in so many ways !

Strongly impressed as my mind is with' the importance of this subject, it is not possible for me to remain silent. The subject is too nearly connected with many of the most essential enjoyments of life not to be highly interesting to all those who feel pleasure in promoting or in contemplating the comfort and happiness of mankind. And without suffering myself to be deterred either by the fear of being thought to give to the subject a degree of importance to which it is not entitled, or by the apprehension of being tiresome to my readers by the prolixity of my descriptions, I shall proceed to investigate the subject in all its parts and details with the utmost care and attention. And first with regard to smoking chimneys.

There are various causes by which chimneys may be prevented from carrying smoke, but there are none that may not easily be discovered and completely removed. This will doubtless be considered as a bold assertion ; but I trust I shall be able to make it appear in a manner perfectly satisfactory to my readers that I have not ventured to give this opinion but upon good and sufficient grounds.

Those who will take the trouble to consider the nature and properties of elastic fluids, of air, smoke, and vapour, and to examine the laws of their motions, and the necessary consequences of their being rarified by heat, will perceive that it would be as much a miracle if smoke should not rise in a chimney, all hindrances to its ascent being removed, as that water should refuse to run in a syphon, or to descend in a river.

The whole mystery, therefore, of curing smoking chimneys, is comprised in this simple direction; *find out and remove those local hindrances which forcibly prevent the smoke from following its natural tendency to go up the chimney;* or rather, to speak more accurately, which prevent its being forced up the chimney by the pressure of the heavier air of the room.

Although the causes by which the ascent of smoke in a chimney *may be* obstructed are various, yet that cause which will most commonly, and I may say almost universally, be found to operate, is one which it is always very easy to discover, and as easy to remove, — the bad construction of the chimney *in the neighbourhood of the fireplace.*

In the course of all my experience and practice in curing smoking chimneys, — and I certainly have not had less than five hundred under my hands, and among them many which were thought to be quite incurable, — I never have been obliged, except in one single instance, to have recourse to any other method of cure than merely reducing the fireplace, and the throat of the chimney, or that part of it which lies immediately above the fireplace, to a proper form and just dimensions.

That my principles for constructing fireplaces are equally applicable to those which are designed for burning coal, as to those in which wood is burned, has lately been abundantly proved by experiments made here in

London; for of above a hundred and fifty fireplaces
which have been altered in this city under my direction,
within these last two months, there is not one which has
not answered perfectly well.* And by several experiments which have been made with great care, and with
the assistance of thermometers, it has been demonstrated, that the saving of fuel, arising from these improvements of fireplaces, amounts in all cases to more
than *half*, and in many cases to more than *two thirds*, of
the quantity formerly consumed. Now as the alterations in fireplaces which are necessary may be made at a
very trifling expense,— as any kind of grate or stove may
be made use of, and as no iron work but merely a few
bricks and some mortar, or a few small pieces of fire-
stone, are required, — the improvement in question is
very important when considered merely with a view to
economy; but it should be remembered, that not only
a great saving is made of fuel by the alterations proposed, but that rooms are made much more comfortable,

---

* Eves and Sutton, bricklayers, Broad Sanctuary, Westminster, have alone altered
above ninety chimneys. The experiment was first made in London at Lord Palmerston's house in Hanover Square; then two chimneys were altered in the house of Sir
John Sinclair, Baronet, President of the Board of Agriculture; one in the room in
in which the Board meets, and the other in the Secretary's room; which last being
much frequented by persons from all parts of Great Britain, it was hoped that circumstance would tend much to expedite the introduction of these improvements in various
parts of the kingdom. Several chimneys were then altered in the house of Sir Joseph
Banks, Baronet, K. B., President of the Royal Society. Afterwards a number were
altered in Devonshire House; in the house of Earl Besborough, in Cavendish Square,
and at his seat at Rockhampton; at Holywell House, near St. Alban's, the seat of
the Countess Dowager Spencer; at Melbourne House; at Lady Templeton's, in Port-
land Place; at Mrs. Montagu's, in Portman Square; at Lord Sudley's, in Dover
Street; at the Marquis of Salisbury's seat, at Hatfield, and at his house in town; at
Lord Palmerston's seat in Broadlands, near Southampton, and at several gentlemen's
houses in that neighborhood; and a great many others; but it would be tiresome to
enumerate them all, and even these are mentioned merely for the satisfaction of
those who may wish to make inquiries respecting the success of the experiments.

and more salubrious ; that they may be more equally warmed, and more easily kept at any required temperature ; that all draughts of cold air from the doors and windows towards the fireplace, which are so fatal to delicate constitutions, will be completely prevented ; that in consequence of the air being equally warm all over the room, or in all parts of it, it may be entirely changed with the greatest facility, and the room completely ventilated when this air is become unfit for respiration, and this merely by throwing open for a moment a door opening into some passage from whence fresh air may be had, and the upper part of a window ; or by opening the upper part of one window and the lower part of another. And as the operation of ventilating the room, even when it is done in the most complete manner, will never require the door and window to be open more than one minute, in this short time the walls of the room will not be sensibly cooled, and the fresh air which comes into the room will, in a very few minutes, be so completely warmed by these walls, that the temperature of the room, though the air in it be perfectly changed, will be brought to be very nearly the same as it was before the ventilation.

Those who are acquainted with the principles of pneumatics, and know why the warm air in a room rushes out at an opening made for it at the top of a window when colder air from without is permitted to enter by the door or by any other opening situated lower than the first, will see that it would be quite impossible to ventilate a room in the complete and expeditious manner here described, where the air in a room is partially warmed, or hardly warmed at all, and where the walls of the room, remote from the fire, are con-

stantly cold; which must always be the case where, in
consequence of a strong current up the chimney, streams
of cold air are continually coming in through all the
crevices of the doors and windows, and flowing into the
fireplace.

But although rooms furnished with fireplaces constructed upon the principles here recommended, may be
easily and most effectually ventilated (and this is certainly a circumstance in favour of the proposed improvements), yet such total ventilations will very seldom, if ever, be necessary. As long as *any fire* is kept
up in the room, there is so considerable a current of air
up the chimney, notwithstanding all the reduction that
can be made in the size of its throat, that the continual
change of air in the room which this current occasions
will, generally, be found to be quite sufficient for keeping the air in the room sweet and wholesome; and, indeed, in rooms in which there is no open fireplace, and
consequently no current of air from the room setting
up the chimney,— which is the case in Germany and all
the northern parts of Europe, where rooms are heated
by stoves, whose fireplaces, opening without, are not supplied with the air necessary for the combustion of
the fuel from the room ; and although in most of the
rooms abroad, which are so heated, the windows and·
doors are double, and both are closed in the most exact
manner possible, by slips of paper pasted over the crevices, or by slips of list or fur, yet when these rooms are
tolerably large, and when they are not very much crowded
by company, nor filled with a great many burning lamps
or candles, the air in them is seldom so much injured
as to become oppressive or unwholesome, and those
who inhabit them show by their ruddy countenances, as

well as by every other sign of perfect health, that they
suffer no inconvenience whatever from their closeness.
There is frequently, it is true, an oppressiveness in the
air of a room heated by a German stove, of which those
who are not so much accustomed to living in those
rooms seldom fail to complain, and indeed with much
reason ; but this oppressiveness does not arise from the
air of the room being injured by the respiration and
perspiration of those who inhabit it ; it arises from a
very different cause, — from a fault in the construction
of German stoves in general, but which may be easily
and most completely remedied, as I shall show more
fully in another place.   In the mean time, I would just
observe here with regard to these stoves, that as they
are often made of iron, and as this metal is a very good
conductor of heat, some part of the stove in contact
with the air of the room becomes so hot as to calcine or
rather to *roast* the dust which lights upon it ; which never
can fail to produce a very disagreeable effect on the air
of the room.   And even when the stove is constructed
of pantiles or pottery-ware, if any part of it in contact
with the air of the room is suffered to become very hot,
which seldom fails to be the case in German stoves constructed on the common principles, nearly the same
effects will be found to be produced on the air as when
the stove is made of iron, as I have very frequently had
occasion to observe.

Though a room be closed in the most perfect manner
possible, yet, as the quantity of air injured and rendered
unfit for further use by the respiration of two or three
persons in a few hours is very small compared to the
immense volume of air which a room of a moderate
size contains ; and as a large quantity of fresh air

always enters the room, and an equal quantity of the warm air of the room is driven out of it every time the door is opened, there is much less danger of the air of a room becoming unwholesome for the want of ventilation than has been generally imagined; particularly in cold weather, when all the different causes which conspire to change the air of warmed rooms act with increased power and effect.

Those who have any doubts respecting the very great change of air or ventilation which takes place each time the door of a warm room is opened in cold weather, need only set the door of such a room wide open for a moment, and hold two lighted candles in the doorway, one near the top of the door and the other near the bottom of it: the violence with which the flame of that above will be driven outwards, and that below inwards, by the two strong currents of air which, passing in opposite directions, rush in and out of the room at the same time, will be convinced that the change of air which actually takes place must be very considerable indeed; and these currents will be stronger, and consequently the change of air greater, in proportion as the difference is greater between the temperatures of the air within the room and of that without. I have been more particular upon this subject, — the ventilation of warmed rooms which are constantly inhabited, — as I know that people in general in this country have great apprehensions of the bad consequences to health of living in rooms in which there is not a continual influx of cold air from without. I am as much an advocate for a *free circulation* of air as anybody, and always sleep in a bed without curtains on that account; but I am much inclined to think, that the currents of cold air which never fail to be produced in rooms heated by fireplaces constructed upon the common principle, — those partial heats on one side of the body, and cold blasts on the other, so often felt in houses in this country, — are infinitely more detrimental to health than the supposed closeness of the air in a room warmed more equally, and by a smaller fire.

All these advantages, attending the introduction of the improvements in fireplaces here recommended, are certainly important, and I do not know that they are counterbalanced by any one disadvantage whatsoever. The only complaint that I have ever heard made against them was that they made the rooms *too* warm; but the remedy to this evil is so perfectly simple and obvious, that I should be almost afraid to mention it, lest it might be considered as an insult to the understanding of the person to whom such information should be given; for nothing surely can be conceived more perfectly ridiculous than the embarrassment of a person on account of the too great heat of his room, when it is in his power to diminish *at pleasure* the fire by which it is warmed; and yet, strange as it may appear, this has sometimes happened!

Before I proceed to give directions for the construction of fireplaces, it will be proper to examine more carefully the fireplaces now in common use; to point out their faults; and to establish the principles upon which fireplaces ought to be constructed.

The great fault of all the open fireplaces, or chimneys, for burning wood or coals in an open fire, now in common use, is, that they are much too large; or, rather, it is *the throat of the chimney*, or the lower part of its open canal, in the neighborhood of the mantle and immediately over the fire, which is too large. This opening has hitherto been left larger than otherwise it probably would have been made, in order to give a passage to the chimney-sweeper; but I shall show hereafter how a passage for the chimney-sweeper may be contrived without leaving the throat of the chimney of such enormous dimensions as to swallow up and devour all the warm air of the room, instead of merely giving a passage to the smoke and heated vapour which rise from the fire, for which last purpose alone it ought to be destined.

Were it my intention to treat my subject in a formal scientific manner, it would doubtless be proper, and even necessary, to begin by explaining in the fullest manner, and upon the principles founded on the laws of nature, relative to the motions of elastic fluids, as far as they have been discovered and demonstrated, the causes of the ascent of smoke; and also to explain and illustrate upon the same principles, and even to measure or estimate by calculations, the precise effects of all those mechanical aids which may be proposed for assisting it in its ascent, or rather for removing those obstacles which hinder its motion upwards; but as it is my wish rather to write a useful practical treatise than a learned dissertation,—being more desirous to contribute in diffusing useful knowledge by which the comforts and enjoyments of mankind may be increased, than to acquire the reputation of a philosopher among learned men,—I shall endeavour to write in such a manner as to be easily understood *by those who are most likely to profit by the information I have to communicate*, and consequently most likely to assist in bringing into general use the improvements I recommend. This being premised, I shall proceed, without any further preface or introduc-

tion, to the investigation of the subject I have undertaken to treat.

As the immoderate size of the throats of chimneys is
the great fault of their construction, it is this fault
which ought always to be first attended to in every
attempt which is made to improve them; for however
perfect the construction of a fireplace may be in other
respects, if the opening left for the passage of the
smoke is larger than is necessary for that purpose, nothing can prevent the warm air of the room from escaping
through it; and whenever this happens, there is not
only an unnecessary loss of heat, but the warm air which
leaves the room to go up the chimney being replaced by
cold air from without, the draughts of cold air, so often
mentioned, cannot fail to be produced in the room,
to the great annoyance of those who inhabit it. But
although both these evils may be effectually remedied
by reducing the throat of the chimney to a proper size,
yet in doing this several precautions will be necessary.
And first of all, the throat of the chimney should be in
its proper place: that is to say, in that place in which it
ought to be, in order that the ascent of the smoke may
be most facilitated; for every means which can be employed for facilitating the ascent of the smoke in the
chimney must naturally tend to prevent the chimney
from smoking; now as the smoke and hot vapour which
rise from a fire naturally tend *upwards*, the proper place
for the throat of the chimney is evidently perpendicularly *over the fire*.

But there is another circumstance to be attended to
in determining the proper place for the throat of a
chimney, and that is to ascertain its distance from the
fire, or *how far* above the burning fuel it ought to be

placed. In determining this point, there are many things to be considered, and several advantages and disadvantages to be weighed and balanced.

As the smoke and vapour which ascend from burning fuel rise in consequence of their being rarefied by heat, and made lighter than the air of the surrounding atmosphere; and as the degree of their rarefaction, and consequently their tendency to rise, is in proportion to the intensity of their heat; and further, as they are hotter near the fire than at a greater distance from it, it is clear that the nearer the throat of a chimney is to the fire, the stronger will be what is commonly called its *draught*, and the less danger there will be of its smoking. But on the other hand, when the draught of a chimney is very strong, and particularly when this strong draught is occasioned by the throat of the chimney being very near the fire, it may so happen that the draught of air into the fire may become so strong as to cause the fuel to be consumed too rapidly. There are likewise several other inconveniences which would attend the placing of the throat of a chimney *very near* the burning fuel.

In introducing the improvements proposed, in chimneys already built, there can be no question in regard to the height of the throat of the chimney, for its place will be determined by the height of the mantle. It can hardly be made lower than the mantle; and it ought always to be brought down as nearly upon the level with the bottom of it as possible. If the chimney is apt to smoke, it will sometimes be necessary either to lower the mantle or to diminish the height of the opening of the fireplace, by throwing over a flat arch, or putting in a straight piece of stone from one side of it to the other, or, which will be still more simple and easy in practice, building a wall of bricks, supported by a flat bar of iron, immediately under the mantle.

Nothing is so effectual to prevent chimneys from smoking as diminishing the opening of the fireplace in the manner here described, and lowering and diminishing the throat of the chimney; and I have always found, except in the single instance already mentioned, that a perfect cure may be effected by *these means alone*, even in the most desperate cases. It is true, that when the construction of the chimney is very bad indeed, or its situation very unfavourable to the ascent of the smoke, and especially when both these disadvantages exist at the same time, it may sometimes be necessary to diminish the opening of the fireplace, and particularly to lower it, and also to lower the throat of the chimney, more than might be wished; but still I think this can produce no inconveniences to be compared with that greatest of all plagues, a smoking chimney.

The position of the throat of a chimney being determined, the next points to be ascertained are its size and form, and the manner in which it ought to be connected with the fireplace below, and with the open canal of the chimney above.

But as these investigations are intimately connected with those which relate to the form proper to be given to the fireplace itself, we must consider them all together.

That these inquiries may be pursued with due method, and that the conclusions drawn from them may be clear and satisfactory, it will be necessary to consider, first, what the objects are which ought principally to be had in view in the construction of a fireplace; and secondly, to see how these objects can best be attained.

Now the design of a chimney fire being simply to

warm a room, it is necessary, first of all, to contrive matters so that the room shall be actually warmed; secondly, that it be warmed with the smallest expense of fuel possible; and, thirdly, that, in warming it, the air of the room be preserved perfectly pure and fit for respiration, and free from smoke and all disagreeable smells.

In order to take measures with certainty for warming a room by means of an open chimney fire, it will be necessary to consider *how*, or *in what manner*, such a fire communicates heat to a room. This question may perhaps, at the first view of it, appear to be superfluous and trifling, but a more careful examination of the matter will show it to be highly deserving of the most attentive investigation.

To determine in what manner a room is heated by an open chimney fire, it will be necessary, first of all, to find out *under what form* the heat generated in the combustion of the fuel exists, and then to see how it is communicated to those bodies which are heated by it.

In regard to the first of these subjects of inquiry, it is quite certain that the heat which is generated in the combustion of the fuel exists under *two* perfectly distinct and very different forms. One part of it is *combined* with the smoke, vapour, and heated air, which rise from the burning fuel, and goes off with them into the upper regions of the atmosphere; while the other part, which appears to be *uncombined*, or, as some ingenious philosophers have supposed, combined only with light, is sent off from the fire in rays in all possible directions.

With respect to the second subject of inquiry, namely, — how this heat, existing under these two different forms, is communicated to other bodies; it is highly probable that the combined heat can only be communicated to other bodies by *actual contact* with the body with which it is combined; and with regard to the rays which are sent off by burning fuel, it is certain that *they* communicate or generate heat only *when* and *where* they are stopped or absorbed. In passing through air, which is transparent, they certainly do not communicate any heat to it; and it seems highly probable that they do not communicate heat to solid bodies by which they are reflected.

In these respects they seem to bear a great resemblance to the solar rays. But n order not to distract the attention of my reader or carry him too far away from the subject more immediately under consideration, I must not enter too deeply into these inquiries respecting the nature and properties of what has been called *radiant heat*. It is certainly a most curious subject of philosophical investigation, but more time would be required to do it justice than we now have to spare. We must, therefore, content ourselves with such a partial examination of it as will be sufficient for our present purpose.

A question which naturally presents itself here is, What proportion does the radiant heat bear to the combined heat? Though that point has not yet been determined with any considerable degree of precision, it is, however, quite certain, that the quantity of heat which goes off combined with the smoke, vapour, and heated air, is much more considerable, perhaps three or four times greater, at least, than that which is sent off from the fire in rays. And yet, small as the quantity is of this radiant heat, it is the only part of the heat generated in the combustion of fuel burned in an open fireplace, which is ever employed, or which can ever be employed, in heating a room.

The whole of the combined heat escapes by the chimney, and is totally lost; and, indeed, no part of it could ever be brought into a room from an open fireplace, without bringing along with it the smoke with which it is combined; which, of course, would render it impossible for the room to be inhabited. There is, however, one method by which combined heat, and even that which arises from an open fireplace, may be made to assist in warming a room; and that is by making it pass through something analogous to a German stove, placed in the chimney above the fire. But of this contrivance I shall take occasion to treat more fully hereafter; in the mean time I shall continue to investigate the properties of open chimney fireplaces, constructed upon the most simple principles, such as are now in common use; and shall endeavour to point out and explain all those improvements of which *they* appear to me to be capable. When fuel is burned in fireplaces upon this simple construction, where the smoke escapes immediately by the open canal of the chimney, it is quite evident that all the combined heat must of necessity be lost; and as it is the radiant heat alone which can be employed in heating a room, it becomes an object of much importance to determine how the greatest quantity of it may be generated in the combustion of the fuel, and how the greatest proportion possible of that generated may be brought into the room.

Now, the quantity of radiant heat generated in the combustion of a given quantity of any kind of fuel depends very much upon the management of the fire, or upon the manner in which the fuel is consumed. When the fire burns bright, much radiant heat will be sent off from it; but when it is *smothered up*, very little will be generated; and indeed very little combined heat, that can be employed to any useful purpose; most of the heat produced will be immediately *expended* in giving elasticity to a thick dense vapour or smoke which will be seen rising from the fire; and the combustion being very incomplete, a great part of the inflammable matter of the fuel being merely rarefied and driven up the chimney without being inflamed, the fuel will be wasted to little purpose. And hence it appears of how much importance it is, whether it be considered with a view to economy, or to cleanliness, comfort, and elegance, to pay due attention to the management of a chimney fire.

Nothing can be more perfectly void of common-sense, and wasteful and slovenly at the same time, than the manner in which chimney fires, and particularly where coals are burned, are commonly managed by servants. They throw on a load of coals at once, through which the flame is hours in making its way; and frequently it is not without much trouble that the fire is prevented from going quite out. During this time, no heat is communicated to the room; and what is still worse, the throat of the chimney, being occupied merely by a heavy dense vapour not possessed of any considerable degree of heat, and consequently not having much elasticity, the warm air of the room finds less difficulty in forcing its way up the chimney and escaping, than when the fire burns bright; and it happens not unfrequently, especially in chimneys and fireplaces ill constructed, that this current of warm air from the room, which presses into the chimney, crossing upon the current of heavy smoke which rises slowly from the fire, obstructs it in its ascent, and beats it back into the room; hence it is that chimneys so often smoke when too large a quantity of fresh coals is put upon the fire. So many coals should never be put on the fire at once, as to prevent the free passage of the flame between. In short, a fire should never be smothered; and when proper attention is paid to the quantity of coals put on, there will be very little use for the poker; and this circumstance will contribute very much to cleanliness and to the preservation of furniture.

Those who have feeling enough to be made miserable by anything careless, slovenly, and wasteful, which happens under their eyes, who know what comfort is, and consequently are worthy of the enjoyments of a *clean hearth* and *cheerful fire*, should really either take the trouble themselves to manage their fires (which, indeed, would rather be an amusement to them than a trouble), or they should instruct their servants to manage them better.

But to return to the subject more immediately under consideration. As we have seen what is necessary to the production or generation of radiant heat, it remains to determine how the greatest proportion of that generated and sent off from the fire in all directions may be made to enter the room, and assist in warming it. Now, as the rays which are thrown off from burning fuel have this property in common with light, that they generate heat only *when* and *where* they are stopped or absorbed, and also in being capable of being reflected *without generating heat* at the surfaces of various bodies, the knowledge of these properties will enable us to take measures, with the utmost certainty, for producing the effect required, — that is to say, for bringing as much radiant heat as possible into the room.

This must be done, first, by causing as many as possible of the rays, as they are sent off from the fire in straight lines, to come *directly* into the room ; which can only be effected by bringing the fire as far forward as possible, and leaving the opening of the fireplace as wide and as high as can be done without inconvenience ; and secondly, by making the sides and back of the fireplace of such form, and constructing them of such materials, as to cause the direct rays from the fire, which strike against them, to be sent into the room *by reflection* in the greatest abundance.

Now it will be found, upon examination, that the best form for the vertical sides of a fireplace, or the *covings* (as they are called), is that of an upright plane, making an angle with the plane of the back of the fireplace of about 135 degrees. According to the present construction of chimneys, this angle is 90 degrees, or forms a right angle ; but as in this case the two sides or covings of the fireplace (A C, B D, Plate VIII., Fig. 1) are parallel to each other, it is evident that they are very ill contrived for throwing into the room by reflection the rays from the fire which fall on them.

To have a clear and perfect idea of the alterations I propose in the forms of fireplaces, the reader need only observe, that, whereas the backs of fireplaces, as they are now commonly constructed, are as wide as the opening of the fireplace in front, and the sides of it are of course perpendicular to it and parallel to each other, — in the fireplaces I recommend, the back (*i k*, Plate IX., Fig. 3) is only about one third of the width of the opening of the fireplace in front (*a b*), and consequently that the two sides or covings of the fireplace (*a i* and *b k*), instead of being perpendicular to the back, are inclined

to it at an angle of about 135 degrees; and in consequence of this position, instead of being parallel to each
other, each of them presents an oblique front towards
the opening of the chimney, by means of which the
rays which they reflect are thrown into the room. A
bare inspection of the annexed drawings (Plate VIII.,
Fig. 1, and Plate IX., Fig. 3) will render this matter
perfectly clear and intelligible.

In regard to the materials which it will be most advantageous to employ in the construction of fireplaces,
so much light has, I flatter myself, already been thrown
on the subject we are investigating, and the principles
adopted have been established on such clear and obvious
facts, that no great difficulty will attend the determination of that point. As the object in view is to bring
radiant heat into the room, it is clear that that material
is best for the construction of a fireplace, which reflects
the most, or which *absorbs the least* of it; for that heat
which is *absorbed* cannot be *reflected*. Now, as bodies
which absorb radiant heat are necessarily heated in consequence of that absorption, to discover which of the
various materials that can be employed for constructing
fireplaces are best adapted for that purpose, we have
only to find out by an experiment, very easy to be
made, what bodies acquire *least heat* when exposed to
the direct rays of a clear fire; for those which are least
heated evidently absorb the least, and consequently
reflect the most radiant heat. And hence it appears
that iron, and, in general, metals of all kinds, which are
well known to *grow very hot* when exposed to the rays
projected by burning fuel, are to be reckoned among the
very worst materials that it is possible to employ in the
construction of fireplaces.

The best materials I have hitherto been able to discover are fire-stone, and common bricks and mortar. Both these materials are, fortunately, very cheap; and as to their comparative merits, I hardly know to which of them the preference ought to be given.

When bricks are used, they should be covered with a thin coating of plaster, which, when it is become perfectly dry, should be whitewashed. The fire-stone should likewise be whitewashed, when that is used; and every part of the fireplace, which is not exposed to being soiled and made black by the smoke, should be kept as white and clean as possible. As *white* reflects more heat, as well as more light, than any other colour, it ought always to be preferred for the inside of a chimney fireplace, and *black*, which reflects neither light nor heat, should be most avoided.

I am well aware how much the opinion I have here ventured to give, respecting the unfitness of iron and other metals to be employed in the construction of open fireplaces, differs from the opinion generally received upon that subject; and I even know that the very reason, which, according to my ideas of the matter, renders them totally unfit for the purpose, is commonly assigned for making use of them; namely,—that they soon grow very hot. But I would beg leave to ask what advantage is derived from heating them?

I have shown the disadvantage of it; namely,—that the quantity of radiant heat thrown into the room is diminished; and it is easy to show that almost the whole of that absorbed by the metal is ultimately carried up the chimney by the air, which, coming into contact with this hot metal, is heated and rarefied by it, and, forcing its way upwards, goes off with the smoke; and as no current of air ever sets from any part of the opening of a fireplace into the room, it is impossible to conceive how the heat existing in the metal composing any part of the apparatus of the fireplace, and situated within its cavity, can come, or be brought, into the room.

This difficulty may be in part removed, by supposing, what indeed seems to be true in a certain degree, that the heated metal sends off in rays the heat it acquires from the fire, even when it is not heated red-hot; but still, as it never can be admitted that the heat absorbed by the metal, and afterwards thrown off by it in rays, is *increased* by this operation, nothing can be gained by it; and as much must necessarily be lost in consequence of the great quantity of heat communicated by the hot metal to the air in contact with it, which, as has already been shown, always makes its way up the chimney, and flies off into the atmosphere, the loss of heat attending the use of it is too evident to require being further insisted on.

There is, however, in chimney fireplaces destined for burning coals, one essential part, the grate, which cannot well be made of anything else but iron; but there is no necessity whatever for that immense quantity of iron which surrounds grates as they are now commonly constructed and fitted up, and which not only renders them very expensive, but injures very essentially the fireplace. If it should be necessary to diminish the opening of a large chimney in order to prevent its smoking, it is much more simple, economical, and better in all respects, to do this with marble, fire-stone, or even with bricks and mortar, than to make use of iron, which, as has already been shown, is the very worst material that can possibly be employed for that purpose; and as to registers, they not only are quite unnecessary where the throat of a chimney is properly constructed, and of proper dimensions, but in that case would do much harm. If they act at all, it must be by opposing their flat surfaces to the current of rising smoke in a manner which cannot fail to embarrass and impede its motion. But we have shown that the passage of the smoke through the throat of a chimney ought to be facilitated as much as possible, in order that it may be enabled to pass by a small aperture.

Register stoves have often been found to be of use; but it is because, the great fault of all fireplaces constructed upon the common principles being the enormous dimensions of the throat of the chimney, this fault has been in some measure corrected by them; but I will venture to affirm that there never was a fireplace so corrected that would not have been much more improved, and with infinitely less expense, by the alterations here recommended, and which will be more particularly explained in the next chapter.

CHAPTER II.

Practical Directions designed for the Use of Workmen, showing how they are to proceed in making the Alterations necessary to improve Chimney Fireplaces, and effectually to cure smoking Chimneys.

ALL chimney fireplaces, without exception, whether they are designed for burning wood or coals, and even those which do not smoke, as well as those which do, may be greatly improved by making the alterations

in them here recommended; for it is by no means
merely to prevent chimneys from smoking that these improvements are recommended, but it is also to make
them better in all other respects as fireplaces; and when
the alterations proposed are properly executed, which
may very easily be done with the assistance of the following plain and simple directions, the chimneys will
never fail to answer, I will venture to say, even beyond
expectation.   The room will be heated much more
equally and more pleasantly with *less than half the fuel*
used before; the fire will be more cheerful and more agreeable, and the general appearance of the fireplace more
neat and elegant; and the chimney *will never smoke.*

The advantages which are derived from mechanical
inventions and contrivances are, I know, frequently
accompanied by disadvantages which it is not always
possible to avoid; but in the case in question, I can say
with truth, that I know of no disadvantage whatever
that attends the fireplaces constructed upon the principles here recommended.   But to proceed in giving
directions for the construction of these fireplaces.

That what I have to offer on this subject may be the
more easily understood, it will be proper to begin by
explaining the precise meaning of all those technical
words and expressions which I may find it necessary or
convenient to use.

By the *throat* of a chimney, I mean the lower extremity of its canal, where it unites with the upper part
of its open fireplace.   This throat is commonly found
about a foot above the level of the lower part of the
mantle, and it is sometimes contracted to a smaller size
than the rest of the canal of the chimney, and sometimes not.

Plate X., Fig. 5, shows the section of a chimney on the common construction, in which *d e* is the throat.

Fig. 6 shows the section of the same chimney altered and improved, in which *d i* is the reduced throat.

The *breast* of a chimney is that part of it which is immediately behind the mantle. It is the wall which forms the entrance from below, into the throat of the chimney in front, or towards the room. It is opposite to the upper extremity of the back of the open fireplace, and parallel to it; in short, it may be said to be the back part of the mantle itself. In the figures 5 and 6, it is marked by the letter *d*. The *width* of the throat of the chimney (*d e*, Fig. 5, and *d i*, Fig. 6) is taken from the breast of the chimney to the back, and its *length* is taken at right angles to its width, or in a line parallel to the mantle (*a*, Figs. 5 and 6).

Before I proceed to give particular directions respecting the exact forms and dimensions of the different parts of a fireplace, it may be useful to make such general and practical observations upon the subject as can be clearly understood without the assistance of drawings; for the more complete the knowledge of any subject is, which can be acquired without drawings, the more easy will it be to understand the drawings when it becomes necessary to have recourse to them.

The bringing forward of the fire into the room, or rather bringing it nearer to the front of the opening of the fireplace, and the diminishing of the throat of the chimney, being two objects principally had in view in the alterations in fireplaces here recommended, it is evident that both these may be attained, merely by bringing forward the back of the chimney. The only question therefore is, how far it should be brought forward.

The answer is short, and easy to be understood, — bring it forward as far as possible, without diminishing too much the passage which must be left for the smoke. Now as this passage, which, in its narrowest part, I have called the *throat of the chimney*, ought, for reasons which are fully explained in the foregoing chapter, to be immediately, or perpendicularly, over the fire, it is evident that the back of the chimney must always be built perfectly upright. To determine therefore the place for the new back, or how far precisely it ought to be brought forward, nothing more is necessary than to ascertain how wide the throat of the chimney ought to be left, or what space must be left between the top of the breast of the chimney, where the upright canal of the chimney begins, and the new back of the fireplace carried up perpendicularly to that height.

In the course of my numerous experiments upon chimneys, I have taken much pains to determine the width proper to be given to this passage, and I have found, that, when the back of the fireplace is of a proper width, the best width for the throat of a chimney, when the chimney and the fireplace are at the usual form and size, is *four inches.* Three inches might sometimes answer, especially where the fireplace is very small, and the chimney good, and well situated; but as it is always of much importance to prevent those accidental puffs of smoke which are sometimes thrown into rooms by the carelessness of servants in putting on suddenly too many coals at once upon the fire, and as I found these accidents sometimes happened when the throats of chimneys were made very narrow, I found that, upon the whole, all circumstances being well considered, and advantages and disadvantages compared and balanced, *four inches* is the best width that can be given to the throat of a chimney; and this, whether the fireplace be destined to burn wood, coals, turf, or any other fuel commonly used for heating rooms by an open fire.

In fireplaces destined for heating very large halls, and where very great fires are kept up, the throat of the chimney, may, if it should be thought necessary, be made four inches and an half, or five inches wide; but I have frequently made fireplaces for halls, which have answered perfectly well, where the throats of the chimneys have not been wider than four inches.

It may perhaps appear extraordinary, upon the first view of the matter, that fireplaces of such different sizes should all require the throat of the chimney to be of the same width; but when it is considered that the *capacity* of the throat of a chimney does not depend on its width alone, but on its width and length taken together, and that in large fireplaces, the width of the back, and consequently the length of the throat of the chimney, is greater than in those which are smaller, this difficulty vanishes.

And this leads us to consider another important point respecting open fireplaces, and that is, the width which it will, in each case, be proper to give to the back. In fireplaces as they are now commonly constructed, the back is of equal width with the opening of the fireplace in front; but this construction is faulty on two accounts. First, in a fireplace so constructed, the sides of the fireplace — or *covings*, as they are called — are parallel to each other, and consequently ill contrived to throw out into the room the heat they receive from the fire in the form of rays; and secondly, the large open corners, which are formed by making the back as wide as the opening of the fireplace in front, occasion eddies of wind which frequently disturb the fire, and embarrass the smoke in its ascent in such a manner as often to bring it into the room. Both these defects may be entirely remedied by diminishing the width of the back of the fireplace. The width which, in most cases, it will be best to give it is *one third* of the width of the opening of the fireplace in front. But it is not absolutely necessary to conform rigorously to this decision, nor will it always be possible. It will frequently happen that the back of a chimney must be made wider than, according to the rule here given, it ought to be. This may be either to accommodate the fireplace to a stove, which, being already on hand, must, to avoid the expense of purchasing a new one, be employed; or for other reasons; and any small deviation from the general rule will be attended with no considerable inconvenience. It will always be best, however, to conform to it as far as circumstances will allow.

Where a chimney is designed for warming a room of a middling size, and where the thickness of the wall of the chimney in front, measured from the front of the mantle to the breast of the chimney, is nine inches, I should set off four inches more for the width of the throat of the chimney, which, supposing the back of the chimney to be built upright, as it always ought to be, will give thirteen inches for the depth of the fireplace, measured upon the hearth from the opening of the fireplace in front to the back. In this case, thirteen inches would be a good size for the width of the back; and three times thirteen inches, or thirty-nine inches, for the width of the opening of the fireplace in front ; and the angle made by the back of the fireplace and the sides of it, or covings, would be just 135 degrees, which is the best position they can have for throwing heat into the room.

But I will suppose that in altering such a chimney it is found necessary, in order to accommodate the fireplace to a grate or stove already on hand, to make the fireplace sixteen inches wide. In that case, I should merely increase the width of the back to the dimensions required, without altering the depth of the chimney or increasing the width of the opening of the chimney in front. The covings, it is true, would be somewhat reduced in their width by this alteration; and their position with respect to the plane of the back of the chimney would be a little changed; but these alterations would produce no bad effects of any considerable consequence, and would be much less likely to injure the fireplace, than an attempt to bring the proportions of its parts nearer to the standard, by increasing the depth of the chimney, and the width of its opening in front; or than an attempt to preserve that particular obliquity of the covings which is recommended as the best (135 degrees), by increasing the width of the opening of the fireplace, without increasing its depth.

In order to illustrate this subject more fully, we will suppose one case more. We will suppose that in the chimney which is to be altered, the width of the fireplace in front is either wider or narrower than it ought to be, in order that the different parts of the fireplace, after it is altered, may be of the proper dimensions. In this case, I should determine the depth of the fireplace, and the width of the back of it, without any regard to the width of the opening of the fireplace in front; and when this is done, if the opening of the fireplace should be only two or three inches too wide, — that is to say, only two or three inches wider than is necessary in order that the covings may be brought into their proper position with respect to the back, — I should not alter the width of this opening, but should accommodate the covings to this width, by increasing their breadth, and increasing the angle they make with the back of the fireplace ; but if the opening of the fireplace should be more than three inches too wide, I should reduce it to the proper width by slips of stone, or by bricks and mortar.

When the width of the opening of the fireplace in front is very great compared with the depth of the fireplace, and with the width of the back, the covings in that case being very wide and consequently very oblique, and the fireplace very shallow, any sudden motion of the air in front of the fireplace (that motion, for instance, which would be occasioned by the clothes of a woman passing hastily before the fire, and very near it) would be apt to cause eddies in the air, *within the opening of the fireplace*, by which puffs of smoke might easily be brought into the room.

Should the opening of the chimney be too narrow, which however will seldom be found to be the case, it will, in general, be advisable to let it remain as it is, and to accommodate the covings to it, rather than to attempt to increase its width, which would be attended with a good deal of trouble, and probably a considerable expense.

From all that has been said, it is evident that the points of the greatest importance, and which ought most particularly to be attended to in altering fireplaces upon the principles here recommended, are, the bringing

forward the back to its proper place, and making it of a
proper width. But it is time that I should mention
another matter upon which it is probable that my reader
is already impatient to receive information. Provision
must be made for the passage of the chimney-sweeper
up the chimney. This may easily be done in the following manner. In building up the new back of the
fireplace, — when this wall (which need never be more
than the width of a single brick in thickness) is
brought up so high that there remains no more than
about ten or eleven inches between what is then the top
of it and the inside of the mantle, or lower extremity
of the breast of the chimney, — an opening, or doorway, eleven or twelve inches wide, must be begun in the
middle of the back, and continued quite to the top of
it, which, according to the height to which it will commonly be necessary to carry up the back, will make the
opening about twelve or fourteen inches high; which
will be quite sufficient to allow the chimney-sweeper to
pass. When the fireplace is finished, this doorway is to
be closed by a few bricks, by a tile, or a fit piece of
stone, placed in it, dry or without mortar, and confined
in its place by means of a rabbet made for that purpose
in the brick-work. As often as the chimney is swept,
the chimney-sweeper takes down this temporary wall,
which is very easily done, and when he has finished his
work he puts it again into its place. The annexed
drawing (Plate X., Fig. 6) will give a clear idea of this
contrivance; and the experience I have had of it has
proved that it answers perfectly well the purpose for
which it is designed.

I observed above, that the new back, which it will
always be found necessary to build in order to bring the

fire sufficiently forward, in altering a chimney constructed on the common principles, need never be
thicker than the width of a common brick. I may say
the same of the thickness necessary to be given to the
new sides, or covings, of the chimney; or if the new
back and covings are constructed of stone, one inch and
three quarters, or two inches, in thickness, will be sufficient. Care should be taken in building up these new
walls to unite the back to the covings in a solid manner.

Whether the new back and covings are constructed
of stone, or built of bricks, the space between them
and the old back and covings of the chimney ought to
be filled up, to give greater solidity to the structure.
This may be done with loose rubbish, or pieces of
broken bricks, or stones, provided the work be strength-
ened by a few layers or courses of bricks laid in mortar;
but it will be indispensably necessary to finish the work,
where these new walls end, that is to say, at the top of
the throat of the chimney, where it ends abruptly in the
open canal of the chimney, by a horizontal course of
bricks well secured with mortar. This course of bricks
will be upon a level with the top of the doorway left for
the chimney-sweeper.

From these descriptions it is clear, that, where the
throat of the chimney has an end, that is to say, where
it enters into the lower part of the open canal of the
chimney, there the three walls which form the two covings and the back of the fireplace all end abruptly. It
is of much importance that they should end in this
manner; for were they to be sloped outward and raised
in such a manner as to swell out the upper extremity of
the throat of the chimney in the form of a trumpet, and
increase it by degrees to the size of the canal of the

chimney, this manner of uniting the lower extremity of the canal of the chimney with the throat would tend to assist the winds which may attempt to blow down the chimney, in forcing their way through the throat, and throwing the smoke backward into the room ; but when the throat of the chimney ends abruptly, and the ends of the new walls form a flat horizontal surface, it will be much more difficult for any wind from above to find and force its way through the narrow passage of the throat of the chimney.

As the two walls which form the new covings of the chimney are not parallel to each other, but inclined, presenting an oblique surface towards the front of the chimney, and as they are built perfectly upright and quite flat, from the hearth to the top of the throat, where they end, it is evident that a horizontal section of the throat will not be an oblong square ; but its deviation from that form is a matter of no consequence ; and no attempts should ever be made, by twisting the covings above, where they approach the breast of the chimney, to bring it to that form. All twists, bends, prominences, excavations, and other irregularities of form, in the covings of a chimney, never fail to produce eddies in the current of air which is continually passing into and through an open fireplace in which a fire is burning ; and all such eddies disturb either the fire, or the ascending current of smoke, or both, and not unfrequently cause the smoke to be thrown back into the room. Hence it appears, that the covings of chimneys should never be made circular, or in the form of any other curve, but always quite flat.

For the same reason, that is to say, to prevent eddies, the breast of the chimney, which forms that side of the throat that is in front, or nearest to the room, should be neatly cleaned off, and its surface made quite regular and smooth.

This may easily be done by covering it with a coat of plaster, which may be made thicker or thinner in different parts as may be necessary in order to bring the breast of the chimney to be of the proper form.

With regard to the form of the breast of a chimney, this is a matter of very great importance, and which ought always to be particularly attended to. The worst form it can have is that of a vertical plane, or upright flat; and next to this, the worst form is an inclined plane. Both these forms cause the current of warm air from the room, which will, in spite of every precaution, sometimes find its way into the chimney, to cross upon the current of smoke, which rises from the fire, in a manner most likely to embarrass it in its ascent, and drive it back. The inclined plane which is formed by a flat register placed in the throat of a chimney produces the same effects; and this is one reason, among many others, which have induced me to disapprove of register stoves.

The current of air, which, passing under the mantle, gets into the chimney, should be made *gradually to bend its course upwards*, by which means it will unite *quietly* with the ascending current of smoke, and will be less likely to check it, or force it back into the room. Now this may be effected with the greatest ease and certainty, merely by *rounding off* the breast of the chimney or back part of the mantle, instead of leaving it flat, or full of holes and corners; and this, of course, ought always to be done.

I have hitherto given no precise directions in regard

to the height to which the new back and covings ought
to be carried. This will depend not only on the height
of the mantle, but also, and more especially, on the
height of the breast of the chimney, or of that part of
the chimney where the breast ends and the upright canal
begins. The back and covings must rise a few inches, 5
or 6 for instance, higher than this part, otherwise the
throat of the chimney will not be properly formed; but I
know of no advantages that would be gained by carrying them up still higher.

I mentioned above, that the space between the walls
which form the new back and covings, and the old back
and sides of the fireplace, should be filled up; but this
must not be understood to apply to the space between
the wall of dry bricks, or the tile which closes the passage
for the chimney-sweeper, and the old back of the chimney; for that space must be left void, otherwise, though
this tile (which at most will not be more than two
inches in thickness) were taken away, there would not
be room sufficient for him to pass.

In forming this doorway, the best method of proceeding is to place the tile or flat piece of stone destined for
closing it in its proper place, and to build round it, or
rather by the sides of it, taking care not to bring any
mortar near it, in order that it may be easily removed
when the doorway is finished. With regard to the rabbet which should be made in the doorway to receive it
and fix it more firmly in its place, this may either be
formed at the same time when the doorway is built, or
it may be made after it is finished, by attaching to its
bottom and sides, with strong mortar, pieces of thin
roof tiles. Such as are about half an inch in thickness
will be best for this use; if they are thicker, they will

diminish too much the opening of the doorway, and will likewise be more liable to be torn away by the chimney-sweeper in passing up and down the chimney.

It will hardly be necessary for me to add, that the tile, or flat stone, or wall of dry bricks, which is used for closing up this doorway, must be of sufficient height to reach quite up to a level with the top of the walls which form the new back and covings of the chimneys.

I ought, perhaps, to apologize for having been so very particular in these descriptions and explanations; but it must be remembered that this chapter is written principally for the information of those who, having had few opportunities of employing their attention in abstruse philosophical researches, are not sufficiently practised in these intricate investigations to seize, with facility, new ideas, and consequently, that I have frequently been obliged to *labour* to make myself understood.

I have only to express my wishes that my reader may not be more *fatigued* with this labour than I have been; for we shall then most certainly be satisfied with each other. But to return once more to the charge.

There is one important circumstance respecting chimney fireplaces destined for burning coals, which still remains to be further examined; and that is the grate.

Although there are few grates that may not be used in chimneys constructed or altered upon the principles here recommended, yet they are not, by any means, all equally well adapted for that purpose. Those whose construction is the most simple, and which, of course, are the cheapest, are beyond comparison the best, *on all accounts.* Nothing being wanted in these chimneys but merely a grate for containing the coals, and in which they will burn with a clear fire, and all additional apparatus being not only useless, but very pernicious, all complicated and expensive grates should be laid aside, and such as are more simple substituted in the room of them. And in the choice of a grate, as in everything else, *beauty* and *elegance* may easily be united with the *most perfect simplicity*. Indeed, they are incompatible with everything else.

In placing the grate, the thing principally to be attended to is to make the back of it coincide with the back of the fireplace; but as many of the grates now in common use will be found to be too large, when the fireplaces are altered and improved, it will be necessary to diminish their capacities by filling them up at the back and sides with pieces of fire-stone. When this is done, it is the front of the flat piece of fire-stone which is made to form a new back to the grate, which must be made to coincide with and mark part of the back of the fireplace. But in diminishing the capacities of grates with pieces of fire-stone, care must be taken not to make them *too narrow*.

The proper width for grates destined for rooms of a middling size will be from 6 to 8 inches, and their length may be diminished more or less, according as the room is heated with more or less difficulty, or as the weather is more or less severe. But where the width of a grate is not more than 5 inches, it will be very difficult to prevent the fire from going out.

It goes out for the same reason that a live coal from the grate that falls upon the hearth soon ceases to be red-hot; it is cooled by the surrounding cold air of the atmosphere. The knowledge of the cause which produces this effect is important, as it indicates the means which may be used for preventing it. But of this subject I shall treat more fully hereafter.

It frequently happens that the iron backs of grates are not vertical, or upright, but inclined backwards. When these grates are so much too wide as to render it necessary to fill them up behind with fire-stone, the inclination of the back will be of little consequence; for by making the piece of stone with which the width of the grate is to be diminished in the form of a wedge, or thicker above than below, the front of this stone, which in effect will become the back of the grate, may be made perfectly vertical, and, the iron back of the grate being hid in the solid work of the back of the fireplace, will produce no effect whatever; but, if the grate be already so narrow as not to admit of any diminution of its width, in that case it will be best to take away the iron back of the grate entirely, and, fixing the grate firmly in the brickwork, cause the back of the fireplace to serve as a back to the grate. This I have very frequently done, and have always found it to answer perfectly well.

Where it is necessary that the fire in a grate should be very small, it will be best, in reducing the grate with fire-stone, to bring its cavity, destined for containing the fuel, to the form of one half of a hollow hemisphere; the two semicircular openings being one above, to receive the coals, and the other in front, or towards the bars of the grate; for when the coals are burned in such a confined space, and surrounded on all sides, except in the front and above, by fire-stone (a substance peculiarly well adapted for confining heat), the heat of the fire will be concentrated, and, the cold air of the atmosphere being kept at a distance, a much smaller quantity of coals will burn than could possibly be made to burn in a grate where they would be more exposed to be cooled by the surrounding air, or to have their heat carried off by being in contact with iron, or with any other substance through which heat passes with greater facility than through fire-stone.

Being persuaded that, if the improvements in chimney fireplaces here recommended should be generally adopted (which I cannot help flattering myself will be the case), it will become necessary to reduce, very considerably, the sizes of grates, I was desirous of showing how this may, with the greatest safety and facility, be done.

Where grates, which are designed for rooms of a middling size, are longer than 14 or 15 inches, it will always be best, not merely to diminish their lengths, by filling them up at their two ends with fire-stone, but, forming the back of the chimney of a proper width, without paying any regard to the length of the grate, to carry the covings through the two ends of the grate in such a manner as to conceal them, or at least to conceal the back corners of them in the walls of the covings.

I cannot help flattering myself that the directions here given in regard to the alterations which it may be necessary to make in fireplaces, in order to introduce the improvements proposed, will be found to be so perfectly plain and intelligible that no one who reads them will be at any loss respecting the manner in which the work is to be performed; but as order and arrangement tend much to facilitate all mechanical operations, I shall here give a few short directions respecting the manner of *laying out the work*, which may be found useful, and particularly to gentlemen who may undertake to be their own architects, in ordering and directing the alterations to be made for the improvement of their fireplaces.

Directions for laying out the Work.

If there be a grate in the chimney which is to be altered, it will always be best to take it away ; and when this is done, the rubbish must be removed, and the hearth swept perfectly clean.

Suppose the annexed figure (Plate VIII., Fig. 1) to represent the ground plan of such a fireplace ; A B being the opening of it in front, A C and B D the two sides or covings, and C D the back.

Figure 2 shows the elevation of this fireplace.

First, draw a straight line, with chalk or with a lead-pencil, upon the hearth, from one jamb to the other, even with the front of the jambs. The dotted line A B (Plate IX., Fig. 3) may represent this line.

From the middle C of this line (A B) another line *c d* is to be drawn perpendicular to it, across the hearth, to the middle *d* of the back of the chimney.

A person must now stand upright in the chimney, with his back to the back of the chimney, and hold a plumb-line to the middle of the upper part of the breast of the chimney (Plate X., Fig. 5, *d*), or where the canal of the chimney begins to rise perpendicularly ; taking care to place the line above in such a manner that the plumb may fall on the line *c d*, drawn on the hearth from the middle of the opening of the chimney in front to the middle of the back, and an assistant must mark the precise place *e,* on that line where the plumb falls.

This being done, and the person in the chimney having quitted his station, 4 inches are to be set off on the line *c d*, from *e* towards *d ;* and the point *f*, where these 4 inches end (which must be marked with chalk, or with a pencil), will show how far the new back is to be brought forward.

Through *f* draw the line *g h*, parallel to the line A B, and this line *g h* will show the direction of the new back, or the ground line upon which it is to be built.

The line *c f* will show the depth of the new fireplace; and if it should happen that *c f* is equal to about *one third* of the line A B, and if the grate can be accommodated to the fireplace instead of its being necessary to accommodate the fireplace to the grate, in that case half the length of the line *c f* is to be set off from *f* on the line *g f h*, on one side to *k*, and on the other to *i*, and the line *i k* will show the ground line of the forepart of the back of the chimney.

In all cases where the width of the opening of the fireplace in front (A B) happens to be not greater, or not more than two or three inches greater, than *three times* the width of the new back of the chimney (*i k*), this opening may be left, and lines drawn from *i* to A and from *k* to B will show the width and position of the front of the new covings; but when the opening of the fireplace in front is still wider, it must be reduced, which is to be done in the following manner.

From *c*, the middle of the line A B, *c a* and *c b* must be set off equal to the width of the back (*i k*), added to half its width (*f i*), and lines drawn from *i* to *a* and from *k* to *b* will show the ground plan of the fronts of the new covings.

When this is done, nothing more will be necessary than to build up the back and covings, and, if the fireplace is designed for burning coals, to fix the grate in its proper place, according to the directions already given.

When the width of the fireplace is reduced, the edges of the covings *a* A and *b* B are to make a finish with the front of the jambs. And in general it will be best, not only for the sake of the appearance of the chimney, but for other reasons also, to lower the height of the opening of the fireplace, whenever its width in front is diminished.

Fig. 4 (Plate IX.) shows a front view of the chimney after it has been altered according to the directions here given. By comparing it with Fig. 2 (which shows a front view of the same chimney before it was altered), the manner in which the opening of the fireplace in front is diminished may be seen. In Fig. 4, the under part of the doorway by which the chimney-sweeper gets up the chimney is represented by white dotted lines. The doorway is represented closed.

I shall finish this chapter with some general observations relative to the subject under consideration; with directions how to proceed where such local circumstances exist as render modifications of the general plan indispensably necessary.

Whether a chimney be designed for burning wood upon the hearth, or wood or coals in a grate, the form of the fireplace is, in my opinion, most perfect when *the width of the back* is equal to *the depth of the fireplace*, and the opening of the fireplace in front equal to *three times* the width of the back, or, which is the same thing, to *three times the depth of the fireplace.*

But if the chimney be designed for burning wood upon the hearth, upon handirons, or dogs, as they are called, it will sometimes be necessary to accommodate the width of the back to the length of the wood; and when this is the case, the covings must be accommodated to the width of the back and the opening of the chimney in front.

When the wall of the chimney in front, measured from the upper part of the breast of the chimney to the front of the mantle, is very thin, it may happen, and especially in chimneys designed for burning wood upon the hearth, or upon dogs, that the depth of the chimney, determining according to the directions here given, may be too small.

Thus, for example, supposing the wall of the chimney in front, from the upper part of the breast of the chimney to the front of the mantle, to be only 4 inches (which is sometimes the case, particularly in rooms situated near the top of a house), in this case, if we take 4 inches for the width of the throat, this will give 8 inches only for the depth of the fireplace, which would be too little, even were coals to be burned instead of wood. In this case I should increase the depth of the fireplace at the hearth to 12 or 13 inches, and should build the back perpendicular to the height of the top of the burning fuel (whether it be wood burned upon the hearth, or coals in a grate), and then, sloping the back by a gentle inclination forward, bring it to its proper place, that is to say, *perpendicularly under the back part of the throat of the chimney.* This slope (which will bring the back forward 4 or 5 inches, or just as much as the depth of the fireplace is increased), though it ought not to be too abrupt, yet it ought to be quite finished at the height of 8 or 10 inches above the fire, otherwise it may perhaps cause the chimney to smoke; but when it is very near the fire, the heat of the fire will enable the current of rising smoke to overcome the obstacle which this slope will oppose to its ascent, which it could not do so easily were the slope situated at a greater distance from the burning fuel.*

Figs. 7, 8, and 9 (Plate X.) show a plan, elevation,

---

* Having been obliged to carry backward the fireplace in the manner here described, in order to accommodate it to a chimney whose walls in front were remarkably thin, I was surprised to find, upon lighting the fire, that it appeared to give out more heat into the room than any fireplace I had ever constructed. This effect was quite unexpected; but the cause of it was too obvious not to be immediately discovered. The flame rising from the fire broke against the part of the back which sloped forward over the fire, and this part of the back being soon very much heated, and in consequence of its being very hot, (and when the fire burned bright it was frequently quite red-hot,) it threw off into the room a great deal of radiant heat. It is not possible that this oblique surface (the slope of the back of the fireplace) could have been heated red-hot *merely* by the radiant heat projected by the burning fuel; for other parts of the fireplace nearer the fire, and better situated for receiving radiant heat, were never found to be so much heated; and hence it appears that the combined heat in the current of smoke and hot vapour which rises from an open fire *may be*, at least *in part*, stopped in its passage up the chimney, changed into radiant heat, and afterwards thrown into the room. This opens a new and very interesting field for experiment, and bids fair to lead to important improvements in the construction of fireplaces. I have of late been much engaged in these investigations, and am now actually employed daily in making a variety of experiments with grates and fireplaces, upon different constructions, in the room I inhabit in the Royal Hotel in Pall Mall; and Mr. Hopkins, of Greek Street, Soho, Ironmonger to his Majesty, and Mrs. Hempel, at her Pottery at Chelsea, are both at work in their different lines of business, under my direction, in the construction of fireplaces upon a principle entirely new, and which, I flatter myself, will be found to be not only elegant and convenient, but very economical. But as I mean soon to publish a particular account of these fireplaces, with drawings and ample directions for constructing them, I shall not enlarge further on the subject in this place. It may, however, not be amiss just to mention here, that these new invented fireplaces not being fixed to the walls of the chimney, but merely set down upon the hearth, may be used in any open chimney; and that chimneys altered or constructed on the principles here recommended are particularly well adapted for receiving them.

The public in general, and more particularly those tradesmen and manufacturers whom it may concern, are requested to observe, that, as the author does not intend to take out himself, or to suffer others to take out, any patent for any invention of his which may be of public utility, all persons are at full liberty to imitate them, and vend them, for their own emolument, when and where and in any way they may think proper; and those who may wish for any further information respecting any of those inventions or improvements will receive (*gratis*) all the information they can require by applying to the author, who will take pleasure in giving them every assistance in his power.

and section of a fireplace constructed or altered upon
this principle. The wall of the chimney in front at *a*
(Fig. 9) being only 4 inches thick, 4 inches more added
to it for the width of the throat would have left the
depth of the fireplace measured upon the hearth *b c* only
8 inches, which would have been too little; a niche *c*
and *e* was therefore made in the new back of the fire-
place for receiving the grate, which niche was 6 inches
deep in the centre of it, below 13 inches wide (or equal
in width to the grate), and 23 inches high; finishing
above with a semicircular arch, which, in its highest
part, rose 7 inches above the upper part of the grate.
The doorway for the chimney-sweeper, which begins
just above the top of the niche, may be seen distinctly
in both the Figs. 8 and 9. The space marked *g* (Fig.
9) behind this doorway may either be filled with loose
bricks, or may be left void. The manner in which the
piece of stone (*f*, Fig. 9) which is put under the mantle
of the chimney to reduce the height of the opening of
the fireplace, is rounded off on the inside, in order to
give a fair run to the column of smoke in its ascent
through the throat of the chimney, is clearly expressed
in this figure.

The plan (Fig. 7) and elevation (Fig. 8) show how
much the width of the opening of the fireplace in front
is diminished, and how the covings in the new fireplace
are formed.

A perfect idea of the form and dimension of the fire-
place in its original state, as also after its alteration, may
be had by a careful inspection of these figures.

I have added the drawing (Fig. 10, Plate XI.) merely
to show how a fault, which I have found workmen in
general whom I have employed in altering fireplaces are

very apt to commit, is to be avoided. In chimneys like
that represented in this figure, where the jambs A and B
project far into the room, and where the front edge of
the marble slab *o,* which forms the coving, does not
come so far forward as the front of the jambs, the workmen in constructing the new covings are very apt to
place them, not in the line *c* A, which they ought to do,
but in the line *c o,* which is a great fault. The covings
of a chimney should never range *behind* the front of the
jambs, however those jambs may project into the room;
but it is not absolutely necessary that the covings should
make a finish with the internal front corners of the
jambs, or that they should be continued from the back
c quite to the front of the jambs at A. They may
finish in front at *a* and *b,* and small corners, A, *o, a,* may
be left for placing the shovels, tongs, etc.

Were the new coving to range with the front edge of
the old coving *o,* the obliquity of the new coving would
commonly be too great; or the angle *d c o* would exceed
135 degrees, *which it never should do,* or at least never by
more than a very few degrees.

No inconvenience of any importance will arise from
making the obliquity of the covings *less* than what is
here recommended; but many cannot fail to be produced by making it much greater; and as I know from
experience that workmen are very apt to do this, I have
thought it necessary to warn them particularly against it.

Fig. 11 shows how the width and obliquity of the
covings of a chimney are to be accommodated to the
width of the back, and to the opening in front and
depth of the fireplace, where the width of the opening
of the fireplace is less than three times the width of the
new back.

As all those who may be employed in altering chimneys may not perhaps know how to set off an angle of any certain number of degrees, or may not have at hand the instruments necessary for doing it, I shall here show how an instrument may be made which will be found to be very useful in laying out the work for the bricklayers.

Upon a board about 18 inches wide and 4 feet long, or upon the floor or a table, draw three equal squares (A, B, C, Fig. 12, Plate XIII.), of about 12 or 14 inches each side, placed in a straight line, and touching each other. From the back corner *c* of the centre square B draw a diagonal line across the square A, to its outward front corner *f*, and the adjoining angle formed by the lines *d c* and *c f* will be equal to 135 degrees, the angle which the plane of the back of a chimney fireplace ought to make with the plane of its covings. And a bevel *m n* being made to this angle with thin slips of hard wood, this little instrument will be found to be very useful in marking out on the hearth, with chalk, the plans of the walls which are to form the covings of fireplaces.

As chimneys which are apt to smoke will require the covings to be placed less obliquely in respect to the back than others which have not that defect, it would be convenient to be provided with several bevels, — three or four, for instance, forming different angles. That already described, which may be called No. 1, will measure the obliquity of the covings when the fireplace can be made of the most perfect form; another, No. 2, may be made to a smaller angle, *d c e*; and another, No. 3, for chimneys which are very apt to smoke, at the still smaller angle *d c i.* Or a bevel may be so contrived, by means of a joint, and an arch, properly graduated, as to serve for all the different degrees of obliquity which it may ever be necessary to give to the covings of fireplaces.

Another point of much importance, and particularly in chimneys which are apt to smoke, is to form the throat of the chimney properly, by carrying up the back and covings to a proper height.

This workmen are apt to neglect to do, probably on account of the difficulty they find in working where the opening of the canal of the chimney is so much reduced. But it is absolutely necessary that these walls should be carried up 5 or 6 inches at least above the upper part of the breast of the chimney, or to that point where the wall which forms the front of the throat begins to rise perpendicularly. If the workman has intelligence enough to avail himself of the opening which is formed in the back of the fireplace to give a passage to the chimney-sweeper, he will find little difficulty in finishing his work in a proper manner.

In placing the plumb-line against the breast of the chimney, in order to ascertain how far the new back is to be brought forward, great care must be taken to place it at the very top of the breast, where the canal of the chimney *begins to rise perpendicularly;* otherwise, when the plumb-line is placed too low, or against the slope of the breast, when the new back comes to be raised to its proper height, the throat of the chimney will be found to be too narrow.

Sometimes, and indeed very often, the top of the breast of a chimney lies very high, or far above the fire (see Figs. 13 and 14, Plate XIII., where *d* shows the top of the breast of the chimney); when this is the case, it must be brought lower, otherwise the chimney will be very apt to smoke. So much has been said, in the first chapter of this essay, of the advantages to be derived from bringing the throat of a chimney near to the burning fuel, that I do not think it necessary to enlarge on them in this place, taking it for granted that the utility and necessity of that arrangement have already been made sufficiently evident; but a few directions for workmen, to show them how the breast (and consequently the throat) of a chimney can most readily be lowered, may not be superfluous.

Where the too great height of the breast of a chimney is owing to the great height of the mantle (see Fig. 13), or, which is the same thing, of the opening of the fireplace in front, which will commonly be found to be the case, the only remedy for the evil will be to bring down the mantle lower; or, rather, to make the opening of the fireplace in front lower, by throwing across the top of this opening, from one jamb to the other, and immediately under the mantle, a very flat arch, a wall of bricks and mortar, supported on straight bars of iron, or a piece of stone (*h*, Fig. 13). When this is done, the slope of the' old throat of the chimney, or of the back side of the mantle, is to be filled up with plaster, so as to form one continued flat, vertical, or upright plane surface with the lower part of the wall of the canal of the chimney, and a new breast is to be formed lower down, care being taken to round it off properly, and make it finish at the lower surface of the new wall built under the mantle; which wall forms, in fact, a new mantle.

The annexed drawing (Fig. 13), which represents the section of a chimney in which the breast has been lowered according to the method here described, will show these various alterations in a clear and satisfactory manner. In this figure, as well as in most of the others in this essay, the old walls are distinguished from the new ones by the manner in which they are shaded; the old walls being shaded by diagonal lines, and the new ones by vertical lines. The additions, which are formed of plaster, are shaded by dots instead of lines.

Where the too great height of the breast of a chimney is occasioned, not by the height of the mantle, but by the too great width of the breast, in that case (which, however, will seldom be found to occur), this defect may be remedied by covering the lower part of the breast with a thick coating of plaster, supported, if necessary, by nails or studs driven into the wall which forms the breast, and properly rounded off at the lower part of the mantle. (See Fig. 14.)

CHAPTER III.

Of the Cause of the Ascent of Smoke. — Illustration of the Subject by familiar Comparisons and Experiments. — Of Chimneys which affect and cause each other to smoke. — Of Chimneys which smoke from Want of Air. — Of the Eddies of Wind which sometimes blow down Chimneys, and cause them to smoke.

THOUGH it was my wish to avoid all abstruse philosophical investigations in this essay, yet I feel that it is necessary to say a few words upon a subject generally considered as difficult to be explained, which is too intimately connected with the matter under consideration to be passed over in silence. A knowledge of the cause of the ascent of smoke being indispensably necessary to those who engage in the improvement of fireplaces, or who are desirous of forming just ideas relative to the operations of fire and the management of heat, I shall devote a few pages to the investigation of that curious and interesting subject. And as many of those who may derive advantage from these inquiries are not much accustomed to philosophical disquisitions, and would not readily comprehend either the language or the diagrams commonly used by scientific writers to explain the phenomena in question, I shall take pains to express myself in the most familiar manner, and to use such comparisons for illustration as may easily be understood.

If small leaden bullets, or large goose-shot, be mixed with peas, and the whole well shaken in a bushel, the shot will separate from the peas, and will take its place at the bottom of the bushel; forcing, by its greater weight, the peas, which are lighter, to move upwards, contrary to their natural tendency, and take their places above.

If water and linseed oil, which is lighter than water, be mixed in a vessel by shaking them together, upon suffering this mixture to remain quiet the water will descend and occupy the bottom of the vessel, and the oil, being forced out of its place by the greater pressure downwards of the heavier liquid, will be obliged to rise and swim on the surface of the water.

If a bottle containing linseed oil be plunged in water with its mouth upwards, and open, the oil will ascend out of the bottle, and, passing upwards through the mass of water, in a continued stream, will spread itself over its surface.

In like manner, when two fluids of any kind, of different densities, come into contact, or are mixed with each other, that which is the lightest will be forced upwards by that which is the heaviest.

And as heat rarefies all bodies, fluids as well as solids, air as well as water or mercury, it follows that two portions of the same fluid, at different temperatures, being brought into contact with each other, that portion which is the hottest, being more rarified, or specifically *lighter* than that which is colder, must be forced upwards by this last. And this is what always happens in fact.

When hot water and cold water are mixed, the hottest part of the mixture will be found to be at the surface above; and when cold air is admitted into a warmed room, it will always be found to take its place at the bottom of the room, the warmer air being in part expelled, and in part forced upwards to the top of the room.

Both air and water being transparent and colourless fluids, their internal motions are not easily discovered by the sight; and when these motions are very slow, they make no impression whatever on any of our senses, consequently they cannot be detected by us without the aid of some mechanical contrivance. But where we have reason to think that those motions exist, means should be sought, and may often be found, for rendering them perceptible.

If a bottle containing hot water tinged with logwood, or any other colouring drug, be immersed, with its mouth open, and upwards, into a deep glass jar filled with cold water, the ascent of the hot water from the bottle through the mass of cold water will be perfectly visible through the glass. Now, nothing can be more evident than that both of these fluids are forced or *pushed*, and not *drawn* upwards. Smoke is frequently said to be drawn up the chimney, and that a chimney draws well or ill; but these are careless expressions, and lead to very erroneous ideas respecting the cause of the ascent of smoke, and consequently tend to prevent the progress of improvements in the management of fires. The experiment just mentioned with the coloured water is very striking and beautiful, and it is well calculated to give a just idea of the cause of the ascent of smoke. The cold water in the jar, which, in consequence of its superior weight or density, forces the heated and rarefied water in the bottle to give place to it, and to move upwards out of its way, may represent the cold air of the atmosphere, while the rising column of coloured water will represent the column of smoke which ascends from a fire.

If smoke required a chimney to *draw* it upwards, how happens it that smoke rises from a fire which is made in the open air, where there is no chimney?

If a tube, open at both ends, and of such a length that its upper end be below the surface of the cold water in the jar, be held vertically over the mouth of the bottle which contains the hot coloured water, the hot water will rise up through it, just as smoke rises in a chimney.

If the tube be previously heated before it is plunged into the cold water, the ascent of the hot coloured water will be facilitated and accelerated, in like manner as smoke is known to rise with greater facility in a chimney which is hot, than in one in which no fire has been made for a long time. But in neither of these cases can it, with any propriety, be said that the hot water is *drawn* up the tube. The hotter the water in the bottle is, and the colder that in the jar, the greater will be the velocity with which the hot water will be forced up through the tube; and the same holds of the ascent of hot smoke in a chimney. When the fire is intense, and the weather very cold, the ascent of the smoke is very rapid; and under such circumstances chimneys seldom smoke.

As the cold water of the jar immediately surrounding the bottle which contains the hot water will be heated by the bottle, while the other parts of the water in the jar will remain cold, this water so heated, becoming specifically lighter than that which surrounds it, will be forced upwards; and if it finds its way into the tube will rise up through it with the coloured hot water. The warmed air of a room heated by an open chimney fireplace has always a tendency to rise (if I may use that inaccurate expression), and, finding its way into the chimney, frequently goes off with the smoke.

What has been said will, I flatter myself, be sufficient to explain and illustrate, in a clear and satisfactory manner, the cause of the ascent of smoke; and just ideas upon that subject are absolutely necessary in order to judge, with certainty, of the merit of any scheme proposed for the improvement of fireplaces, or to take effectual measures, in all cases, for curing smoking chimneys. For, though the perpetual changes and alterations which are produced by accident, whim, and caprice, do sometimes lead to useful discoveries, yet the progress of improvement under such guidance must be exceedingly slow, fluctuating, and uncertain.

As to the causes of the smoking of chimneys, they are very numerous and various; but as a general idea of them may be acquired from what has already been said upon that subject in various parts of this essay, and as they may, in all cases (a very few only excepted), be completely remedied by making the alterations in fireplaces here pointed out, I do not think it necessary to enumerate them all in this place, or to enter into those long details and investigations which would be required to show the precise manner in which each of them operates, either alone or in conjunction with others.

There is, however, one cause of smoking chimneys which I think it is necessary to mention more particularly. In modern-built houses, where the doors and windows are generally made to close with such accuracy that no crevice is left for the passage of the air from without, the chimneys in rooms adjoining to each other, or connected by close passages, are frequently found to affect each other; and this is easy to be accounted for. When there is a fire burning in one of the chimneys, as the air necessary to supply the current up the chimney where the fire burns cannot be had in sufficient quantities from without, through the very small crevices of the doors and windows, the air in the room becomes rarefied, not by heat, but by subtraction of that portion of air which is employed in keeping up the fire, or supporting the combustion of the fuel, and, in consequence of this rarefaction, its elasticity is diminished, and being at last overcome by the pressure of the external air of the atmosphere, this external air rushes into the room by the only passage left for it, namely, by the open chimney of the neighbouring room; and the flow of air into the fireplace, and up the chimney where the fire is burning, being constant, this expense of air is supplied by a continued current down the other chimney.

If an attempt be made to light fires in both chimneys

at the same time, it will be found to be very difficult to get the fires to burn, and the rooms will both be filled with smoke.

One of the fires — that which is made in the chimney where the construction of the fireplace is best adapted to facilitate the ascent of the smoke; or, if both fireplaces are on the same construction, that which has the wind most favourable, or in which the fire happens to be soonest kindled — will overcome the other, and cause its smoke to be beat back into the room by the cold air which descends through the chimney. The most obvious remedy in this case is to provide for the supply of fresh air necessary for keeping up the fires by opening a passage for the external air into the room by a shorter road than down one of the chimneys; and when this is done, both chimneys will be found to be effectually cured.

But chimneys so circumstanced may very frequently be prevented from smoking, even without opening any new passage for the external air, merely by diminishing the draught (as it is called) up the chimneys; which can best be done by altering both fireplaces upon the principles recommended and fully explained in the foregoing chapters of this essay.

Should the doors and windows of a room be closed with so much nicety as to leave no crevices by which a supply of air can enter sufficient for maintaining the fire, *after the current of air up the chimney has been diminished as much as possible by diminishing the throat of the fireplace,* in that case there would be no other way of preventing the chimney from smoking but by opening a passage for the admission of fresh air from without; but this, I believe, will very seldom be found to be the case.

A case more frequently to be met with is, where currents of air set down chimneys in consequence of a diminution and rarefaction of the air in a room, occasioned by the doors of the room opening into passages or courts where the air is rarefied by the action of some particular winds. In such cases the evil may be remedied, either by causing the doors in question to close more accurately, or (which will be still more effectual) by giving a supply of air to the passage or court which wants it by some other way.

Where the top of a chimney is commanded by high buildings, by cliffs, or by high grounds, it will frequently happen, in windy weather, that the eddies formed in the atmosphere by these obstacles will blow down the chimney, and beat down the smoke into the room. This, it is true, will be much less likely to happen when the throat of the chimney is contracted and properly formed than when it is left quite open, and the fireplace badly constructed; but as it is *possible* that a chimney may be so much exposed to these eddies in very high winds as to be made to smoke sometimes when the wind blows with violence from a certain quarter, it is necessary to show how the effects of those eddies may be prevented.

Various mechanical contrivances have been imagined for preventing the wind from blowing down chimneys, and many of them have been found to be useful; there are, however, many of these inventions, which, though they prevent the wind from blowing down the chimney, are so ill-contrived on other accounts as to obstruct the ascent of the smoke, and do more harm than good.

Of this description are all those chimney-pots with flat horizontal plates or roofs placed upon supporters just above the opening of the pot; and most of the caps which turn with the wind are not much better. One of the most simple contrivances that can be made use of, and which in most cases will be found to answer the purpose intended as well or better than more complicated machinery, is to cover the top of the chimney with a hollow truncated pyramid or cone, the diameter of which above, or opening for the passage of the smoke, is about 10 or 11 inches. This pyramid, or cone (for either will answer), should be of earthenware or of cast-iron; its perpendicular height may be equal to the diameter of its opening above, and the diameter of its opening below equal to three times its height. It should be placed upon the top of the chimney, and it may be contrived so as to make a handsome finish to the brick-work. Where several flues come out near each other, or in the same stack of chimneys, the form of a pyramid will be better than that of a cone for these covers.

The intention of this contrivance is, that the winds and eddies which strike against the oblique surface of these covers may be reflected upwards, instead of blowing down the chimney. The invention is by no means new, but it has not hitherto been often put in practice. As often as I have seen it tried, it has been found to be of use; I cannot say, however, that I was ever obliged to have recourse to it, or to any similar contrivance; and if I forbear to enlarge upon the subject of these inventions, it is because I am persuaded that when chimneys are properly constructed *in the neighbourhood of the fireplace,* little more will be necessary to be done at the top of the chimney than to leave it open.

I cannot conclude this essay without again recommending, in the strongest manner, a careful attention to the management of fires in open chimneys; for not only the quantity of heat produced in the combustion of fuel depends much on the manner in which the fire is managed, but even of the heat actually generated a very small part only will be saved, or usefully employed, when the fire is made in a careless and slovenly manner.

In lighting a coal fire, more wood should be employed than is commonly used, and fewer coals; and as soon as the fire burns bright, and the coals are well lighted, and *not before*, more coals should be added to increase the fire to its proper size.*

The enormous waste of fuel in London may be estimated by the vast dark cloud which continually hangs over this great metropolis, and frequently overshadows

---

* *Kindling-balls*, composed of equal parts of coal, charcoal, and clay, the two former reduced to a fine powder, well mixed and kneaded together with the clay moistened with water, and then formed into balls of the size of hens' eggs, and thoroughly dried, might be used with great advantage instead of wood for kindling fires. These *kindling-balls* may be made so inflammable as to take fire in an instant, and with the smallest spark, by dipping them in a strong solution of nitre and then drying them again; and they would neither be expensive nor liable to be spoiled by long keeping. Perhaps a quantity of pure charcoal, reduced to a very fine powder and mixed with the solution of nitre in which they are dipped, would render them still more inflammable.

I have often wondered that no attempts should have been made to improve the fires which are made in the open chimneys of elegant apartments, by preparing the fuel; for nothing surely was ever more dirty, inelegant, and disgusting than a common coal fire.

Fire-balls, of the size of goose-eggs, composed of coal and charcoal in powder, mixed up with a due proportion of wet clay, and well dried, would make a much more cleanly, and in all respects a pleasanter, fire than can be made with crude coals; and I believe would not be more expensive fuel. In Flanders and in several parts of Germany, and particularly in the Duchies of Juliers and Bergen, where coals are used as fuel, the coals are always prepared before they are used, by pounding them to a powder, and mixing them up with an equal weight of clay, and a sufficient quantity of water to form the whole into a mass which is kneaded together and formed into cakes; which cakes are afterwards well dried and kept in a dry place for use. And

the whole country, far and wide; for this dense cloud is certainly composed almost entirely of *unconsumed coal,* which, having stolen wings from the innumerable fires of this great city, has escaped by the chimneys, and continues to sail about in the air, till, having lost the heat which gave it volatility, it falls in a dry shower of extremely fine black dust to the ground, obscuring the atmosphere in its descent, and frequently changing the brightest day into more than Egyptian darkness.

I never view from a distance, as I come into town, this black cloud which hangs over London, without wishing to be able to compute the immense number of caldrons of coals of which it is composed; for, could this be ascertained, I am persuaded so striking a fact would awaken the curiosity and excite the astonishment of all ranks of the inhabitants, and *perhaps* turn their

it has been found by long experience, that the expense attending this preparation is amply repaid by the improvement of the fuel. The coals, thus mixed with clay, not only burn longer, but give much more heat than when they are burned in their crude state.

It will doubtless appear extraordinary to those who have not considered the subject with some attention, that the quantity of heat produced in the combustion of any given quantity of coals should be increased by mixing the coals with clay, which is certainly an incombustible body; but the phenomenon may, I think, be explained in a satisfactory manner.

The heat generated in the combustion of any small particle of coal existing under two distinct forms, namely, in that which is *combined* with the flame and smoke which rise from the fire, and which, if means are not found to stop it, goes off immediately by the chimney and is lost, and the *radiant heat* which is sent off from the fire, in all directions, in right lines; I think it reasonable to conclude, that the particles of clay, which are surrounded on all sides by the flame, arrest a part at least of the combined heat, and prevent its escape; and this combined heat so arrested, heating the clay red-hot, is retained in it, and, being changed by this operation to radiant heat, is afterwards emitted, and may be directed and employed to useful purposes.

In composing *fire-balls,* I think it probable that a certain proportion of chaff—of straw cut very fine, or even of saw-dust—might be employed with great advantage. I wish those who have leisure would turn their thoughts to this subject, for I am persuaded that very important improvements would result from a thorough investigation of it.

minds to an object of economy to which they have hitherto paid little attention.

Conclusion.

Though the saving of fuel which will result from the improvements in the forms of *chimney fireplaces*, here recommended, will be very considerable, yet I hope to be able to show in a future essay that still greater savings may be made, and more important advantages derived, from the introduction of improvements I shall propose in *kitchen fireplaces*.

I hope, likewise, to be able to show in an essay on *cottage fireplaces*, which I am now preparing for publication, that *three quarters*, at least, of the fuel which cottagers now consume in cooking their victuals and in warming their dwellings, may with great ease, and without any expensive apparatus, be saved.

PLATES

EXPLANATION OF THE FIGURES.

PLATE VIII.

Fig. 1.

The plan of a fireplace on the common construction.
A B, the opening of the fireplace in front.
C D, the back of the fireplace.
A C and B D, the covings.
See page 261.

Fig. 2.

This figure shows the elevation, or front view, of a fireplace on the common construction. See page 261.

PLATE VIII.

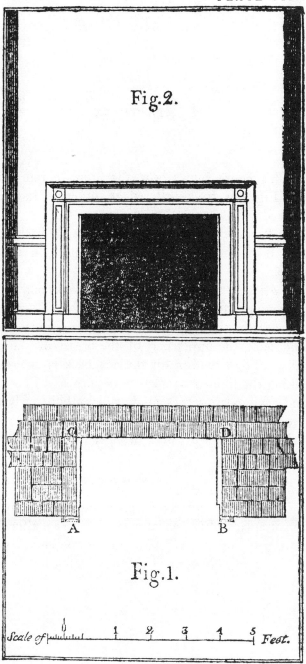

Fig.2.

Fig.1.

Scale of 1 2 3 4 5 Feet.

PLATE IX.

Fig. 3.

This figure shows how the fireplace represented by the Fig. 1 is to be altered, in order to its being improved.

A B is the opening in front, C D the back, and A C and B D the covings of the fireplace in its original state.

a b its opening in front, *i k* its back, and *a i* and *b k* its covings after it has been altered ; *e* is a point upon the hearth upon which a plumb suspended from the middle of the upper part of the breast of the chimney falls. The situation for the new back is ascertained by taking the line *e f* equal to four inches. The new back and covings are represented as being built of bricks, and the space between these and the old back and covings as being filled up with rubbish. See page 261.

Fig. 4.

This figure represents the elevation or front view of the fireplace (Fig. 3) after it has been altered. The lower part of the doorway left for the chimney-sweeper is shown in this figure by white dotted lines. See page 263.

PLATE IX.

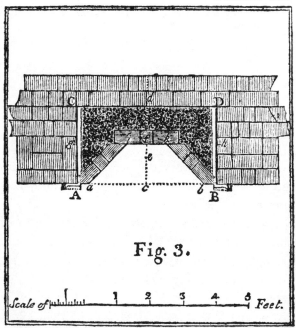

Fig. 3.

Scale of [1] 1 2 3 4 5 Feet.

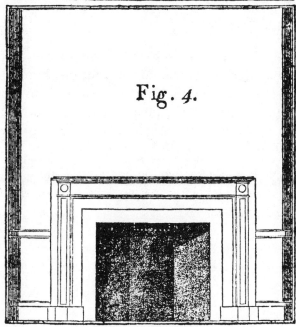

Fig. 4.

PLATE X.

Fig. 5.

This figure shows the section of a chimney fireplace and of a part of the canal of the chimney on the common construction.

a b is the opening in front; *b c* the depth of the fireplace at the hearth; *d* the breast of the chimney.

d e, the throat of the chimney, and *d f*, *g e*, a part of the open canal of the chimney.

Fig. 6.

Shows a section of the same chimney after it has been altered.

k l is the new back of the fireplace; *l i* the tile or stone which closes the doorway for the chimney-sweeper; *d i* the throat of the chimney, narrowed to four inches; *a*, the mantle, and *b*, the new wall made under the mantle, to diminish the height of the opening of the fireplace in front.

N. B.—These two figures are sections of the same chimney which is represented in each of the four preceding figures.

PLATE X.

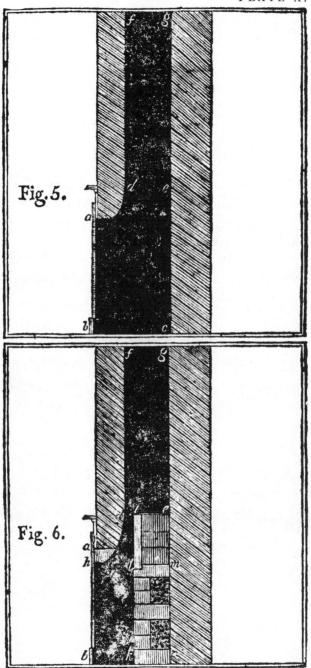

PLATE XI.

Fig. 7.

This figure represents the ground plan of a chimney fireplace in which the grate is placed in a niche, and in which the original width A B of the fireplace is considerably diminished.

a b is the opening of the fireplace in front after it has been altered, and *d* is the back of the niche in which the grate is placed. See page 265.

Fig. 8.

Shows a front view of the same fireplace after it has been altered; where may be seen the grate, and the doorway for the chimney-sweeper. See page 265.

Fig. 9.

Shows a section of the same fireplace, *c d e* being a section of the niche, *g* the doorway for the chimney-sweeper, closed by a piece of firestone, and *f* the new wall under the mantle, by which the height of the opening of the fireplace in front is diminished. See page 265.

PLATE XI.

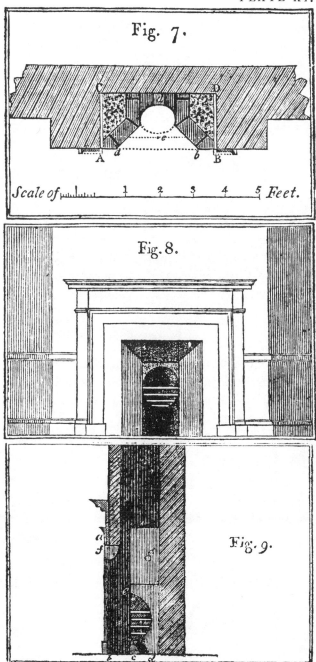

Fig. 7.

Scale of ‖‖‖‖‖‖‖‖‖ 1 2 3 4 5 *Feet.*

Fig. 8.

Fig. 9.

PLATE XII.

Fig. 10.

This figure shows how the covings are to be placed when the front of the covings (*a* and *b*) do not come so far forward as the front of the opening of the fireplace, or the jambs (A and B). See page 266.

Fig. 11.

This figure shows how the width and obliquity of the covings are to be accommodated to the width of the back of a fireplace, in cases where it is necessary to make the back very wide. See page 267.

PLATE XII.

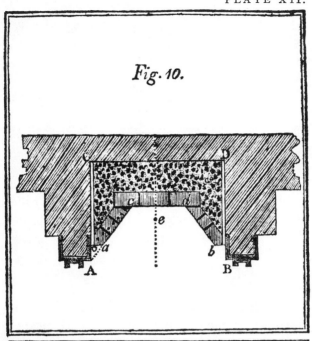

Fig. 10.

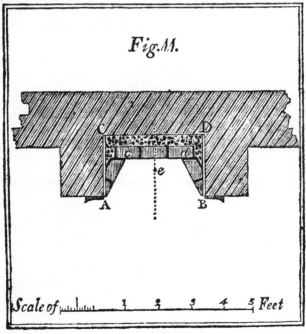

Fig.M.

Scale of 1 2 3 4 5 Feet

PLATE XIII.

Fig. 12.

This figure snows how an instrument called a bevel (*m n*), useful in laying out the work, in altering chimney fireplaces, may be constructed. See page 268.

Fig. 13.

This shows how, when the breast of a chimney (*d*) is too high, it may be brought down by means of a wall (*h*) placed under the mantle, and a coating of plaster, which in this figure is represented by the part marked by dots. See page 270.

Fig. 14.

This shows how the breast of a chimney may be brought down merely by a coating of plaster. See page 271.

PLATE XIII.

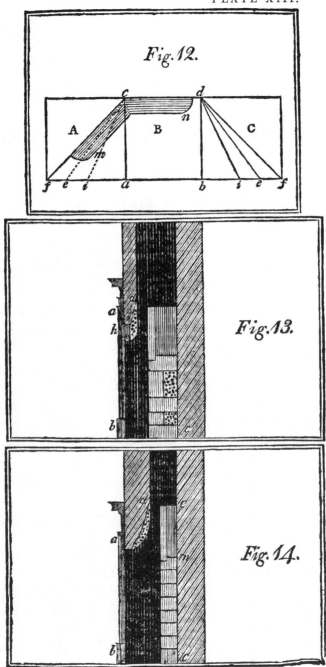

Fig. 12.

Fig. 13.

Fig. 14.

SUPPLEMENTARY OBSERVATIONS

CONCERNING

CHIMNEY FIREPLACES.

OBSERVATIONS CONCERNING OPEN CHIMNEY FIRE-PLACES.

An Account of various Faults that have been committed by Workmen, in England, who have been employed in altering Chimney Fireplaces, and fitting them up according to the Method recommended by the Author, in his Fourth Essay. — Consequences which have resulted from these Mistakes. — Necessity of adhering strictly, and without Deviation, to the Directions which have been given. — Those Particulars are pointed out in which Workmen are most liable to fail.

I WAS much flattered on my return to England, in September, 1798, after an absence of two years, to find that the improvements in the construction of chimney fireplaces, which I had recommended in my Fourth Essay, published in London in the beginning of the year 1796, were coming into use in various parts of the country; and I have since taken a good deal of pains to find out how they have answered, and what faults and imperfections have been discovered in them. And as the information I have obtained by these inquiries has enabled me to make several remarks and observations relative to the construction and management of these fireplaces, that may be of use to those who have introduced them, or may be desirous of introducing them, I feel it to be my duty to lay them before the public.

It has been objected to these fireplaces, that they sometimes occasion dust and ashes to come into the room when the fire is stirred. I have examined several fireplaces said to have been fitted up on my principles, that have certainly had that fault; but I have commonly, I might say invariably, found, that their imperfections have arisen from faults in their construction. Either the grate has been brought out *too far* into the room, or the opening of the fireplace in front has been left too wide or too high, or the workman has neglected to lower and to round off the breast of the chimney, or, what I have often found to be the case, several of these faults have existed together, in the same fireplace.

When the throat of a chimney is situated very high up above the mantle, and especially when the mantle and breast of the chimney, or the wall that reposes on the mantle, are very thin, workmen who are employed to alter chimneys, setting about the work with their minds strongly prepossessed with what they consider as the *leading principle* in the construction of these fireplaces, namely, that the throat of the chimney should not be more than four inches wide, they are very apt to bring the grate too far forward. In dropping their plumbline from the breast of the chimney, they do not reach up high enough into the chimney, but take a part of the breast, where it still goes on to slope backwards, for the bottom of the perpendicular canal of the chimney. They also very often commit another fault, not less essential, and that has the same tendency, in neglecting

to *bring down the throat of the chimney nearer to the fire,*
when it happens to be situated too high.

This I have not only recommended in my Essay on
Chimney Fireplaces, but have given the most particular
directions how it is to be done (see page 531), and, to
mark the importance of the object still more strongly,
have accompanied those directions by an engraving.

It is indeed a very important point, that the throat
of the chimney should be near the fire, and it should
always be carefully attended to. It is likewise very important to "*round off the breast of the chimney,*" though
this, I find, is very often entirely neglected, even by
workmen who have had much practice in the construction of the fireplaces I have recommended.

The breast of a chimney should always be rounded
off in the neatest manner possible, beginning from the
very front of the lower part of the mantle, and ending
at the narrowest part of the throat of the chimney,
where the breast ends in the front part of the perpendicular canal of the chimney. If the under surface of
the mantle is flat and wide, it will be impossible to
round off the breast properly; and that circumstance
alone renders it indispensably necessary, in those cases,
to alter the mantle, or to run under it a thinner piece of
stone, or a thin wall of bricks, supported on an iron
bar, in order that the breast of the chimney may be
brought to be of the proper form, and the throat of the
chimney may be brought into its proper situation.

If the under side of the mantle be left broad and flat,
it is easy to perceive that the cloud of dust or light
ashes that rises from a coal fire nearly burned out when
it is violently stirred about with a poker, striking perpendicularly against this flat part of it, must unavoid-

ably be beat back into the room ; but when the breast
of the chimney is properly rounded off, the ascending
cloud of dust and smoke more easily finds its way into
the throat of the chimney, and is even directed and
assisted in some measure by the warm air of the room
that gets under the mantle, and is going the same way.

Another very common fault that I have observed in
chimney fireplaces, that have been altered on what have
been called my principles, and which has a direct tendency to bring dust, and even smoke, into the room, is
the sloping of the covings too much, and leaving the
opening of the fireplace in front too wide. I have said,
in my Essay on Chimney Fireplaces,[11] that where chimneys are well constructed and well situated, and have
never been apt to smoke, in altering them the covings
may be placed at an angle of 135 degrees with the back;
but I have expressly said that they should never exceed
that angle, and have stated at large the bad consequences
that must follow from making the opening of a fireplace
very wide, when its depth is very shallow (see page
510). I have also expressly said (page 530), that,
for chimneys that are apt to smoke, the covings should
be placed *less obliquely*, in respect to the back, than in
others that have not that fault. But most of the workmen who have altered chimneys seem to have paid little
attention to these distinctions, and I have frequently
found, and sometimes in fireplaces that have been remarkably shallow, that the covings have been placed at
an angle even more oblique than that above mentioned.

Another cause that sometimes has considerable effect
in bringing dust and smoke into rooms, from the fires
that are made in them, is the great nicety with which the
doors and windows are fitted in their frames, which pre-

vents a sufficient quantity of fresh air from coming into
the room to supply a brisk current up the chimney.   It
is, however, evident, that all the alterations in fireplaces
on the common construction, that have been recommended in order to improve them, must tend directly
and very powerfully to lessen this evil; but nothing will
so completely remedy it as lowering the mantle, and
diminishing the width of the fireplace.

How many fireplaces in close rooms have been cured
completely of throwing puffs of smoke and dust into
the room, merely by placing a register stove in them!
But there is surely nothing peculiar to a register-stove
that could enable it to perform such a cure, but merely
as it serves to diminish the width and height of the
opening of the fireplace; and how much easier could
this be done with marble, or other stone, or with bricks
and mortar, plastered over and incrusted in front with
proper ornaments in stucco, or in artificial stone!

I am the more anxious that something of this sort
should be introduced, as the openings of chimney fire-
places are in general certainly too wide and too high,
and as I am convinced that there is no way of reducing
them to a proper size, that would be so cheap, or more
effectual, or that could be made more ornamental.

Those who are fond of the glitter of polished steel,
and have no objection to the expense of it, or to the
labour that is required to keep it bright, may surround
their fireplaces *in front* with a border of it, for *there* it
will do no harm, and may use grates and fenders of the
most exquisite workmanship; but if they wish to have
a pleasant, cheerful, and economical fire, the covings of
their fireplaces must be placed obliquely, and they must
not be constructed of metal; and if the sides and back

of the grate be constructed of fire-bricks instead of iron, the fire will burn still brighter, and will send off considerably more radiant heat into the room.

I have abundant reason to think, that if, in constructing or altering chimney fireplaces, the rules laid down in my essay on that subject are *strictly* adhered to, chimneys so fitted up will very seldom be found either to smoke, or to throw out dust into the room ; and should they be found to have either of these faults, there is a remedy for the evil, as effectual as it is simple and obvious : *Bring down the mantle and the throat of the chimney lower ; and if it should be found necessary, reduce the width of the opening of the fireplace in front, and diminish obliquity of the covings.*

These alterations will certainly be effectual to prevent either smoke or dust from coming into the room *when there is a fire burning in the grate ;* but it sometimes happens, and indeed not unfrequently, that dust and soot are drawn down a chimney in which there is no fire, to the great annoyance of those who are in the room, and to the great damage of the furniture. When this happens, it is commonly occasioned by a very strong draught up *another chimney*, in which there *is a fire*, in an adjoining room ; and when that is the case, the most simple remedy is to alter that other chimney, and, constructing its fireplace on good principles, to reduce its throat to reasonable dimensions. But if the passage of the air down a chimney in which there is no fire is occasioned by strong eddies of wind, there is no remedy for that evil but placing a chimney-pot, of a peculiar construction, on the top of the chimney, which shall counteract the effect of those eddies ; or by closing up the throat of the chimney occasionally, by a door made for that purpose of sheet-iron.

If the doorway that is left in the back of the fireplace for giving a passage to the chimney-sweeper, instead of being closed with a tile, or with a flat piece of stone, set in a groove made to receive it, according to the directions given in my Fourth Essay,[11] be closed with a flat piece of cast-iron, or of plate-iron, fixed at its lower end, to the lower end of the doorway, by a hinge, or movable on two gudgeons, — this plate may easily be so contrived as to serve occasionally as a register or door for diminishing or closing the throat of the chimney.

As this plate, situated at the *back part* of the chimney, could not produce any of those bad effects that have with reason been attributed to the registers of common register-stoves (which are placed on the breast of the chimney), it appears to me to be very probable, that it would be found useful as a register for occasionally altering the size of the throat of the chimney, and regulating its draught, as well as for occasionally closing up that passage entirely. It would certainly be worth while to try the experiment.*

Before I quit this subject, I must mention another fault, which workmen employed in altering chimney fireplaces that are furnished with grates or stoves with sloping backs are very apt to make. They leave the back of the grate in its place, and instead of carrying up the back of the fireplace perpendicularly *from the bottom of the grate,* they first begin to carry it up perpendicularly from the top of the iron plate that forms the back of the grate; and as this plate not only slopes backwards considerably, but rises several inches above the

---

* Since the introduction of the cottage and gridiron grates, this contrivance has come into very general use, and experience has shown it to be extremely useful. I would strongly recommend it to those who fit up chimney fireplaces on these principles, never to omit this register; it costs a mere trifle, and is very useful on many accounts.

level of the upper bar of the grate, this necessarily throws the fire very far into the room. This tends to bring both smoke and dust into the room, not only because it brings the fire too far forward, but also because it occasions the air of the room, that slips in by the sides of the covings, to get behind the current of smoke that rises perpendicularly from the fire, which air frequently crowds the smoke forward, and causes it to strike against the mantle. This is a great fault, and I am sorry to say that I have found it very common in many parts of England, where attempts have been made to introduce the fireplaces I have recommended. Where grates *with sloping backs* are used in fitting up these fireplaces, these backs must either be taken quite away or bricked up, and the new back part, or back wall of the fireplace, must be made to serve as a back for the grate, against which the burning fuel is laid.

As I am giving an account of the mistakes that have been made by some of those who have been employed in fitting up chimney fireplaces on the principles I have publicly recommended, it will naturally be expected that I should take some notice of those numerous *improvements* that have been announced to the public, said to have been made in stoves, grates, etc., to which advertisers in the newspapers have thought proper to affix my name. As I am extremely anxious not to injure any man, either in his reputation for ingenuity, or in his trade, or in any other way, I shall not say one word more on this subject than what I feel it to be my duty to the public to declare, namely, that I am not the inventor of any of those stoves or grates that have been offered to the public for sale under my name.

Having mentioned the inconveniences that sometimes

arise from doors and windows being fitted to their frames with so much nicety as not to give a sufficient passage to air from without to get into the room to supply the current up the chimney, which must always exist when a fire is burning in the room, I embrace this opportunity of mentioning a contrivance for remedying this defect, which I am persuaded would not only be found most effectual for that purpose, but would at the same time contribute very essentially to rendering dwelling-houses more salubrious and more comfortable, by facilitating the means of warming them more equally and ventilating them more easily and more effectually.

In building a house, an *air-canal*, about twelve or fifteen inches square, in the clear, and open at both ends, may be constructed in or near the centre of each stack of chimneys; and two branches from this air-canal, both furnished with registers, may open into each of the adjoining rooms, — one of these branches opening into the fireplace, just under the grate, and the other over the fireplace, and near the top of the room, or just under the ceiling. Each of these branches should be about four inches square, in the clear; and to prevent the uncouth appearance of the open mouth of that which opens into the room over the fireplace, it may be masked by a medallion, a picture, or any other piece of ornamental furniture proper for that use, placed before it at the distance of one or two inches from the side or wall of the room.

The bottom of this *air-tube* should reach to the ground, where it should communicate freely with the open air of the atmosphere; but it should not rise quite so high as the chimneys (or canals for carrying off the smoke) are carried up, but should end (by lateral openings, communicating with the air of the atmosphere) immediately above the roof of the house.

If this air-tube be situated in the middle of a building, it is evident that a horizontal canal or tube of communication must be carried from its lower orifice to some open place without the building, in order to establish a free circulation of fresh air, both upwards and downwards, in the *air-tube*. I say both *upwards* and *downwards*, for sometimes the current of air in the tube will be found to set upwards, and sometimes downwards. Its direction will depend on the winds that happen to prevail, or rather on the eddies they occasion in the air out of doors in the neighbourhood of the buildings ; and it is no small advantage that will arise from leaving both ends of the air-tube open, that the tube will always be supplied with a sufficiency of air, whatever eddies the winds may occasion. It is easy to perceive how powerfully this must operate to prevent those puffs of smoke which, in high winds, are frequently thrown into some rooms by the eddies, and the partial rarefactions of the air that they occasion ; but this is far from being the only or the most important of the advantages that will be derived from this air-tube. Those who consider what an immense quantity of air is required to supply the current that sets up the chimney of an open fireplace, where there is a fire burning, must perceive what an enormous loss of heat there must be, when all this expense of air is supplied by the warmed air of the room, and that all this warmed air is necessarily and constantly replaced by the cold air from without, which finds its way into the room by the crevices of the doors and windows. But all this waste of heat, or any part of it, at pleasure, may be prevented by the scheme proposed ; for if the air necessary to the combustion of the fuel, and to the supplying of the current up the chimney, be furnished by the air-tube, the warmed air in the room will remain in its place ; and as this will in a great measure prevent the cold currents from the crevices of the door and windows, the heat in the room will be the more equable, and consequently the more wholesome and agreeable on that account.

But there are, I am told, persons in this country, who are so fond of seeing what is called a great roaring fire, that even with its attendant inconveniences, of roasting and freezing opposite sides of the body at the same time, they prefer it to the genial and equable warmth which a smaller fire, properly managed, may be made to produce, even in an open chimney fireplace. To recommend the air-tubes to persons of that description, I would tell them, that, by closing up, by means of its register, the lower branch of communication (that which ends just under the grate) and setting that situated near the top of the room wide open, they may indulge themselves with having a very large fire in the room *with little heat*, and this with much less inconvenience from currents of cold air from the doors and windows than they now experience.

It is easy to perceive that by a proper use of the two registers, together with a judicious management of the fire, the air in the room may either be made hotter or colder, or may be kept at any given temperature, or the room may be most effectually ventilated ; and that this change of air may be effected either gradually or more suddenly. And here it may perhaps be the proper place to observe, that in all our reasonings and speculations relative to the heating of rooms by means of open chimney fires, we must never forget that it is the *room that heats the air*, and not the air that heats the room.

The rays that are sent off from the burning fuel generate heat only *when* and *where* they are *stopped* or *absorbed;* consequently they generate no heat in the air in the room in passing through it, because they *pass through it*, and are not *stopped* by it, but, striking against the walls of the room, or against any solid body in the room, these rays are *there* stopped and absorbed, and it is *there* that the heat found in the room is *generated.* The air in the room is afterwards heated by coming into contact with these solid bodies. Many capital mistakes have arisen from inattention to this most important fact.

It is really astonishing how little attention is paid to events which happen frequently, however interesting they may be as objects of curious investigation, or however they may be connected with the comforts and enjoyments of life. Things near us, and which are familiar to us, are seldom objects of our meditations. How few persons are there who ever took the trouble to bestow a thought on the subject in question, though it is, in the highest degree, curious and interesting!

OF THE MANAGEMENT OF FIRE AND THE ECONOMY OF FUEL.

CHAPTER I.

The Subject of this Essay curious and interesting in a very high Degree. — All the Comforts, Conveniences, and Luxuries of Life are procured by the Assistance of FIRE *and of* HEAT. — *The Waste of Fuel very great. — Importance of the Economy of Fuel to Individuals, and to the Public. — Means used for estimating the Amount of the Waste of Fuel. — An Account of the first Kitchen of the House of Industry at Munich, and of the Expense of Fuel in that Kitchen compared with the Quantity consumed in the Kitchens of private Families. — An Account of several other Kitchens constructed on various Principles at Munich, under the Direction of the Author. — Introduction to a more scientific Investigation of the Subject under Consideration.*

NO subject of philosophical inquiry within the limits of human investigation is more calculated to excite admiration and to awaken curiosity than fire; and there is certainly none more extensively useful to mankind. It is owing, no doubt, to our being acquainted with it from our infancy, that we are not more struck with its appearance, and more sensible of the benefits we derive from it. Almost every comfort

and convenience which man by his ingenuity procures
for himself is obtained by its assistance ; and he is not
more distinguished from the brute creation by the use
of speech, than by his power over that wonderful
agent.

Having long been accustomed to consider the
management of heat as a matter of the highest importance to mankind, a habit of attending carefully
to every circumstance relative to this interesting subject that occasionally came under my observation soon
led me to discover how much this science has been
neglected, and how much room there is for very essential improvements in almost all those various operations in which heat is employed for the purposes of
human life.

The great waste of fuel in all countries must be
apparent to the most cursory observer; and the uses
to which fire is employed are so very extensive, and the
expense for fuel makes so considerable an article in the
list of necessaries, that the importance of the subject
cannot be denied.

And with regard to the economy of fuel, it has this
in particular to recommend it, that whatever is saved
by an individual is at the same time a positive saving
to the whole community ; for the less demand there is
for any article in the market, the lower will be its price ;
and as all the subjects of useful industry — all the arts
and manufactures, without exception — depend directly
or indirectly on operations in which fire is necessary,
it is of much importance to a manufacturing and com-
mercial country to keep the price of fuel as low as
possible ; and even in countries where there are no
manufactures, and where the inhabitants subsist entirely

by agriculture, if wood be used as fuel, — as the proportion of woodland to arable must depend in a great measure on the consumption of fire-wood, — any saving of fuel will be attended with a proportional diminution of the forests reserved for fire-wood, consequently with an increase of the lands under cultivation, with an increase of inhabitants and of national wealth, strength and prosperity.

But what renders this subject peculiarly interesting is the great relief to the poor in all countries, and particularly in all cold climates, and in all great cities in every climate, that would result from any considerable diminution of the price of fuel, or from any simple contrivance by which a smaller quantity of this necessary article than they now are obliged to employ to make themselves comfortable might be made to perform the same services. Those who have never been exposed to the inclemencies of the seasons — who have never been eye-witnesses to the sufferings of the poor in their miserable habitations, pinched with cold and starving with hunger — can form no idea of the importance *to them* of the subject which I propose to treat in this Essay.

To all those who take pleasure in doing good to mankind by promoting useful knowledge, and facilitating the means of procuring the comforts and conveniencies of life, these investigations cannot but be very interesting.

Though it is generally acknowledged that there is a great waste of fuel in all countries, arising from ignorance and carelessness in the management of fire, yet few — very few, I believe — are aware of the real amount of this waste.

From the result of all my inquiries upon this subject, I have been led to conclude that not less than *seven eighths* of the heat generated, or which *with proper management might be generated*, from the fuel actually consumed, is carried up into the atmosphere with the smoke, and totally lost.   And this opinion has not been formed hastily; on the contrary, it is the result of much attentive observation, and of many experiments.   But in a matter of so much importance I feel it to be my duty not merely to give the public my *opinions*, but to lay before them the grounds upon which those opinions have been founded, in order that every one may judge for himself of the certainty or probability of my deductions.

It would not be difficult, merely from a consideration of the nature of heat, — of the manner in which it is generated in the combustion of fuel, and the manner in which it exists when generated, — to show that, as the process of boiling is commonly performed, there must of necessity be a very great loss of heat; for when the vessel, in which the fluid to be boiled is contained, is placed over an open or naked fire, not only by far the greater part of the radiant heat is totally lost, but also of that which exists in the flame, smoke, and hot vapour, a very small proportion only enters the vessel; the rest going off with great rapidity, by the chimney, into the higher regions of the atmosphere.   But, without insisting upon these reasonings (though they are certainly incontrovertible), I shall endeavour to establish the facts in question upon still more solid ground, — that of actual experiment.

In the prosecution of the experiments necessary in this investigation, I proceeded in the following manner: As the quantity of heat which any given quantity of any given kind of fuel is capable of generating is not known, there is no fixed standard with which the result of an experiment can be compared, in order to ascertain exactly the proportion of the heat saved, or usefully employed, to that lost. Instead therefore of being able to determine this point *directly*, I was obliged to have recourse to *approximations*. Instead of determining the quantity of heat lost in any given operation, I endeavoured to find out with how much less fuel the same operation might be performed, by a more advantageous arrangement of the fire and disposition of the machinery: and several extensive public establishments, which have been erected in Bavaria within these last six or seven years, under my direction, by order of His Most Serene Highness, the ELECTOR PALATINE, — particularly an establishment for the poor of Munich (of which an account has been given to the public in my First Essay), and the establishment of a Public Academy for the education of one hundred and eighty young men, destined for the service of the State in the different civil and military departments, — the economical arrangements of these establishments afforded me a most favorable opportunity of putting into practice all my ideas relative to the management of fire; and of ascertaining, by numerous experiments made upon a large scale, and often varied and repeated, the real importance of the improvements I have introduced.

That many experiments have been actually made in these two establishments, during the seven years they have existed, will not be doubted by those who are informed that the kitchen, or rather the fire-place of the kitchen of the House of Industry, has been pulled down and built entirely anew no less than *three times*, and that of the Military Academy *twice*, during that period; and that the forms of the boilers, and the internal construction of the fire-places, have been changed still oftener.

The importance of the improvements in the management of heat employed in culinary operations, which have resulted from these investigations, will appear by comparing the quantity of fuel now actually used in those kitchens to that consumed in performing the same operations in kitchens on the common construction. And this will at the same time show, in a clear and satisfactory manner, what I proposed to prove, — namely, that in all the common operations in which fire is employed there is a very great waste of fuel.

The waste of fuel in boiling water or any other liquid over an open fire, in the manner in which that process is commonly performed, and the great saving of fuel which will result from a more advantageous disposition and management of the fire, will be evident from the results of the following experiments, all of which were made by myself, and with the utmost care.

Experiment No. 1. — A copper boiler belonging to the kitchen of the Military Academy in Munich, 22 Rhinland inches in diameter above, 19¼ inches in diameter below, and 24 inches in depth, and which weighed 50 lbs. weight of Bavaria (=61.92 lbs. Avoirdupois), being fixed in its fire-place, was filled with 95 Bavarian measures (= 28 English wine-gallons) of water, which weighed 187 Bavarian pounds (= 232.58 lbs. Avoirdupois); and this water being at the temperature of 58° F., a fire was lighted under the boiler with dry beech-wood, and the water was made to boil, and was continued boiling two hours. The time employed and wood consumed in this experiment were as follows : —

	Time employed. h. m.		Wood consumed. lbs.
To make the water boil . .	1	1	11
To keep the water boiling . .	2	0	2½
Total	3	1	13½

Experiment No. 2. — The same boiler, containing the same quantity of water at the same temperature, being now removed to the kitchen of a private gentleman in the neighbourhood (Baron de Schwachheim, a brother of the Commandant of the Academy), and placed upon a tripod, a quantity of the same kind of wood used in the former experiment being provided, a fire was lighted under it by the gentleman's cook (directions having been given to be as sparing as possible of fuel), and it was made to boil and continued boiling two hours.

The result of the experiment was as follows : —

	Time employed. h. m.		Wood consumed. lbs.
To make the water boil . .	1	31	45
To keep it boiling	2	0	17½
Total	3	31	62½

As in these two experiments the same boiler was employed; as the quantity of water was the same, as also its temperature at the beginning of the experiments; and as it was made to continue boiling during the same length of time, it is evident that the quantities of wood consumed show the relative advantages of the different methods employed in the management of the fire. The difference of these quantities of fuel is very great (the one being only 13½ lbs. and the other

amounting to no less than 62½ lbs.). And this shows
how very considerable the waste of fuel really is, in the
manner in which it is commonly employed for culinary
purposes, and how important the savings are which may
be made by introducing a more advantageous arrangement for the management of fire. But great as these
savings may appear to be, as shown by the results of
the foregoing experiments, yet they are in fact still
more considerable, as will be abundantly proved in the
sequel. In the Experiment No. 2, in which the boiler
was put over an open fire, great care was taken to
place the fuel in the most advantageous manner; but
in general little attention is paid to that circumstance,
and the waste of fuel is greatly increased by such negli-
gence. But in closed fire-places, upon a good construction, as the *proper place* for the fuel cannot be
mistaken, and as it is fixed and bounded on all sides
by a wall, the ignorance or inattention of those who
take care of the fire can never be productive of any
great waste of fuel; and this is an advantage of no
small importance attending these fire-places.

Experiment No. 3. — A large copper sauce-pan or
casserole, 11¼ inches in diameter above, 10¾ in diameter
below, and 3¾ inches deep, containing 4 measures of
water weighing 7$\frac{15}{16}$ lbs., and at the temperature of 58° F.,
being placed in its closed fire-place, and a fire being
made under it with small pieces of dry beech-wood cut
in lengths of about 4 inches, the water was made to
boil, and was continued boiling two hours.

The result of the experiment was as follows: —

	Time employed. h.	m.	Wood consumed. lbs.
To make the water boil . .	0	12	1
To keep it boiling	2	0	0¾
Total	2	12	1¾

Experiment No. 4. — The same sauce-pan, containing the same quantity of water, and at the same temperature as in the last experiment, was now taken from its proper fire-place, and placed upon a tripod; and a fire being made under it with dry beech-wood, the result of the experiment was as follows: —

	Time employed.		Wood consumed.
	h.	m.	lbs.
To make the water boil . .	o	28	6
To keep it boiling	2	o	5½
Total	2	28	11½

The difference in the results of these two experiments is nearly the same as that in the results of those before mentioned, and they all tend to show that, in cooking or boiling over an open fire, nearly *five times* as much fuel is required as when the heat is confined in a closed fire-place, and its operation properly directed.

But I must again repeat, what I have already observed with respect to the two former experiments, as the Experiments No. 2 and No. 4 were both made with the utmost care, the results of them, compared with those which were made with the same boilers placed in closed fire-places, can give no adequate idea of the real loss of heat and waste of fuel which take place in the common operations of cookery.

From several estimates which I have made with great care relative to this subject, founded upon the quantity of fuel actually consumed in the kitchens of several private families, compared with the quantities of different kinds of food prepared for the table, it appears that at least *nine tenths* of the wood actually consumed in common kitchens, where cooking is carried on over an open fire, might be saved, by introducing the various

improvements I have brought into use in the kitchens
which have been constructed under my directions.

But it is not alone in kitchens, in which cooking is
carried on over open fires, that useful alterations may be
made: kitchens with closed fire-places, and indeed all
the kitchens which have yet been contrived (as far as
my knowledge extends), are susceptible of great improvement.

The various improvements that may be made in
mechanical arrangements for the economy of fuel will
appear in a striking manner from a detail of the different alterations which have from time to time been made
in the kitchen of the House of Industry at Munich,
and in that of the Military Academy, and of the effects
produced by those progressive improvements.

The House of Industry being an establishment of
public charity, and the number of those fed from the
kitchen amounting from 1000 to 1500 persons daily,
the economy of fuel, in a kitchen upon so large a scale,
became an object of serious consideration; and I attended to this matter with peculiar pleasure, as it so
completely coincided with my favorite philosophical
pursuits.

The investigation of heat, and of the laws of its
operations, had long occupied my attention, and I had
been so fortunate, in the course of my experiments upon
that subject, as to make some discoveries which were
thought worthy of being inserted in the Philosophical
Transactions of the Royal Society of London;[1, 5, 12, 13, 14]
and for my last paper upon that subject, published in the
Transactions for the year 1792,[15] I had the honour to
receive the annual medal of the Society. I hope my mentioning this circumstance will not be attributed to osten-

tation. My motive in doing it is merely to show that, when I undertook to make the arrangements of which I am about to give an account, the subject was by no means new to me; but, on the contrary, that I was prepared, and in some measure qualified, for such investigation.

I conceive it to be the duty of those who propose useful improvements for the benefit of mankind not only to *merit*, but also to do every thing in their power to *obtain* the confidence of those to whom their proposals are submitted; and there appears to me to be a much greater degree of pride and arrogance displayed by an author *in taking it for granted* that the world is already sufficiently acquainted with his merit and his qualifications to treat the subject he undertakes to investigate, than in modestly pointing out the grounds upon which the confidence of the public in his knowledge of his subject and in his integrity may be founded.

But to return from this digression. In the first arrangement of the kitchen in the House of Industry at Munich, which was finished in the beginning of the year 1790, eight large copper boilers, each capable of containing about 38 English wine-gallons, were placed in such a manner in two rows, in a solid mass of brickwork, 3 feet high, 9 feet wide, and 18 feet long, built in the middle of the kitchen, that, from a single fireplace, situated at one end of this brick-work, by means of canals (furnished with valves or dampers) going from it through the solid mass of the brick-work to all the different boilers, these boilers were all heated, and made to boil with one single fire; and though none of them were in actual contact with the fire-place, and some of them were distant from it near 15 feet, yet they were all heated with great facility, and in a short space of time, by the heat which, upon opening the valves (which were of iron), was made to pass through the canals.

Each boiler having its separate canal and its separate valves, any single boiler, or any number of them, might be heated at pleasure, without heating the rest; and by opening the valves of any boiler more or less, more or less heat, as the occasion required, might be made to pass under the boiler; and when no more heat was wanting for any of the boilers, or when the fire was too strong, by opening a particular valve a communication with a waste canal was formed, by which all the heat, or any part of it at pleasure, might be made to pass off directly into the chimney, without going near any of the boilers.

The fire was regulated by a register in the door of the ash-pit, by which the air was admitted into the fire-place; and, when no more heat was wanted, the fire was put out by closing this register entirely, and by closing at the same time all the valves or dampers in the canals leading from the fire-place.

The fire-place was of an oval form, 3 feet long, 2 feet 3 inches wide, and about 18 inches high, vaulted above with *a double vault*, 4 inches of air being left between the two vaults; and the fuel was introduced into the fire-place by a passage closed by a *double* iron door, which door was kept constantly shut; and the fuel was burned upon an iron grate, the air which supplied the fire coming up from below the grate through the ash-pit.

The loss of heat in its passage from the fire-place

to the boilers was prevented by making the canals of communication *double*, one within the other; the internal canal by which the heat passed, and which was 5 inches wide internally, and 6 inches high, being itself placed, and, as it were *insulated*, in a canal still larger, in such a manner that the canal by which the heat passed (which was constructed of very thin bricks, or rather tiles) was *surrounded on every side* with a wall, 2 inches thick, of *confined air*. The surrounding canal being formed in the solid body of the mass of brick-work, this arrangement of the double canals was entirely concealed. The double canals and the double vault over the fire-place were intended to serve the same purpose; namely, *to confine more effectually the heat*, and prevent its escape into the mass of brick-work, and its consequent loss.

Having found, in the course of my experiments, that confined air is the best barrier[15] that can be opposed to heat, to confine it, I endeavoured to avail myself of that discovery in these economical arrangements, and my attempts were not unsuccessful.

Not only the fire-place itself, and the canals of communication between the fire-place and the boilers, were surrounded by confined air, but it was also made use of for confining the heat in the boilers, and preventing its escaping into the atmosphere. This was done by making the covers of the boilers *double*. These covers (see the Figures 1 and 2, Plate I.) which were made of tin, or rather of thin iron plates tinned, were in the form of a hollow cone. The height of the cone was equal to about one third of its diameter, and the air which it contained was entirely shut up, the bottom of

the cone being closed by a circular plate or thin sheet
of tinned iron. The bottom of the cone was accurately fitted to the top of the boiler, which it completely
closed, by means of a rim about 2 inches wide, which
entered the boiler; which rim was soldered to the flat
sheet of tinned iron which formed the bottom of the
cover. The steam generated by the boiling liquid
was carried off by a tube about half an inch in diameter, which passed through the hollow conical cover,
and which was attached to the cover, both above and
below, with solder, in such a manner that the air with
which the hollow cone was filled remained completely
confined, and cut off from all communications with the
external air of the atmosphere, as well as with the
steam generated in the boiler.

In some of the covers I filled the hollow of the cone
with fur, but I did not find that these were sensibly
better for confining the heat than those in which the
cone was filled simply with air.

To convince the numerous strangers, who from curiosity visited this kitchen, of the great advantage of
making use of double covers to confine the heat in the
boilers, instead of using single covers for that purpose,
a single cover was provided, which, as it was externally
of the same form as the others, when it was placed
upon a boiler, could not be distinguished from them;
but as its bottom was wanting, and consequently there
was no confined air interposed between the hot steam
in the boiler and the external surface of the cover, on
being placed upon a kettle actually boiling, this cover
instantaneously became so exceedingly hot as actually
to burn those who ventured to touch it; while a *double
cover*, formed of the same materials, and placed in the

same situation, was so moderately warm that the naked hand might be held upon it for any length of time without the least inconvenience.

As it was easy to conceive that what was so exceedingly hot as to burn the hand in an instant, upon touching it, could not fail to communicate a great deal of heat to the cold atmosphere which continually lay upon it, this experiment showed in a striking and *convincing* manner the utility of my double covers; and I have since had the satisfaction to see them gradually finding their way into common use.

It is perhaps quite unnecessary that I should inform my readers that one principal motive which induced me to take so much pains in the arrangement of this kitchen was a desire to introduce useful improvements, relative to the management of heat and the economy of fuel, into common practice. An establishment so interesting in all respects, so important in its consequences, and so perfectly new in Bavaria, as a public House of Industry upon a liberal and extensive plan, — where almost every trade and manufacture is carried on under the same roof, where the poor and indigent of both sexes, and of all ages, find a comfortable asylum, and employment suited to their strength and to their talents, and where industry is excited *not by punishments*, but by *the most liberal rewards*, and by the kindest usage, — such an establishment, I thought, could not fail to excite the curiosity of the public, and to draw together a great concourse of visitors; and as this appeared to me a favourable opportunity to draw the public attention to useful improvements, all my measures were taken accordingly; and not only the kitchen, but also the bake-house, the stoves for heating the rooms, the lamps, the various utensils and machines made use of in the different manufactories, all the different economical arrangements and contrivances for facilitating the operations of useful industry, were so many models expressly made for imitation.

But in the arrangements relative to the economy of fuel, besides a view to immediate public utility, another motive, not much less powerful, contributed to induce me to pay all possible attention to the subject; namely, a desire to acquire a more thorough knowledge relative to the nature of heat and of the laws of its operations; and with this view several parts were added to the machinery, which I suspected at the time to be too complicated to be really useful in common practice.

The steam, for instance, which arose from the boiling liquids, instead of being suffered to escape into the atmosphere, was carried up by tubes into a room immediately over the kitchen, where it was made to pass through a spiral worm placed in a large cask full of cold water, and condensed, giving out its heat to the water in the cask; which water thus warmed, without any new expense of fuel, was made use of next day, instead of cold water, for filling the boilers. That this water, so warmed, might not be cooled during the night, the cask that contained it was put into another cask still larger; and the space between the two casks was filled with wool. The cooling of the steam, in its passage from the boiler to the cask where it was condensed, was prevented by warm coverings of sheepskins with the wool on them, by which the tubes of communication, which were of tin, were defended from the cold air of the atmosphere.

By this contrivance, the heat, which would otherwise

have been carried off by the steam into the atmosphere and totally lost, was arrested in its flight, and brought back into the boiler, and made to work the second day.

By other contrivances, the smoke also was laid under contribution. After it had passed under the boilers, and just as it was about to escape by the chimney, it was stopped, and, by being made to pass under a large copper filled with cold water, was deprived of the greater part of the heat it still retained; and thinking it probable that considerable advantages would be derived from drying the wood very thoroughly, and even heating it, before it was made use of for fuel, the smoke from two of the boilers was made to pass under a plate of iron which formed the bottom of an oven, in which the wood, necessary for the consumption of the kitchen for one day (having previously been cut into billets of a proper size), was dried during 24 hours, previous to its being used.

In a smaller kitchen (adjoining to that I have been describing), which was constructed merely as a model for imitation, and which was constantly open for the inspection·of the public, five boilers of different sizes, all heated by the same fire, were placed in a semicircular mass of brick-work, and the smoke, after having passed under all these five boilers, was made to heat, at pleasure, either an oven, or water which was contained in a wooden cask set upright upon the brick-work. A tube of copper, tinned on the outside, which went through the cask, gave a passage to the smoke, and this tube was connected with the bottom of the cask by means of a circular plate of copper through which the tube passed, which plate closed a circular opening in the bottom of the cask somewhat larger in diameter than the tube.

This circular plate was nailed to the bottom of the cask, and the joining made water-tight by interposing between the metallic plate and the wood a sheet of pasteboard; and the tube was fastened to the plate with solder. This tube (which was about 6 inches in diameter), as soon as it had passed the circular plate and entered the barrel, branched out into three smaller tubes, each about 4 inches in diameter, which, running parallel to each other through the whole length of the cask, went out of it above, by three different holes in the upper head of the cask, and ended in a canal which led to the chimney.

This tube, by which the smoke passed through the cask, was branched out into a number of branches in order to increase the surface, by which the heat of the smoke was communicated to the water in the cask. The cask was supplied with water from a reservoir placed in the upper part of the building, by means of a leaden pipe of communication from the one to the other; and the machinery was so contrived that, when any water was drawn out of the cask for use, it was immediately replaced from the reservoir; but as soon as the water in the cask had regained its proper height, the cold water from the reservoir ceased to flow in it.

Nothing more generally excited the surprise and curiosity of those who visited this kitchen, than to see water actually boiled in a wooden cask, and drawn from it boiling hot, by a brass cock. I have been the more particular in describing the manner in which this was done, as I have reason to think that a contrivance of this kind, or something similar to it, might, in many cases, be applied to useful purposes. No contrivance can possibly be invented by which heat can be communicated to fluids with so little loss; and as wood is not only an excellent non-conductor of heat itself, but may easily be surrounded by confined air, by furs, and other like bodies which are known to be useful in confining heat, the loss of heat, by the sides of a containing vessel composed of wood, might be almost entirely prevented.

Why should not the boilers for large salt-works and breweries, and those destined for other similar processes, in which great quantities of water are heated or evaporated, be constructed of wood, with horizontal tubes of iron or of copper, communicating with the fire-place, and running through them, for the circulation of the smoke? But this is not the place to enlarge upon this subject: I shall therefore leave it for the present, and return to my kitchens.

To prepare the soup furnished to the poor from the kitchen of the House of Industry, it was found necessary to keep up the fire near *five hours;* the soup, in order to its being good, requiring to be kept actually boiling above three hours.

The fuel made use of in this kitchen was dry beech-wood; a cord of which (or *klafter*, as it is called), 5 English feet $8\frac{9}{10}$ inches long, 5 feet $8\frac{9}{10}$ inches high, and 3 feet $1\frac{1}{3}$ inches wide, and which weighed at an average about 2200 Bavarian pounds (= 2724 lbs. avoirdupois), cost at an average about $5\frac{1}{4}$ florins (= 9s. $6\frac{1}{2}d.$ sterling) in the market.

Of this wood the daily consumption, when soup was provided for 1000 persons, was about 300 lbs. Bavarian weight, or about $\frac{1}{7}$, or more exactly $\frac{3}{23}$ of a cord or klafter, which cost 43 kreutzers (60 kreutzers making a florin), or about 1s. 3½d. sterling; and this gives $\frac{2}{23}$ of a kreutzer, or $\frac{1}{20}$ of a farthing, for the daily expense for fuel in cooking for each person.

To make an estimate of the daily expense for fuel in cooking the same quantity of the same kind of soup in private kitchens, we will suppose these 1000 persons, who were fed from the public kitchen of the House of Industry, to be separated into families of 5 persons each.

This would make just 200 families; and the quantity of wood consumed in the public kitchen daily for feeding 1000 persons (= 300 lbs.), being divided among 200 families, gives 1½ lbs. of wood for the daily consumption of each family; and, according to this estimate, 1 cord of wood, weighing 2200 lbs., ought to suffice for cooking for such a family 1466 days, or 4 years and 6 days.

But upon the most careful inquiries relative to the real consumption of fuel in private families in operations of cookery, as they are now generally performed over an open fire, I find that 5 Bavarian pounds of good peas-soup can hardly be prepared at a less expense of fuel than 15 lbs. of dry beech-wood of the best quality; consequently, a cord of such wood, instead of sufficing for preparing a soup daily for a family of 5 persons for 4 years, would hardly suffice for so long a time as 5 months.

And hence it appears that the consumption of fuel in the kitchens of private families is to that consumed in the first kitchen of the House of Industry at Munich, *in preparing the same quantity of the same kind of food* (peas-soup), as 10 to 1.* But it must be remembered that this difference in the quantities of fuel expended is not occasioned *entirely* by the difference between the two methods of managing the fire; for, exclusive of the effect produced by a given arrangement of the machinery, with the same arrangement, the greater the quantity of food prepared at once, or the larger the boiler (within certain limits, however, as will be seen hereafter), the less in proportion will be the quantity of fuel required; and the saving of fuel which arises from cooking upon a large scale is very considerable. But I shall take occasion to treat this part of my subject more fully elsewhere.

The kitchen in the House of Industry was finished in the beginning of the year 1790. And much about the same time, two other public kitchens upon a large scale were erected at Munich, under my directions; namely, the kitchen belonging to the Military Academy, and that belonging to the Military Hall (as it is called) in the English garden, in which building near 200 military officers messed daily during the annual encampments, for which purpose this building was erected.

There is likewise in the garden (which is 6 English miles in circumference) an inn, a farm-house, and a large dairy; and these establishments gave me an opportunity of constructing no less than four other kitchens, — namely, two for the inn, one for the farm-house, and one for the use of the dairy. And the uses for which these different kitchens were designed, and to which they were applied, were so various as not only

---

* Afterwards, on altering the kitchen of the House of Industry, and fitting it up on better principles, the economy of fuel was carried still farther, as will be seen in the sequel of this Essay.

to include almost every process of cookery, but also to afford opportunities of performing the same operations upon very different scales, and consequently of making many interesting experiments relative to the management of heat and the economy of fuel.

That I did not neglect these opportunities of pursuing with effect a subject which had long engaged my attention, and to which I was much attached, will readily be believed by those who know what ardour a curious subject of philosophical investigation is capable of inspiring in an inquisitive mind.

As the experiments I have made, or caused to be made, in the different establishments before mentioned, during the six or seven years that they have existed, are extremely numerous, it would take up too much time to give an account of them in detail: I shall therefore content myself with merely noticing the general results of them, and mentioning more particularly only such of them as appear to me to be most important. And in regard to the peculiar construction of the different kitchens above mentioned, as most of them have undergone many alterations, and as no one of them remains exactly in the same state in which it was first constructed, I do not think it necessary to be very particular in my account of them: I shall occasionally mention the principles on which they were constructed, and the faults I discovered in them; but when I shall come to speak of those improvements which have stood the test of actual experience, and which I can recommend as being worthy of imitation, I shall take care to be very exact and particular in my descriptions.

It will not be found very difficult, I fancy, from what has been said, to form a pretty just idea of the construction of the kitchen in the House of Industry above described, even without the help of a plan or drawing of it. That in the Military Academy was constructed upon a different principle. Instead of heating all the boilers from one and the same fire-place, almost every boiler had its own separate fire-place; and though the boilers were all furnished with double covers, similar to those made use of in the kitchen of the House of Industry, yet there was no attempt made to recover the heat carried off by the steam, but it was suffered to escape without hindrance into the atmosphere; it having been found, by the experiments made in the kitchen of the House of Industry, that when the fire is properly managed, — that is to say, when the heat is but just sufficient to keep the liquid boiling hot, or *very gently boiling*, — the quantity of steam generated is inconsiderable, and the heat carried off by it not worth the trouble of saving. Each fire-place was furnished with an iron grate, upon which the wood was burnt; and the opening into the fire, as well as that which communicated with the ash-pit, had in each its separate iron door.

Finding afterwards that the iron door which closed the opening by which the wood was introduced into the fire-place was much heated, and consequently that it caused a considerable loss of heat by communicating it to the cold atmosphere with which it was in contact; in order to remedy this evil without incurring the expense of double doors, the iron door was removed, and in its stead was placed a hollow cylinder, or rather truncated cone, of burnt clay or common earthen ware, which cone was 4 inches long, 6 inches in diameter internally, and 8 inches in diameter externally, at its

larger end or base; and 5½ inches in diameter internally, and 7½ inches in diameter externally, at its smaller
end; and being firmly fixed, with its axis in a horizontal position, and its larger end or base outwards, in
the middle of the opening leading to the fire-place, and
being well united with the solid brick-work by means
of mortar, the cavity of this cone formed the opening
by which the wood was introduced into the fire-place.
This cavity being closed with a fit stopper of earthen
ware, as earthen ware is a non-conductor of heat, or as
heat cannot pass through it but with great difficulty
and very slowly, the external surface of this cone and its
stopper were never much heated, consequently the quantity of heat they could communicate to the atmosphere
was but very trifling.   This contrivance was afterwards
rendered much more simple by substituting, instead of the
hollow cone, a tile, 10 inches square, and about 2½ inches
thick, with a conical hole in its centre, 6 inches in diameter externally, and 5¾ inches in diameter within, provided with a fit baked earthen stopper.   (See the Figures
No. 6, 7, and 8, Plate I.)

A perforated square tile is preferable to a hollow cylinder for forming a passage into the fire-place, not only
because it is cheaper, stronger, and more durable, but
also because it may, on account of its form, be more
easily and more firmly fixed in its place, and united with
the rest of the brick-work.

If proper moulds be provided for forming these perforated tiles and their stoppers, they may be afforded for
a mere trifle.   In Munich they are made of the very best
earth, by the Elector's potter; and they cost no more than
24 kreutzers, or something less than 9*d.* sterling, for a
tile with its stopper.   I had several made of sandstone

by a stone-cutter, but they cost me 1 florin and 30 kreutzers, or about 2*s*. 9*d*. sterling each.

Though those made of stone answered perfectly well, yet I found them not better than those made of earthen ware; and as these last are much cheaper, and I believe equally durable, they ought certainly to be preferred. That the stopper may be made to fit with accuracy the hole it is intended to close (which is necessary, as will be seen hereafter), they may be ground together with fine sand moistened with water.

Sensible from the beginning of the great importance of being absolutely master of the air which is admitted into the fire-place to feed the fire, so as to be able to admit more or less at pleasure, or to exclude it entirely, I took care, in all my fire-places, to close very exactly the passage into the ash-pit by a door carefully fitted to its frame, the air being admitted through a semicircular opening furnished with a register in the middle of this door. This contrivance (which admits of no further improvement) is indispensably necessary in all well-constructed fire-places, great or small. (See the Figures from Fig. 9 to Fig. 16, Plate II.)

Having occasion, in the course of my arrangements, to make use of a great number of boilers, and often of several boilers of the same dimensions, I availed myself of that circumstance to determine, by actual experiments, the best form for boilers, or that form which, with any given capacity, shall be best adapted for saving fuel.

Two or more boilers of the same capacity, but of different forms, constructed of sheet-copper of the same thickness, were placed in closed fire-places, constructed as nearly as possible upon the same principles, and were

used for a length of time in the same culinary processes ;
and the quantity of fuel consumed by each being noted,
the comparative advantages of their different forms were
ascertained.   Some of these boilers were made deep and
narrow, others wide and shallow ; there were some with
flat bottoms, others of a globular form, and others again
with their bottoms drawn inward like the bottom of
a common glass bottle.   The results of these inquiries
were very curious, and led me to a most interesting discovery.   They taught me not only what forms are
best for boilers, but also (what is still more interesting)
why one form is preferable to another.   They gave me
much new light with respect to the *manner* in which
flame and hot vapour part with their heat ; and suggested to me the idea of a very important improvement
in the internal construction of fire-places, which I have
since put in practice with great success.

But in order to be able to explain this matter in a
clear and satisfactory manner, and to render it easier to
be understood by those who have not been much conversant in inquiries of this kind, it will be necessary to
go back a little, and to treat the subject under consideration in a more regular and scientific manner.

Though it was not my intention originally to write an
elementary treatise on heat, yet, as the first or fundamental principles of that science are necessary to be
known, in order to establish upon solid grounds the
practical rules and directions relative to the management of heat which will hereafter be recommended, it
will not, I trust, be deemed either improper or superfluous to take a more extensive view of the subject, and
to treat it methodically, and at some length.

I have perhaps already exposed myself to criticism by

paying so little attention to method in this Essay, as to postpone so long the investigation of the elementary principles of the science I have undertaken to treat. It may be thought that the part of the subject I am now about to consider should have preceded all other investigation; that instead of occupying the middle of my book, it ought to have been discussed in the Introduction, or at least to have been treated in the beginning of the first chapter. But if I have been guilty of a fault in the arrangement of my subject, it has arisen not from inattention, but from an error of judgment. Desirous rather of writing a *useful book*, than of being the author of a *splendid performance*, I have not scrupled to transgress the established rules of elegant composition in all cases where I thought it would contribute to my main design, *public utility;* and well aware that my book, in order to its being really useful, must be read by many who have neither time nor patience to labour through an elementary treatise upon so abstruse a subject, I have endeavoured to *decoy* my reader into the situation in which I wish him to be placed, in order to his having a complete view of the prospect I have prepared for him, rather than to force him into it. If I have used art in doing this, he must forgive me; my design was not only innocent, but such as ought to entitle me to his thanks and to his esteem. I wished to entice him on as far as possible, without letting him perceive the difficulties of the road; and now that we have come on together so far, and are so near our journey's end, I hope and trust that he will not leave me. To proceed, therefore —

CHAPTER II.

Of the GENERATION OF HEAT *in the* COMBUSTION OF FUEL. — *Without knowing what Heat really is, the Laws of its Action may be investigated.* — *Probability that the Heat generated in the Combustion of Fuel is furnished by the Air, and not by the Fuel.* — *Effects of blowing a Fire explained.* — *Of Fire-places in which the Fire is made to blow itself.* — *Of Air-furnaces.* — *These Fire-places illustrated by a Lamp on* ARGAND'S *Principle.* — *Great Importance of being able to regulate the Quantity of Air which enters a closed Fire-place.* — *Utility of Dampers in the Chimneys of closed Fire-places.* — *General Rules and Directions for constructing closed Fire-places; with a full Explanation of the Principles on which these Rules are founded.*

WITHOUT entering into those abstruse and most difficult investigations respecting the nature of fire, which have employed the attention and divided the opinions of speculative philosophers in all ages; without even attempting to determine whether there be such a thing as an *igneous* fluid or not, — whether what we call *heat* be occasioned by the accumulation, or by the increased action of such a fluid, or whether it arises merely from an increased motion in the component particles of the body heated, or of some elastic fluid by which those particles are supposed to be surrounded, and upon which they are supposed to act, or by which they are supposed to be acted upon: in short, without bewildering myself and my reader in this endless labyrinth of darkness and uncertainty, I shall confine my inquiries to objects more useful, and which are clearly within the reach of human investigation; namely, the discovery of the sensible properties of heat, and of the most advantageous methods of generating it, and of directing it with certainty and effect in those various processes in which it is employed in the economy of human life.

Though I do not undertake to determine *what heat really is,* nor even to offer any opinions or conjectures relative to that subject; yet as heat is evidently something capable of being excited or generated, increased or accumulated, measured and transferred from one body to another, — in treating the subject I shall speak of it as being *generated, confined, directed, dispersed,* etc., it being necessary to use these terms in order to make myself understood.

Though it is not known exactly *how much* heat it is possible to produce in the combustion of any given quantity of any given kind of fuel, yet it is more than probable that the quantity depends in a great measure on the management of the fire.   It is likewise probable — I might say certain — that the heat produced is furnished not merely by the fuel, but in a great measure, if not entirely, by the *air* by which the fire is fed and supported.   It is well known that air is necessary to combustion; it is likewise known that the pure part of common atmospheric air, or that part of it (amounting to about $\frac{1}{5}$ of its whole volume) which alone is capable of supporting the combustion of inflammable bodies, undergoes a remarkable change, or is actually *decomposed* in that process; and as in this decomposition of pure air a great quantity of heat is known to be set loose, or to become redundant, it has been supposed by many (and with much appearance of probability) that by far the greater part, if not all the heat produced in the combustion of inflammable bodies, is derived from this source.

But whether it be the air or the fuel which furnishes the heat, it seems to be quite certain that the quantity furnished depends much upon *the management of the fire*, and that the quantity is greater as the combustion or decomposition of the fuel is more complete. In all probability, the decomposition of the air keeps pace with the decomposition of the fuel.

It is well known that the consumption of fuel is much accelerated, and the intensity of the heat augmented, by causing the air by which the combustion is excited to flow into the fire-place in a continued stream, and with a certain degree of velocity. Hence, blowing a fire, when the current of air is properly directed and when it is not too strong, serves to accelerate the combustion and to increase the heat; but when the blast is improperly directed, it will rather serve to derange and to impede the combustion than to forward it; and when it is too strong, it will blow the fire quite out, or totally extinguish it. There is no fire, however intense, but may be blown out by a blast of air, provided it be sufficiently strong, and that as infallibly as by a stream of cold water. Even gunpowder, the most inflammable perhaps of known substances, may be actually on fire at its surface, and yet the fire may be blown out and extinguished before the grain of powder has had time to be entirely consumed.

This fact, however extraordinary and incredible it may appear, I have proved by the most unexceptionable and conclusive experiments.

Fire-places may be so constructed that the fire may be made to blow itself, or — which is the same thing — to cause a current of air to flow into the fire; and this is an object to which the greatest attention ought to be paid in the construction of all fire-places where it is not intended to make use of an artificial blast from bellows for blowing the fire. Furnaces constructed upon this principle have been called *air-furnaces;* but every fire-place, and particularly every closed fire-place, ought to be an air-furnace, and that even were it intended to serve only for the smallest saucepan, otherwise it cannot be perfect.

An Argand's lamp is a fire-place upon this construction; for the glass tube which surrounds the wick (and which distinguishes this lamp from all others) serves merely as a blower. The circular form of the wick is not essential; for by applying a flatted glass tube as a blower to a lamp with a flat or riband wick, it may be made to give as much light as an Argand's lamp, or at least quite as much in proportion to the size of the wick, and to the quantity of oil consumed, as I have found by actual experiment.

But it is not the light alone that is increased in consequence of the application of these blowers: the heat also is rendered much more intense; and as the heat of any fire may be increased by a similar contrivance, on that account it is that I have had recourse to these lamps to assist me in explaining the subject under consideration. In these lamps the fire-place is closed on all sides, and the current of air which feeds the fire

rises up perpendicularly from below the fire-place into
the fire.   By surrounding the fire on all sides by a
wall, the cold atmosphere is prevented from rushing in
laterally from all quarters to supply the place of the
heated air or vapour, which, in consequence of its increased elasticity from the heat, continually rises from
the fire, and this causes the current of air below (the
only quarter from which it can with advantage flow
into the fire) to be very strong.

But in order that a fire-place may be perfect, it
should be so contrived that the combustion of the fuel
and the generation of the heat may occasionally be
accelerated or retarded, *without adding to or diminishing the quantity of fuel;* and, when the fire-place is
closed, this may easily be done by means of a *register*
in the door which closes the passage leading to the ash-
pit; for, as the rapidity of the combustion depends
upon the quantity of air by which the fire is fed, by
opening the register more or less, more or less air will
be admitted into the fire-place, and consequently more
or less fuel will be consumed, and more or less heat
generated in any given time, though the quantity of
fuel in the fire-place be actually much greater than what
otherwise would be sufficient.   Fig. 9 shows the form
of the register I commonly use for this purpose.

In order that this register may produce its proper
effect, a valve, or a *damper*, as it is commonly called,
should be placed in the chimney or canal by which the
smoke is carried off; which damper should be opened
more or less, as the quantity of air is greater or less
which is admitted into the fire-place.   This register
and this damper will be found very useful in another
respect, and that is, in putting out the fire when there

is no longer an occasion for it; for, upon closing them both entirely, the fire will be immediately extinguished, and the half-consumed fuel, instead of being suffered to burn out to no purpose, will be saved.

Nearly the same effects as are produced by a damper may be produced without one, by causing the smoke, after it has quitted the fire-place, to descend several feet below the level of the grate on which the fuel is burned before it is permitted to go up the chimney.

There is another circumstance of much importance which must be attended to in the construction of fire-places, and that is, the proper disposition of the fuel; for in order that the combustion may go on well, it is necessary not only that the fuel be in its proper place, but also that it be properly disposed; that is to say, that the solid parts of the fuel be of a just size, and that they be not placed too near each other, so as to prevent the free passage of the air between them, nor too far asunder; and if the fire-place can be so contrived that solid pieces of the inflamed fuel, as they go on to be diminished in size as they burn, may naturally fall together in the centre of the fire-place without any assistance, it will be a great improvement, as I have found by experience. This may be done, in small fire-places (and in these it is more particularly necessary), by burning the fuel upon a grate in the form of a segment of a hollow sphere, or of a dish. (See the Figures 3 and 4, Plate I.) All those I now use, except it be for fire-places which are very large indeed, are of this form; and where wood is made use of for fuel, it is cut into small billets from 4 to 6 inches in length. Instead of a grate of iron, I have lately introduced grates, or rather hollow dishes or pans of earthen-ware, perforated with a great number of holes for giving a passage to the air.

These perforated earthen pans, which are made very thick and strong, are incomparably cheaper than iron grates; and judging from the experience I have had of them, I am inclined to think they answer even better than the grates; indeed it appears to me not difficult to assign a reason why they ought to be better.

For large fire-places I have sometimes used grates, the bars of which were common bricks placed edgewise, and these have been found to answer very well.

As only *that part of the air* which, entering the fire-place in a proper manner and in a just quantity, and coming into actual contact with the burning fuel, *is decomposed*, contributes to the generation of heat, it is evident that all the air that finds its way into the fire-place, *and out of it again*, without being decomposed, is a thief; that it not only *contributes nothing* to the heat, but being itself heated at the expense of the fire, and going off *hot* into the atmosphere by the chimney, occasions an actual loss of heat; and this loss is often very considerable, and the prevention of it is such an object, that too much attention cannot be paid to it in the construction of fire-places.

When the fire-place is closed on all sides by a wall, and when the opening by which the fuel is introduced is kept closed, no air can press in laterally upon the fire; but yet, when the grate is larger than the heap of burning fuel, which must often be the case, a great quantity of air may insinuate itself by the sides of the grate into the fire-place, without going through the fire. But when, instead of an iron grate, a perforated hollow earthen pan is used, by making the bottom of the pan of a certain thickness, 2, 3, or 4 inches, for instance, and making all the air-holes point to one common centre (to the focus or centre of the fire), this furtive entrance of cold air into the fire-place will in a great measure be prevented.

This evil may likewise be prevented when circular hollow iron grates are used, by narrowing the fire-place immediately under the grate in the form of an inverted, truncated, hollow cone, the opening or diameter of which above being equal to the internal diameter of the circular rim of the grate, and that below (by which the air rises to enter the fire-place) about *one third* of that diameter. (See the Figure 5, Plate I.) This opening below, through which the air rises, must be immediately under the centre of the grate, and as near to it as possible; care must be taken, however, that a small space be left between the outside, or underside of the iron bars which form the hollow grate and the inside surface of this inverted hollow cone, in order that the ashes may slide down into the ash-pit.

As to the form and size of the ash-pit, these are matters of perfect indifference, provided, however, that it be large enough to give a free passage to the air necessary for feeding the fire, and that the only passage into it by which air can enter is closed by a good door furnished with a register. The necessity of being completely master of the passage by which the air enters the fire-place has already been sufficiently explained.

It is perhaps unnecessary for me to observe that, where perforated earthen pans are used instead of iron grates, the air-holes in the pans ought to be rather smaller above than below, in order that they may not be choked up by the small pieces of coal and the ashes which occasionally fall through them into the ash-pit.

One great advantage attending fire-places on the construction here proposed is, that they serve equally well for every kind of fuel. Wood, pit-coal, charcoal, turf, etc., may indifferently be used, and all of them with the same facility, and with the same advantages; or any two, or more, of these different kinds of fuel may be used at the same time without the smallest inconvenience; or the fire having been lighted with dry wood, or any other very inflammable material, the heat may afterwards be kept up by cheaper or more ordinary fuel of a more difficult and slow combustion. Some kinds of fuel will perhaps be found most advantageous for making the pot boil, and others for keeping it boiling; and a very considerable saving will probably be found to result from paying due attention to this circumstance. When the fire-place is so contrived as to serve equally well for all kinds of fuel, this may be done without the least difficulty or trouble.

I have just shown that narrowing that part of the fire-place which lies below the grate serves to make the air enter the fire in a more advantageous manner. This construction has another advantage, perhaps still more important: the heat which is projected downwards through the openings between the bars of the grate, instead of being permitted to escape into the ash-pit (where it would be lost), striking against the sides of this inverted hollow *cone*, it is there stopped, and afterwards rises into the fire-place again with the current of air which feeds the fire, or it is immediately reflected by this conical surface, and, after two or three bounds from side to side, is thrown up against the bottom of the boiler.

But in order to be able to form a clear and distinct idea upon this subject, it is necessary to examine with care all the circumstances attending the generation of heat in the combustion of inflammable bodies, and to see in what manner or under what form the heat generated manifests itself and how it may be collected, accumulated, confined, and directed.

This opens a wide field for philosophical inquiry; but as these investigations are not only curious and entertaining, but also useful and important in a high degree, I trust my reader will pardon me for requesting his particular attention while I endeavour to do justice to this most interesting, but, at the same time, most abstruse and most difficult part of the subject I have undertaken to treat.

The heat generated in the combustion of fuel manifests itself in two ways; namely, in the hot vapour which rises from the fire, with which it may be said to be *combined*, and in the calorific rays which are thrown off from the fire in all directions. These rays may, with greater propriety, be said to be *calorific*, or *capable of generating heat*, in any body by which they are *stopped*, than to be called hot; for when they pass freely through any medium (as through a mass of air, for instance), they are not found to communicate any heat whatever to such medium; neither do they appear to excite any considerable degree of heat in bodies from whose surfaces they are reflected; and in these respects they bear a manifest resemblance to the rays emitted by the sun.

What proportion this *radiant heat* (if I may be allowed to use so inaccurate an expression) bears to that which goes off from burning bodies in the smoke and heated vapour, is not exactly known; it is certain, however, that the quantity of heat which goes off in the heated elastic fluids, visible and invisible, which rise from a fire, is much greater than that which all the calorific rays united would be capable of producing. But though the quantity of *radiant heat* is less than that existing in the hot vapour (and which, for the sake of distinction, may be called *combined heat*), the former is still much too considerable to be neglected.

That the heat generated, or excited, by the calorific rays which proceed from burning bodies is in fact considerable, is evident from the heat which is felt in a room warmed by a chimney fire; for as all the heat, combined with the smoke and hot vapour, goes up the chimney, it is certain that the increase of heat in the room, occasioned by the fire, is entirely owing to the calorific rays thrown into it from the burning fuel.

The activity of these rays may be shown in various ways, but in no way in a more striking manner than by the following simple experiment: When the fire burns bright upon the hearth, let the arm be extended in a straight line towards the centre of the fire, with the hand open, and all the fingers extended and pointing to the fire. If the hand is not nearer the fire than the distance of two or three yards, except the fire be very large indeed, the heat will scarcely be perceptible; but if, without moving the arm, the wrist be bent upwards so as to present the inside or flat of the hand perpendicular to the fire, the heat will not only be very sensibly felt, but if the fire be large, and if it burns clear and bright, it will be found to be so intense as to be quite insupportable.

It is not, however, burning bodies alone that emit calorific rays. All bodies — those which are fixed and incombustible as well as those which are inflammable, fluids as well as solids — are found to throw off these rays in great abundance, as soon as they are heated to that degree which is necessary to their becoming luminous in the dark, or till they are red-hot.

Bodies even which are heated to a less degree than that which is necessary to their emitting *visible* light send off calorific rays in all directions. This is a matter of fact, which has been proved by experiment. Do all bodies, at all temperatures, — freezing mercury as well as melting iron, — continually emit these rays in greater or less quantities, or with greater or less velocities? Are bodies cooled in consequence of their emitting these rays? Do these calorific rays always generate heat, even when the body by which they are stopped or absorbed is hotter than that from which the rays proceeded? But I forget that I promised not to involve myself in abstruse speculation. To return, then. Whatever may be the nature of the rays emitted by burning fuel, as *one* of their *known properties* is to generate heat, they ought certainly to be very particularly attended to in every arrangement in which the economy of heat, or of fuel, is a principal object in view.

As these calorific rays generate heat in the body by which they are *stopped or absorbed*, and not in the medium through which they pass, it is necessary to dispose those bodies which are designed for stopping them in such a manner that they may easily and *necessarily* communicate the heat they thus acquire to the body upon which it is intended that it should operate.

The closed fire-places which I have recommended, and which will hereafter be more particularly described, will answer this purpose completely. The fire being closed in these fire-places on every side, as well below the grate as laterally, and in short everywhere, except where the bottom of the boiler presents itself to the fire, none of these rays can possibly escape; and as the materials of which the fire-place is constructed (bricks and mortar) are bad conductors of heat, but a small part of the heat generated in the combustion of the fuel will be absorbed and transmitted by them into the interior parts of the wall, there to be dispersed and lost. But the confining of heat is a matter of sufficient importance to deserve being treated in a separate chapter.

CHAPTER III.

Of the Means of CONFINING HEAT, *and* DIRECTING ITS OPERATIONS. — *Of Conductors and Non-conductors of Heat.* — *Common Atmospheric Air a good Non-conductor of Heat, and may be employed with great Advantage for confining it; is employed by Nature for that Purpose, in many Instances; is the principal Cause of the Warmth of Natural and Artificial Clothing; is the sole Cause of the Warmth of Double Windows.* — *Great Utility of Double Windows and Double Walls: they are equally useful in Hot Countries as in Cold.* — ALL ELASTIC FLUIDS *Non-conductors of Heat.* — STEAM *proved by Experiment to be a Non-conductor of Heat.* — FLAME *is also a Non-conductor of Heat.*

THAT heat passes more freely through some bodies than through others, is a fact well known; but the cause of this difference in the conducting powers of bodies with respect to heat has not yet been discovered.

The utility of giving a wooden handle to a tea-pot or coffee-pot of metal, or of covering its metallic handle with leather, or with wood, is well known. But the difference in the conducting powers of various bodies with regard to heat may be shown by a great number of very simple experiments, such as are in the power of every one to make at all times and in all places, and almost without either trouble or expense.

If an iron nail and a pin of wood, of the same form and dimensions, be held successively in the flame of a candle, the difference in the conducting powers of the metal and of wood will manifest itself in a manner in which there will be no room left for doubt. As soon as the end of the nail which is exposed in the flame of the candle begins to be heated, the other end of it will grow so hot as to render it impossible to hold it in the hand without being burned; but the wood may be held any length of time in the same situation without the least inconvenience; and, even after it has taken fire, it may be held till it is almost entirely consumed, for the uninflamed wood will not grow hot, and, till the flame actually comes in contact with the fingers, they will not be burned. If a small slip or tube of glass be held in the flame of the candle in the same manner, the end of the glass by which it is held will be found to be more heated than the wood, but incomparably less so than the pin or nail of metal ; and among all the various bodies that can be tried in this manner, no two of them will be found to give a passage to heat through their substances with exactly the same degree of facility.*

To confine heat is nothing more than to prevent its escape out of the hot body in which it exists, and in which it is required to be retained ; and this can only be done by surrounding the hot body by some covering composed of a substance through which heat cannot pass, or through which it passes with great difficulty. If a covering could be found perfectly impervious to heat, there is reason to believe that a hot body, completely surrounded by it, would remain hot for ever ; but we are acquainted with no such substance, nor is it probable that any such exists.

Those bodies in which heat passes freely or rapidly are called *conductors* of heat; those in which it makes its way with great difficulty or very slowly, *non-conductors*, or bad conductors of heat. The epithets, good, bad, indifferent, excellent, etc., are applied indifferently to *conductors* and to *non-conductors*. A good conductor, for instance, is one in which heat passes very freely; a good non-conductor is one in which it passes with great difficulty; and an indifferent conductor may likewise be called, without any impropriety, an indifferent non-conductor.

* To show the relative conducting power of the different metals, Doctor Ingenhouz contrived a very pretty experiment. He took equal cylinders of the different metals (being straight pieces of stout wire, drawn through the same hole, and of the same length), and, dipping them into melted wax, covered them with a thin coating of the wax. He then held one end of each of these cylinders in boiling water, and observed how far the coating of wax was melted by the heat communicated through the metal, and with what celerity the heat passed.

Those bodies which are the worst conductors, or rather the best non-conductors of heat, are best adapted for forming coverings for confining heat.

All the metals are remarkably good conductors of heat; wood, and in general all light, dry, and spongy bodies are non-conductors. Glass, though a very hard and compact body, is a non-conductor. Mercury, water, and liquids of all kinds, are conductors; but air, and in general all elastic fluids, *steam* even not excepted, are non-conductors.

Some experiments which I have lately made, and which have not yet been published, have induced me to suspect that water, mercury, and all other non-elastic fluids, do not permit heat to pass through them from particle to particle, as it undoubtedly passes through solid bodies, but that their apparent conducting powers depend essentially upon the extreme mobility of their parts; in short, that they rather *transport* heat than allow it a passage. But I will not anticipate a subject which I propose to treat more fully at some future period.

The conducting power of any solid body in one solid mass is much greater than that of the same body reduced to a powder, or divided into many smaller pieces. An iron bar, or an iron plate, for instance, is a much better conductor of heat than iron filings; and sawdust is a better non-conductor than wood. Dry wood-ashes is a better non-conductor than either; and very dry charcoal reduced to a fine powder is one of the best non-conductors known; and as charcoal is perfectly incombustible when confined in a space where fresh air can have no access, it is admirably well calculated for forming a barrier for confining heat, where the heat to be confined is intense.

But among all the various substances of which coverings may be formed for confining heat, none can be employed with greater advantage than common atmospheric air. It is what nature employs for that purpose; and we cannot do better than to imitate her.

The warmth of the wool and fur of beasts, and of the feathers of birds, is undoubtedly owing to the air in their interstices; which air, being strongly attracted by these substances, is confined, and forms a barrier which not only prevents the cold winds from approaching the body of the animal, but which opposes an almost insurmountable obstacle to the escape of the heat of the animal into the atmosphere. And in the same manner the air in snow serves to preserve the heat of the earth in winter. The warmth of all kinds of artificial clothing may be shown to depend on the same cause; and were this circumstance more generally known, and more attended to, very important improvements in the management of heat could not fail to result from it. A great part of our lives is spent in guarding ourselves against the extremes of heat and of cold, and in operations in which the use of fire is indispensable; and yet how little progress has been made in that most useful and most important of the arts, — the management of heat!

Double windows have been in use many years in most of the northern parts of Europe, and their great utility, in rendering the houses furnished with them warm and comfortable in winter, is universally acknowledged; but I have never heard that anybody has thought of employing them in hot countries to keep their apartments cool in summer; yet how easy and natural is this application of so simple and so useful an invention! If a double window can prevent the heat which is *in* a room from passing *out of it*, one would imagine it could require no great effort of genius to discover that it would be equally efficacious for preventing the heat *without* from coming *in*. But natural as this conclusion may appear, I believe it has never yet occurred to anybody; at least I am quite certain that I have never seen a double window either in Italy or in any other hot country I have had occasion to visit.*

But the utility of double windows and double walls, in hot as well as in cold countries, is a matter of so much importance that I shall take occasion to treat it more fully in another place. In the mean time, I shall only observe here that it is the *confined air* shut up between the two windows, and not the double glass plates, that renders the passage of heat through them so difficult. Were it owing to the increased thickness of the glass, a single pane of glass twice as thick would answer the same purpose; but the increased thickness of the glass of which a window is formed is not found to have any sensible effect in rendering a room warmer.

But air is not only a non-conductor of heat, but its non-conducting power may be greatly increased. To be able to form a just idea of the manner in which air may be rendered a worse conductor of heat, or, which is the same thing, a better non-conductor of it than it is in its natural unconfined state, it will be necessary to consider *the manner* in which heat passes through air.

---

* When double windows are used in hot countries to keep dwelling-houses cool, great care must be taken to screen those windows from the sun's direct rays, and even from the strong light of day, otherwise they will produce effects directly contrary to those intended. This may easily be done either by Venetian blinds or by awnings. In all cases where rooms are to be kept cool in hot weather, the less light that is permitted to enter them the cooler they will be.

Now it appears, from the result of a number of experiments which I made with a view to the investigation of this subject, and which are published in a paper read before the Royal Society,[15] that though the particles of air, *each particle for itself*, can receive heat from *other bodies*, or communicate it to them, yet there is no communication of heat *between one particle of air and another particle of air.* And from hence it follows that though air may, and certainly does, *carry off* heat and *transport it* from one place or from one body to another, yet a mass of air in a quiescent state, or with all its particles at rest, *could it remain in that state*, would be totally impervious to heat, or such a mass of air would be a perfect non-conductor.

Now if heat passes in a mass of air merely in consequence of the motion it occasions in that air; if it be *transported, — not suffered to pass, —* in that case, it is clear that whatever can obstruct and impede the internal motion of the air must tend to diminish its conducting power. And this I have found to be the case in fact. I found that a certain quantity of heat which was able to make its way through a wall, or rather a sheet of confined air, $\frac{1}{2}$ an inch thick in $9\frac{3}{5}$ minutes, required $21\frac{2}{5}$ minutes to make its way through the same wall, when the internal motion of this air was impeded by mixing with it $\frac{1}{56}$ part of its bulk of eider-down, of very fine fur, or of fine silk, as spun by the worm.

But in mixing bodies with air, in order to impede its internal motion and render it more fit for confining heat, such bodies only must be chosen as are themselves non-conductors of heat, otherwise they will do more harm than good, as I have found by experience.

When, instead of making use of eider-down, fur, or fine silk for impeding the internal motion of the confined air, I used an equal volume of exceedingly fine silver-wire flatted (being the ravellings of gold or silver lace), the passage of the heat through the barrier, so far from being impeded, was remarkably facilitated by this addition, — the heat passing through this compound of air and fine threads of metal much sooner than it would have made its way through the air alone.

Another circumstance to be attended to in the choice of a substance to be mixed with air, in order to form a covering or barrier for confining heat, is the fineness or subtilty of its parts; for the finer they are, the greater will be their surface in proportion to their solidity, and the more will they impede the motions of the particles of the air. Coarse horse-hair would be found to answer much worse for this purpose than the fine fur of a beaver, though it is not probable that there is any essential difference in the chemical properties of those two kinds of hair.

But it is not only the fineness of the parts of a substance, and its being a non-conductor, which render it proper to be employed in the formation of covering to confine heat; there is still another property, more occult, which seems to have great influence in rendering some substances better fitted for this use than others: and this is a certain attraction which subsists between certain bodies and air. The obstinacy with which air adheres to the fine fur of beasts and to the feathers of birds is well-known; and it may easily be proved that this attraction must assist very powerfully in preventing the motion of the air concealed in the interstices of those substances, and consequently in impeding the passage of heat through them.

Perhaps there may be another still more hidden cause which renders one substance better than another for confining heat. I have shown by a direct and unexceptionable experiment that heat can pass through the Torricellian vacuum,[13] though with rather more difficulty than in air (the conducting power of air being to that of a Torricellian vacuum as 1000 to 604, or as 10 to 6, very nearly); but if heat can pass where there is no air, it must in that case pass by a medium more subtile than air, — a medium which most probably pervades all solid bodies with the greatest facility, and which must certainly pervade either the glass or the mercury employed in making a Torricellian vacuum.

Now, if there exists a medium more subtile than air by which heat may be conducted, is it not possible that there may exist a certain affinity between that medium and sensible bodies? a certain attraction or cohesion, by means of which bodies in general, or some kinds of bodies in particular, may, somehow or other, impede this medium in its operations in conducting or transporting heat from one place to another? It appeared from the result of several of my experiments, of which I have given an account in detail in my paper before mentioned, published in the year 1786, in vol. lxxvi. of the Philosophical Transactions, that the conducting power of a Torricellian vacuum is to that of air as 604 to 1000; but I found by a subsequent experiment (see my second Paper on Heat, published in the Philosophical Transactions for the year 1792) that 55 parts in bulk of air, with 1 part of fine raw silk, formed a covering for confining heat, the conducting power of which was to that of air as 576 to 1284, or as 448 to 1000. Now, from the result of this last-mentioned experiment, it should seem that the introduction into the space through which the heat passed of so small a quantity of raw silk as $\frac{1}{56}$ part of the volume or capacity of that space, rendered that space (which now contained 55 parts of air and 1 part of silk) more impervious to heat than even a Torricellian vacuum. The silk must therefore not only have completely destroyed the conducting power of the air, but must also at the same time have very sensibly impaired that of the ethereal fluid which probably occupies the interstices of air, and which serves to conduct heat through a Torricellian vacuum : for a Torricellian vacuum was a better conductor of heat than this medium, in the proportion of 604 to 448. But I forbear to enlarge upon this subject, being sensible of the danger of reasoning upon the properties of a fluid whose existence even is doubtful, and feeling that our knowledge of the nature of heat, and of the manner in which it is communicated from one body to another, is much too imperfect and obscure to enable us to pursue these speculations with any prospect of success or advantage.

Whatever may be the *manner* in which heat is communicated from one body to another, I think it has been sufficiently proved that it passes with great difficulty through confined air; and the knowledge of this fact is very important, as it enables us to take our measures with certainty and with facility for confining heat, and directing its operations to useful purposes.

But atmospheric air is not the only non-conductor of

heat. All kinds of air, artificial as well as natural, and in general all elastic fluids, *steam not excepted*, seem to possess this property in as high a degree of perfection as atmospheric air.

That steam is not a conductor of heat I proved by the following experiment: A large globular bottle being provided, of very thin and very transparent glass, with a narrow neck, and its bottom drawn inward so as to form a hollow hemisphere about 6 inches in diameter; this bottle, which was about 8 inches in diameter externally, being filled with cold water, was placed in a shallow dish, or rather plate, about 10 inches in diameter, with a flat bottom formed of very thin sheet brass, and raised upon a tripod, and which contained a small quantity (about $\frac{2}{10}$ of an inch in depth) of water; a spirit-lamp being then placed under the middle of this plate, in a very few minutes the water in the plate began to boil, and the hollow formed by the bottom of the bottle was filled with clouds of steam, which, after circulating in it with surprising rapidity 4 or 5 minutes, and after forcing out a good deal of air from under the bottle, began gradually to clear up. At the end of 8 or 10 minutes (when, as I supposed, the air remaining with the steam in the hollow cavity formed by the bottom of the bottle had acquired nearly the same temperature as that of the steam) these clouds totally disappeared; and though the water continued to boil with the utmost violence, the contents of this hollow cavity became so perfectly invisible, and so little appearance was there of steam, that had it not been for the streams of water which were continually running down its sides I should almost have been tempted to doubt whether any steam was actually generated.

Upon lifting up for an instant one side of the bottle, and letting in a smaller quantity of cold air, the clouds instantly returned, and continued circulating several minutes with great rapidity, and then gradually disappeared as before. This experiment was repeated several times, and always with the same result; the steam always becoming visible when cold air was mixed with it, and afterwards recovering its transparency when, part of this air being expelled, that which remained had acquired the temperature of the steam.

Finding that cold air introduced under the bottle caused the steam to be partially condensed, and clouds to be formed, I was desirous of seeing what visible effects would be produced by introducing a cold solid body under the bottle. I imagined that if steam was a conductor of heat, some part of the heat in the steam passing out of it into the cold body, clouds would of course be formed; but I thought if steam was a *non-conductor* of heat, — that is to say, *if one particle of steam could not communicate any part of its heat to its neighbouring particles*, — in that case, as the cold body could only affect the particles of steam *actually in contact with it*, no cloud would appear; and the result of the experiment showed that steam is in fact a *non-conductor of heat*. For, notwithstanding the cold body used in this experiment was very large and very cold, being a solid lump of ice nearly as large as a hen's egg, placed in the middle of the hollow cavity under the bottle, upon a small tripod or stand made of iron wire; yet as soon as the clouds which were formed in consequence of the unavoidable introduction of cold air in lifting up the bottle to introduce the ice were dissipated, which soon happened, the steam became so perfectly transparent and invisible that *not the smallest appearance of cloudiness was to be seen anywhere,* not even about the ice, which, as it went on to melt, appeared as clear and as transparent as a piece of the finest rock crystal.

This experiment, which I first made at Florence, in the month of November, 1793, was repeated several times in the presence of Lord Palmerston, who was then at Florence, and M. de Fontana.*

In these experiments the air was not entirely expelled from under the bottle; on the contrary, a considerable quantity of it remained mixed with the steam even after the clouds had totally disappeared, as I found by a particular experiment made with a view to ascertain that fact. But that circumstance does not render the result of this experiment less curious; on the contrary, I think it tends to make it more surprising. It should seem that neither the mass of steam, nor that of air, were at all cooled by the body of ice which they surrounded; for

---

* The bottle made use of in this experiment, though it appeared very large externally, contained but a very small quantity of water, owing to its bottom being very much drawn inwards. As the hollow cavity under the bottom of the bottle (which, as I just observed, was nearly in the form of a hemisphere, and 6 inches in diameter) served as a receiver for confining the steam which rose from the boiling water in the plate, it may perhaps be imagined that a common glass receiver in the form of a bell, such as are used in pneumatical experiments, might answer as well as this bottle ; I thought so myself, but upon making the experiment I found my mistake. A common receiver will answer perfectly well for confining the steam, but the glass soon becomes so hot that the drops of water which are formed upon its internal surface, in consequence of the condensation of the steam, instead of running down the sides of the receiver in clear transparent streams, form blotches and streaks, which render the glass so opaque that nothing can be seen distinctly through it ; and this of course completely frustrates the main design of the experiment. But cold water in the bottle keeping the glass cool, the condensation of the steam upon the sides of the hollow cavity formed by the bottom of the bottle goes on more regularly, and the streams of water which are continually running down the sides of the glass, uniting together, form one transparent sheet of water, by which means every thing that goes on under the bottle may be distinctly seen.

if the air had been cooled (in mass), it seems highly probable that the clouds would have returned.

The results of these experiments compared with those formerly alluded to, in which I had endeavoured to ascertain the most advantageous forms for boilers, opened to me an entirely new field for speculation and for improvement in the management of fire. They shewed me that not only cold air, but also hot air and hot steam, and hot mixtures of air and steam, are non-conductors of heat; consequently that the hot vapour which rises from burning fuel, and even the *flame itself, is a non-conductor of heat.*

This may be thought a bold assertion; but a little calm reflection, and a careful examination of the phenomena which attend the combustion of fuel, and the communication of heat by flame, will show it to be well-founded; and the advantages which may be derived from the knowledge of this fact are of very great importance indeed. But this subject deserves to be thoroughly investigated.

CHAPTER IV.

Of the MANNER *in which* HEAT *is* COMMUNICATED *by* FLAME *to other Bodies. — Flame acts on Bodies in the same Manner as a hot Wind. — The Effect of a Blowpipe in increasing the Activity of Flame explained, and illustrated by Experiments. — A Knowledge of the Manner in which Heat is communicated by Flame necessary in order to determine the most ad-*

vantageous Forms for Boilers. — *General Principles on which Boilers of all Dimensions ought to be constructed.*

IF flame be merely vapour, or a mixture of air and steam heated red-hot, as air and steam are both non-conductors of heat, there seems to be no difficulty in conceiving that flame may, notwithstanding its great degree of heat, still retain the properties of its component fluids, and remain *a non-conductor of heat.* The non-conducting power of air does not appear to be at all impaired by being heated to the temperature of boiling water; and I see no reason why that property in air, or in any other elastic fluid, should be impaired by any augmentation of temperature, however great. If steam, or if air, at the temperature of 212 degrees of Fahrenheit's thermometer, be a non-conductor of heat, why should it not remain a non-conductor at that of 1000 degrees, or when heated red-hot? I confess I do not see how a body *could* be deprived of a property so essential, without being at the same time totally changed; and I believe nobody will imagine that either air or steam undergoes any chemical change merely by being heated to the temperature of red-hot iron. But without insisting upon these reasonings, however conclusive I may think them, I shall endeavour to show, from experiment and observation, in short to *prove*, that flame is in fact a non-conductor of heat.

Taking it for granted — what I imagine will not be denied — that air is a non-conductor of heat, at least in the sense I have used that appellation, I shall endeavour to show that flame acts precisely in the same manner as a hot wind would do in communicating heat, and in no other way; and if I succeed in this, I fancy I may consider the proposition as sufficiently proved.

The effect of a blast of cold air in cooling any hot body exposed to it is well known, and the causes of this effect may easily be traced to that property of air which renders it a non-conductor of heat; for if the particles of cold air in contact with a hot body could, with perfect facility, give the heat they acquire from the hot body to other particles of air by which they are immediately surrounded, and these again to others, and so on, the heat would be carried off *as fast as the hot body could part with it*, and any motion of the particles of the air, any wind or blast, would not sensibly facilitate or hasten the cooling of the body ; and by a parity of reasoning it may be shown that, if flame were in fact a perfect conductor of heat, any cold body plunged into it would always be heated *as fast as that body could receive heat;* and neither any motion of the internal parts of the flame, nor the velocity with which it impinged against the cold body, could have any sensible effect either to facilitate or accelerate the heating of the body. But if flame be a non-conductor of heat, its action will be exactly similar to that of a hot wind, and consequently much will depend upon the manner in which it is applied to any body intended to be heated by it. Those particles of it *only* which are in *actual contact* with the body will communicate heat to it; and the greater the number of different particles of the flame which are brought into contact with it, the greater will be the quantity of heat communicated. Hence the importance of causing the flame to impinge with force against the body to be heated, and to strike it in such a manner that its current may be broken, and that whirlpools may be formed in it; for the rapid motion of the flame causes a quick succession of hot particles; and, admitting our assumed principles to be true, it is quite evident that every kind of internal motion among the particles of the flame by which it can be agitated must tend very powerfully to accelerate the communication of the heat.

The effect of a blowpipe is well known, but I do not think that the *manner* in which it increases the *action* of *flame* has ever been satisfactorily explained. It has generally been imagined, I believe, that the current of fresh air which is forced through the flame by a blowpipe actually increases the quantity of heat; I rather suppose it does little more than direct the heat *actually existing in the flame* to a given point. A current of air cannot *generate* heat without at the same time being decomposed; and, in order to its being decomposed in a fire, it must be brought into actual contact with the burning fuel, or at least with the uninflamed inflammable vapour which rises from it. But can it be supposed that there can be any thing inflammable, and not actually inflamed, in the clear, bright, .and perfectly transparent flame of a wax candle? A blowpipe has however as sensible an effect, when directed against the clear flame of a wax candle, as when it is employed to increase the action of a common glass-worker's lamp.

Conceiving that the discovery of the *manner* in which the current of air from a blowpipe serves to increase the intensity of the action of the flame could not fail to throw much light upon the subject under consideration, — namely, the investigation of *the man-*

ner in which heat is communicated to bodies by flame, — I made the following experiments, the results of which I conceive to be decisive.

Concluding that the current of air from a blowpipe, directed against the flame of any burning body, could tend to increase the intensity of the action of the flame only in one or both of these two ways, — namely, by increasing its *action* upon the body against which it is directed, or by actually increasing the *quantity* of heat generated in the combustion of the fuel, — a method occurred to me by which I thought it possible to determine, by actual experiment, to which of these causes the effect in question is owing, or how much each of them might contribute to it. To do this, I filled a large bladder, containing above a gallon, with *fixed air*, which, as is well known, is totally unfit for supporting the combustion of inflammable bodies, and which, of course, could not be suspected of *adding* any heat to a flame against which a current of it should be directed. I imagined therefore that if a blowpipe supplied with this air, on being directed against the flame of a candle, should be found to produce nearly the same effect as when common air is used for the same purpose, it would prove to a demonstration that the augmentation of the intensity of the action, or activity of the flame which arises from the use of a blowpipe, is owing to the agitation of the flame, to its being directed to a point, to the impetuosity with which it is made to strike against the body which is heated by it, and to the rapid succession of fresh particles of this hot vapour, and not to any *positive increase of heat.*

A blowpipe being attached to the bladder containing fixed air, the end of this pipe was directed to the clear brilliant flame of a wax candle, which had just been snuffed; and, by compressing the bladder, the flame was projected against a small tube of glass, which was very soon made red-hot, and even melted.

Having repeated this experiment several times, and having found how long it required to melt the tube when the flame of the candle was forced against it by a blast of *fixed air*, I now varied the experiment, by making use of common atmospheric air instead of fixed air; taking care to employ the same candle and the same blowpipe used in the former experiments, and even making use of the bladder, in order that, the experiments being exactly similar and differing only in the kinds of air made use of, the effect of that difference might be discovered and estimated.

The results of these experiments were most perfectly conclusive, and proved in a decisive manner that the effect of a blowpipe, *when applied to clear flame*, arises not from any real augmentation of heat, but merely from the increased activity of the flame, in consequence of its being impelled with force, and broken in eddies on the surface of the body against which it is made to act; the effect of the blowpipe on these experiments being to all appearance quite as great when fixed air was made use of (which *could not* increase the quantity of heat), as when atmospheric air was used.

But, conceiving the determination of this question relative to the manner in which flame communicates heat to be a matter of much importance, I did not rest my inquiries here. I repeated the experiments very often, and varied them in a great number of different ways, sometimes making use of fixed air, sometimes of atmospheric air, and at other times using dephlogisticated air, and common air rendered unfit for the support of animal life and of combustion, by burning a candle in it till the candle went out.

It would take up too much time to give an account in detail of all these experiments. I shall therefore content myself with merely observing that they all tended to show that the effect of a blowpipe *used in the manner here described* is owing to the direction and velocity it gives to the flame against which it is employed, and not to any real increase of heat.

It must be remembered that the principal object I had in view in these experiments was to discover the *manner* in which flame communicates heat to other bodies, and by what means that communication may be facilitated. Were it required to increase the intensity of the heat by *blowing the fire*, the current of air must be applied in such a manner as to expedite the combustion: it must be directed to the inflamed surface of the burning fuel, and not to the red-hot vapour or flame which rises from it, and in which the combustion is most probably already quite complete ; and in this case there is no doubt but the effect produced by blowing would depend much upon the quality of the air made use of.

The results of the foregoing experiments with the blowpipe will, I am confident, be thought quite conclusive by those who will take the trouble to consider them attentively; and the advantages that may be derived from the knowledge of the fact established by them are very obvious. If flame, or the hot vapour which arises from burning bodies, be a non-conductor of heat; and if, in order to communicate its heat to any other body, it be necessary that its particles *individually* be brought into actual contact with that body, it is evident that the form of a boiler, and of its fire-place, must be matters of much importance; and that *that form* must be most advantageous which is best calculated to produce an internal motion in the flame, and to bring alternately as many of its particles as possible into contact with the body which is to be heated by it. The boiler must not only have as large a surface as possible, but it must be of such a form as to cause the flame which embraces it to impinge against it with force, to break against it, and to play over its surface in eddies and whirlpools.

It is therefore against the *bottom* of a boiler, and not against its sides, that the principal efforts of the flame must be directed; for when the flame, or hot vapour, is permitted to rise freely by the vertical sides of a boiler, it slides over its surface very rapidly, and, there being no obstacle in the way to break the flame into eddies and whirlpools, it glides quietly on like a stream of water in a smooth canal; and the same hot particles of this vapour which happen to be in immediate contact with the sides of the boiler at its bottom or lower extremity, being continually pressed against the surface of the boiler as they are forced upwards by the rising current, prevent other hot particles from approaching the boiler; so that by far the greatest part of the heat in the flame and hot vapour which rise from the fire, instead of entering the boiler, goes off into the atmosphere by the chimney, and is totally lost.

The amount of this loss of heat, arising from the faulty construction of boilers and their fire-places, may be estimated from the results of the experiments recorded in the following chapter.

CHAPTER V.

*An Account of Experiments made with Boilers and
Fire-places of various Forms and Dimensions; together with Remarks and Observations on their
Results, and on the Improvements that may be derived from them. — An Account of some Experiments made on a very large Scale in a Brewhouse
Boiler. — An Account of a Brewhouse Boiler constructed and fitted up on an improved Plan. —
Results of several Experiments which were made
with this new Boiler. — Of the Advantage in regard
to the Economy of Fuel in boiling Liquids, which
arises from performing that Process on a large
Scale. — These Advantages are limited. — An Ac-
count of an Alteration which was made in the new
Brewhouse Boiler, with a view to the* SAVING OF
TIME *in causing its Contents to boil. — Experiments showing the Effects produced by these Alterations. — An Estimate of the* RELATIVE QUANTITIES
OF HEAT *producible from* COKES, PIT-COAL, CHAR-
COAL, *and* OAK. — *A Method of Estimating the
Quantity of Pit-coal which would be necessary to
perform any of the Processes mentioned in this
Essay, in which Wood was used as Fuel. — An
Estimate of the* TOTAL QUANTITIES *of Heat produ-
cible in the Combustion of different Kinds of Fuel;
and of the real Quantities of Heat which are lost,
under various Circumstances, in culinary Processes.*

W̤HAT has been said in the foregoing chapter
will, I trust, be sufficient to give my reader a

clear and distinct idea of the subject under consideration in all its various details and connections, and enable him to comprehend without the smallest difficulty every thing I have to add on this subject; and particularly to discover the different objects I had in view in the experiments of which I am now about to give an account, and to judge with facility and certainty of the conclusions I have drawn from their results.

These experiments, though they occupy so many pages in this Essay, are but a small part of those I have made, and caused to be made under my direction, on the subject of heat, during the last seven years. Were I to publish them all, with all their details as they are recorded in the register that has been kept of them, they would fill several volumes.

It was most fortunate for me that this register is very voluminous; for, had it not been so, I should in all probability have taken it with me to England last year, and in that case I should have lost it, with the rest of my papers, in the trunk of which I was robbed in passing through St. Paul's churchyard, on my arrival in London after an absence of eleven years.*

As I foresaw, when I first began my inquiries respecting heat, that I should have occasion to make many experiments on boiling liquids, to facilitate the registering of them I formed a table (which I had printed), in which, under various heads, every circumstance relative to any common experiment of the kind in question could be entered with much regularity, and with little trouble.

---

* I have many reasons to think that these papers are still in being. What an everlasting obligation should I be under to the person who would cause them to be returned to me !

As this table may be useful to others who may be engaged in similar pursuits, and as the publishing of it will also tend to give my reader a more perfect idea of the manner in which my experiments were conducted, I shall (as an example) give an account of one experiment *in the same form* in which it was registered in one of these printed tables.

These tables, as they are printed for use (on detached sheets), occupy one side of half a sheet of common folio writing paper.

Every thing in this table, except such figures and words as are printed between crotchets, is contained in the printed forms. Hence it is evident how much these tables tend to diminish the trouble of registering the results of experiments of this kind, and also to prevent mistakes.

The example I have here given is an account of an experiment in which a very large quantity of water, equal to 15,590 lbs. avoirdupois in weight, or 1866 wine gallons of 231 cubic inches each; but it is evident that these tables answer equally well for the small quantity contained by the smallest saucepan.

The height of the barometer is expressed in Paris inches; that of the thermometer, in degrees of Fahrenheit's scale. The other measures, as well of length as of capacity, are the common measures of the country (Bavaria); and the weight is expressed in Bavarian pounds, of which 100 make 123.84 lbs. avoirdupois.

What is entered under the head of GENERAL RESULTS OF THE EXPERIMENT requires no explanation; but what I have called the PRECISE RESULT must be explained.

Having frequent occasion to compare the results of

An Experiment on the Management of Fire in Boiling Liquids; made at [Munich], in [a Brewhouse belonging to the Elector], the [15th] of [April, 1795].

Time of the day. Hour.	Minutes.	Fuel put into the fire-place. No. of pieces.	In weight. lbs.	Temperature of the liquid.	Contents of the boiler. Kind of liquid, &c.	Measures.	In weight. lbs.
9	15	29	100	60°	[Water].	6984	12,508
—	30	6	50	70°			
—	41	6	50	—			
10	5	7	50	92°			
—	25	7	50	105°			
—	46	7	50	120°			
11	0	7	50	130°			
—	15	7	50	145°			
—	26	8	50	155°			
—	43	7	50	163°			
—	50	8	50	173°			
12	5	7	50	183°			
—	17	15	100	192°			
—	31			[boiled].			
—	55			—			
1	38	7	50	—			
2	35	7	50	—			
3	38			[ceased boiling].			

Height of the barometer, 26 5/9 inches; of thermometer, 58°.

DIMENSIONS OF THE BOILER.

Diameter { above ———— or long ———— 10 feet.
        { below ———— and wide ———— 8 feet.
Deep, 4 feet. Was constructed of [copper], and weighed [not known]; contained of water, 8176 measures, weighing 14,643 lbs.

KIND OF FUEL USED. — [Pine wood, moderately dry, in lengths of 6 feet.]

GENERAL RESULTS OF THE EXPERIMENT.

Time employed to make the liquid boil, [3 40] h. m.
Fuel consumed to make the liquid boil, [800] lbs.

Time the liquid continued to boil . . . 2 43 h. m.
Fuel added to keep the liquid boiling . 100 lbs.
Quantity of the liquid evaporated [not observed].

PRECISE RESULT.

With the heat generated in the combustion of 1 lb. of the fuel, [13.23 lbs.] of ice-cold water made to boil; or [339.80 lbs.] of boiling-hot water kept boiling 1 hour.

experiments made at different times and in different
seasons of the year, as the temperature of the water
in the boiler when the fire is lighted under it is seldom
the same in any two experiments, and as the boiling
heat varies with the variations of the pressure of the
atmosphere, or of the height of the mercury in the
barometer, it became necessary to make proper allow-
ances for these differences. This I thought could best
be done by determining, by computation, from the
number of degrees the water was *actually heated*, and
the quantity of fuel consumed in heating it that number of degrees, how much fuel would have been
required to have it heated 180 degrees, or from the
point of freezing to that of boiling water (the boiling
point being taken equal to the temperature indicated
by 212° of Fahrenheit's thermometer, which is the
boiling point under the mean pressure of the atmosphere at the surface of the sea). Then, by dividing
the weight of the water used in the experiment (expressed in pounds) by the weight of the fuel expressed
in pounds necessary to heat it 180 degrees, or from
the temperature of freezing to that of boiling water:
this gives the number of pounds of ice-cold water
which (according to the result of the given experiment) *might have been* made to boil, with the heat
generated in the combustion of 1 lb. of the fuel, under
the mean pressure of the atmosphere at the level of the
surface of the sea.

The city of Munich, where all the experiments were
made of which I am about to give an account, being
situated almost in the centre of Germany, lies very high
above the level of the sea. The mean height of the
mercury in the barometer is only about 28 English

inches, consequently water boils at Munich at a lower temperature than at London. The difference is even too considerable to be neglected: it amounts to 2½ degrees of Fahrenheit's scale, being 209½ degrees at a medium at Munich, and 212 degrees in all places situated near the level of the sea. To render the results of my experiments and computations more simple and more generally useful, I shall always make due allowance for this difference.

Having, from the actual result of each experiment, made a computation on the principles here described, showing what (for the want of a better expression) I have called the *precise result* of the experiment, it is evident that these computations show very accurately the comparative merit of the mechanical arrangements, and the management of the fire in conducting the experiments, in as far as relates to the economy of fuel; for the more ice-cold water that can be made to boil with the heat generated in the combustion of any given quantity (1 lb. for instance) of fuel, the more perfect of course (other things being equal) must be the construction of the fire-place.

Under the head of PRECISE RESULT I have sometimes added another computation, showing how much "*boiling-hot water*" might, according to the result of the given experiment, be *kept boiling* "*one hour*" with the heat generated in the combustion of "1 lb. of the fuel." Though I have called this a *precise result*, it is evident that in most cases it cannot be considered as being very exact, owing to the difficulty of estimating the quantity of fuel in the fire-place, which is *unconsumed at the moment when the water begins to boil.*

In the foregoing example, in making this computation I supposed that, when the water began to boil, there was wood enough in the fire-place *unconsumed* to keep the water boiling 43 minutes, and that the wood added afterwards (100 lbs.) kept the water boiling the remainder of the time it boiled, or just 2 hours.

In most cases, however, to save trouble in making these computations, I have supposed that all the wood employed in making the water boil is entirely consumed in that process, and that all the heat expended in *keeping the water boiling* is furnished by the fuel which is added *after the water had begun to boil*. This supposition is evidently erroneous; but, as the computation in question can at best give but an inaccurate and doubtful result, labour bestowed on it would be thrown away. But, imperfect as these rough estimates are, they will however in many cases be found useful.

In giving an account of the following experiments, I shall not place them exactly in the order in which they were made, but shall arrange them in such a manner as I shall think best, in order that the information derived from their results may appear in a clear point of view.

For greater convenience in referring to them, I shall number them all; and as I have already given numbers to the four I mentioned in the first chapter of this Essay, I shall proceed in regular order with the rest.

Experiment No. 5. — The first kitchen of the House of Industry at Munich has already been described in the first chapter of this Essay; and it was there mentioned that the daily expense of fuel in that kitchen, when food (peas-soup) was prepared for 1000 persons, amounted to 300 lbs. in weight of dry beech-wood.

Now as each portion of soup consisted of 1 lb., this gives 0.3 of a pound of wood for each pound of soup.

Experiment No. 6. — The first kitchen of the House of Industry having been pulled down, it was afterwards rebuilt on a different principle. Instead of copper boilers, iron boilers of a hemispherical form were now used, and each of these boilers had its own separate closed fire-place; the boiler being suspended by its rim in the brick-work, and room being left for the flame to play all round it. The smoke went off into the chimney by an horizontal canal, 5 inches wide and 5 inches high, which was concealed in the mass of brick-work, and which opened into the fire-place on the side opposite to the opening by which the fuel was introduced.

The fire was made on a flat iron grate placed directly under the boiler, and distant from its bottom about 12 inches. The ash-pit door was furnished with a register; but there was no damper to the canal by which the smoke went off into the chimney, which was a very great defect. The opening into the fire-place was closed by an iron door. Each of these iron boilers weighed about 148 lbs. avoirdupois, was $25\frac{3}{4}$ English inches in diameter, and 14.935 inches deep, and contained $190\frac{1}{2}$ lbs. Bavarian weight of water, equal to 235.91 lbs. avoirdupois, or about $28\frac{1}{4}$ English wine-gallons.

From this account of the manner in which these iron boilers were fitted up, it is evident that the arrangement was not essentially different from that of kitchens for hospitals as they are commonly constructed.

From experiments made with care, and often repeated, I found that to prepare 89 portions (or 89 lbs.

Bavarian weight) of peas-soup in one of these boilers, 43 lbs. of *dry beech-wood* were required as fuel, and that the process lasted four hours and a half. This gives 0.483 of a pound of wood for each pound of the soup.

In the first arrangement of this kitchen, only 0.3 of a pound of wood was required to prepare 1 lb. of soup. Hence it appears that the kitchen had not been improved, considered with a view to the economy of fuel, by the alterations which had been made in it. This was what I expected; for the object I had in view in constructing this kitchen was not to save fuel, but to find out how much of it is wasted in culinary processes, as they are commonly performed on a large scale in hospitals and other institutions of public charity. Till I knew this, it was not in my power to estimate, with any degree of precision, the advantages of any improvements I might introduce in the construction of kitchen fire-places.

To determine in how far the quantity of fuel necessary in any given culinary process depends on the form of the *fire-place* (the boiler and every other circumstance remaining the same), I made the following experiments.

Experiments Nos. 7 and 8. — Two of the iron boilers in the kitchen of the House of Industry (which, as they were both cast from the same model, were as near alike as possible) being chosen for this experiment, one of them (No. 8) being taken out of the brick-work, its fire-place was altered and fitted up anew on improved principles. The grate was made circular and concave, and its diameter was reduced to 12 inches; the fire-place was made cylindrical above the grate, and only

12 inches in diameter; and the boiler being seated on the top of the wall of this cylindrical fire-place, the flame, passing through a small opening on one side of the fire-place, at the top of it, made one complete turn about the boiler before it was permitted to go off into the canal by which the smoke passed off into the chimney.

Though there was no damper in this canal, yet as its entrance or opening, where it joined the canal which went round the boiler, was considerably reduced in size, this answered (though imperfectly) the purpose of a damper. This fire-place being completed, and a small fire having been kept up in it for several days to dry the masonry, the experiment was made by preparing the same quantity of the same kind of soup in this and in a neighbouring boiler whose fire-place had not been altered.

The food cooked in each was 89 lbs. of peas-soup; and the experiment was begun and finished in both boilers at the same time.

The wood employed as fuel was pine; and it had been thoroughly dried in an oven the day before it was used.

The boilers were both kept constantly covered with their double covers, except only when the soup was stirred about to prevent its burning to the bottoms of the boilers.

The result of this interesting experiment was as follows : —

	Experiment No. 7.	Experiment No. 8.
	In the boiler No. 1.	In the boiler No. 8, with the improved fire-place.
Quantity of wood consumed in cooking 89 lbs. Bavarian weight of peas-soup . . .	37 lbs.	14 lbs.

These experiments were made on the 7th of November, 1794. On repeating them the next day with pine-wood, which had not been previously dried in an oven, the result was as follows : —

Experiments Nos. 9 *and* 10.

	Experiment No. 9.	Experiment No. 10.
	In the Boiler No. 1.	In the Boiler No. 8, with the improved Fire-place.
Quantity of wood consumed in cooking 89 lbs. of peas-soup	39 lbs.	16 lbs.

The first remark I shall make on the results of these experiments is the proof they afford, by comparing them with that which preceded them (No. 6), of the important fact that pine-wood affords more heat in its combustion than beech. This fact is the more extraordinary, as it is directly contrary to the opinion generally entertained on that subject; and it is the more important, as the price of pine-wood is in most places only about half as high as that of beech, when the quantities, *estimated by weight*, are equal.

In the Experiment No. 6 it was found that 43 lbs. of dry beech-wood were necessary when used as fuel, to prepare 89 lbs. of peas-soup. In the Experiment No. 7, the same process was performed with 37 lbs., and in the Experiment No. 9 with 39 lbs., of dry pine. But I shall have occasion to treat this subject more at length in another place. In the mean time I would, however, just observe, that all my experiments have uniformly tended to confirm the fact that dry pine-wood affords more heat in combustion than dry beech. I have reason to think the difference is in fact greater than the experiments before us indicate; but the *apparent* amount of it will always depend in a great measure on the circumstances under which the fuel is consumed, or, in other words, on the construction of the fire-place; and it is no small advantage attending the fire-places I shall recommend, that they are so contrived as to increase as much as it is possible the superiority of the most common and cheapest fire-wood over that which is more scarce and costly.

By comparing the results of these two sets of Experiments (Nos. 7 and 8, Nos. 9 and 10), an estimate may be made of the advantage of using *very dry wood* for fuel, instead of making use of wood that has been less thoroughly dried; but, as I mean to take an opportunity of investigating that matter also more carefully hereafter, I shall not at present enlarge on it farther than just to observe that as the wood, which was dried in an oven, was weighed for use after it had been dried, and as it certainly weighed more before it was put into the oven, the real saving arising from using it in this dried state is not so great as the difference in the weights of the quantities of wood used in the two experiments. To estimate that saving with precision, the wood should be weighed before it is dried, or in the same state in which the other parcel of wood, which is used without being dried, is weighed.

But to proceed to the principal object I had in view in these experiments, — the determination of the effects of the difference in the construction of the two fire-places, — the difference in the quantity of fuel expended in the two fire-places, in performing the same process, shows, in a manner which does not stand in need of any illustration, how much had been gained by the improvements which had been introduced.

Conceiving it to be an object of great importance to ascertain by actual experiment, and with as much precision as possible, the real amount of the advantages, in regard to the economy of fuel, that may be derived from improvements in the forms of fire-places, I did not content myself with improving from time to time the kitchens I had constructed, but I took pains to determine how much I had gained by each alteration that was made. This was necessary, not only to furnish myself with more forcible arguments to induce others to adopt my improvements, but also to satisfy myself with regard to the progress I made in my investigations.

In the first arrangement of the kitchen of the Military Academy, the boilers were suspended by their rims in the brick-work in such a manner that the flame could pass freely all round them, and the smoke went off in horizontal canals which led to the chimney, but which were not furnished with dampers.

The fire was made on a flat square iron grate; and the internal diameter of the fire-place was 2 or 3 inches larger than the diameter of the boiler which belonged to it. The bottom of the boiler was from 6 to 10 or 12 inches (according to its size) above the level of the grate; and the door of the opening into the fire-place by which the fuel was introduced was kept constantly closed. The ash-pit door was furnished with a register, and the boilers were all furnished with double covers.

Having, in consequence of the progress I had made in my inquiries respecting the management of heat and the economy of fuel, come to a resolution to pull down this kitchen, and rebuild it on an improved principle; previous to its being demolished, I made several very accurate experiments to determine the real expense of fuel in the fire-places as they *then existed*, with all their faults; and when the new arrangement of the kitchen was completed, I repeated these experiments *with the same boilers;* and by comparing the results of these two sets of experiments, I was able to estimate with great precision the real amount of the saving of time as well as of fuel, which was derived from the improvements I had introduced.

After all that has been said (and perhaps already too often repeated in different parts of this Essay) on the construction of fire-places, my reader will be able to form a clear and just idea of the construction of those of which I am now speaking (those of the kitchen of the Military Academy, in its *present* improved state), when he is told that the fire burns on a circular concave iron grate, about half the diameter of the circular boiler which belongs to the fire-place ; that the fire-place, properly so called, is a cylindrical cavity in the solid brick-work which supports the boiler, equal in diameter to the circular grate, and from 6 to 10 inches high, more or less according to the size of the boiler; that the boiler is *set down* on the top of the circular wall which forms this fire-place, — a small opening from 3 to 4 or 5 inches in length taken horizontally, and about 2 or 3 inches high, being left on one side of this wall at the top of it, that the flame which burns up under the middle of the bottom of the boiler may afterwards pass round (in a spiral canal constructed for that purpose) under that part of the bottom of the boiler which lies *without* the top of the wall of the fire-place on which the boiler reposes. The flame having made one complete turn *under* the boiler in this spiral canal, it rises upwards, and, going once *round the sides of the boiler*, goes off by a horizontal canal, furnished with a damper, into the chimney.

In order that the top of the circular wall of the fire-place on which the boiler is seated may not cover too much of the bottom of the boiler, its thickness is suddenly reduced *in that part* (that is to say, just where it touches the boiler) to about half an inch.

The opening by which the fuel is introduced into the fire-place is a conical hole in a piece of fire-stone, which hole is closed by a fit stopper made of the same kind of stone. The ash-pit door and its register are finished with so much nicety that, when they are quite closed, the fire almost instantaneously goes out.

The dimensions of the boiler, in which the experiments of which I am about to give an account were made, are as follows: —

$$\text{Diameter} \begin{cases} \text{above} & . \quad 14.935 \\ \text{below} & . \quad 13.39 \end{cases} \text{inches, English measure.}$$
$$\text{Depth} \quad . \quad . \quad . \quad . \quad 14.52$$

It weighs 37 lbs. avoirdupois; and it contains, when quite full, about 73 lbs. avoirdupois, equal to 8¾ gallons (wine-measure) of water.

In two experiments with this boiler, which were both made by myself, and in which attention was paid to every circumstance that could tend to render them perfect, the results were as follows: —

	Experiment No. 11.	Experiment No. 12.
	The first fire-place.	The improved fire-place.
Quantity of water in the boiler, in *Bavarian pounds*	43.63 lbs.	43.63 lbs.
Temperature of the water in the boiler at the beginning of the experiment . . .	59°	60°
Time employed in making the water boil .	67 m.	30 m.
Wood consumed in making the water boil, in *Bavarian pounds*	9 lbs.	3 lbs.
Time the water continued boiling	2 h. 2 m.	3 h.
Wood added to keep the water boiling . .	5 lbs.	2½ lbs.
Kind of wood used	Pine.	Pine.
Precise Results.		
Ice-cold water heated 180 degrees, or made to boil, with 1 lb. of wood	4.02 lbs.	11.93 lbs.
Boiling-hot water kept boiling 1 hour, with 1 lb. of wood	17.74 lbs.	52.36 lbs.

The following experiments were made with two copper boilers (Nos. 1 and 2) nearly of the same dimensions, in the kitchen of the Military Academy at Munich, in the present improved state of that kitchen. These boilers are round and deep, and weigh each about 62 lbs. avoirdupois. They belonged originally to the kitchen of the House of Industry, being two of the eight boilers which, in the first arrangement of that kitchen, were heated by the same fire.

Their exact dimensions, measured in English inches, are as follows : —

	The boiler No. 1.	The boiler No. 2.
	Inches.	Inches.
Diameter { above	22.66	22.66
Diameter { below	19.82	20.85
Depth	24.72	22.04

At the beginning of each of the following experiments, each of these boilers contained just 95 measures (or *Bavarian maasse*) of water, weighing 187 lbs. Bavarian weight (equal to 232.58 lbs. avoirdupois), or a trifle less than 28 gallons.

The grate on which the fire was made under each of these boilers is circular and concave, and 11 inches in diameter; and their fire-places are in all respects similar to that just described (Experiment No. 11). Both boilers are furnished with double covers.

The experiments made with the boiler No. 1, and their results, were as follows: —

	Exp. No. 13.	*Exp. No. 14.*	*Exp. No. 15.*	*Exp. No. 16.*
Quantity of water in the boiler in the beginning of the experiment	lbs. 187	lbs. 187	lbs. 187	lbs. 187
Temperature of the water in the boiler at the beginning of the experiment	61°	59°	64°	55½°
Time employed in making the water boil	m. 78	m. 61	m. 61	m. 62
Wood consumed in making the water boil	lbs. 12	lbs. 11	lbs. 9	lbs. 8
Time the water continued to boil	m. 17	m. 28	m. 6	h. m. 2 19
Quantity of fuel added to keep it boiling this time .	—	—	—	lbs. 4
Kind of wood used as fuel	Beech.	Beech.	Pine.	Pine.
Precise Results of the Experiments.				
Ice-cold water heated 180°, or made to boil with the heat generated in the combustion of 1 lb. of the fuel	lbs. 12.89	lbs. 14.15	lbs. 16.89	lbs. 20
Boiling water kept boiling one hour, with the heat generated in the combustion of 1 lb. of the wood .	—	—	—	lbs. 108.40

All the foregoing experiments were made on the same day (the 13th of October, 1794), and in the same order in which they are numbered.

The following are the results of the experiments made with the boiler No. 2 : —

	Exp. No. 17.	*Exp. No.* 18.	*Exp. No.* 19.	*Exp. No.* 20.	*Exp. No.* 21.
Quantity of water in the boiler at the beginning of the experiment, in *Bavarian pounds* . .	lbs. 187	lbs. 187	lbs. 187	lbs. 187	lbs. 187
Temperature of the water in the boiler at the beginning of the experiment	61°	58°	60°	55°	212°
Time employed in making the water boil. . . .	m. 75	m. 55	m. 57	m. 60	—
Wood consumed in making the water boil . .	lbs. 11	lbs. 11	lbs. 9	lbs. 8	—
Time the water continued to boil	m. 21	m. 17	m. 8	h. m. 2 29	h. m. 1 10
Wood added to keep the water boiling	lb. 1	—	—	lbs. 3½	lbs. 1½
Kind of wood used . .	Beech.	Beech.	Pine.	Pine.	Beech.
Precise Results.					
Ice-cold water heated 180°, or made to boil, with 1 lb. of wood . .	lbs. 13.92	lbs. 14.33	lbs. 17.59	lbs. 20.10	—
Boiling-hot water kept boiling one hour with 1 lb. of wood	—	—	—	lbs. 132.68	lbs. 145.44

This set of experiments was made at the same time with the foregoing set, namely, on the 13th October, 1794, and they were made in the order in which they are here registered. In the last but one (No. 20), the economy of fuel in the process of heating water was carried farther than in any other experiment I have ever made.

In the following experiments, which were made in a large copper boiler fitted up on my most improved principles, belonging to the kitchen of the House of Industry, the economy of fuel was carried nearly as far.

This boiler, which is circular, is 42½ English inches in diameter above, 42.17 inches in diameter below, and 18.54 inches deep. It weighs 78½ lbs. avoirdupois; and contains, when quite full, 714 lbs. *Bavarian weight* (= 884 lbs. avoirdupois, or 106 gallons) of water, at the temperature of 55°.

It is surrounded above by a wooden ring about 2 inches in thickness, into which it is fitted; and in this ring, in a groove about ¾ of an inch deep, is fitted a circular wooden flat cover. This cover is formed in three pieces, united by iron hinges; and one of these pieces being fastened down by hooks to the boiler, the other two are so contrived as to be folded back upon it occasionally. From the upper surface of the part of the cover which is fastened down on the boiler, a tin tube 2 inches in diameter, furnished with a damper, is fixed, by which the steam is carried off into a narrow wooden tube, which conducts it through an opening in the roof of the house into the open air.

To prevent still more effectually the escape of the heat through the wooden cover of the boiler, the upper surface of it is protected from the cold atmosphere by a thick circular blanket covered on both sides by strong canvas, which is occasionally thrown over it.

Though the diameter of this boiler below is more than 40 inches, the diameter of its fire-place (which is just under its centre) is only 11 inches; but as the flame makes two complete turns under the bottom of the boiler in a spiral canal, and one turn round it, the time required to heat it is not so great as, from the smallness of its fire-place, might have been expected.

It has ever been, and still continues to be, the decided favorite of the cook-maids.

The wood used as fuel in the following experiment was pine moderately dried. The billets were 6 inches long, and from 1 to 2 inches in diameter.

The following table shows the results of five experiments that were made with this boiler by myself, just after it was fitted up: —

	Exp. No. 22.	*Exp. No. 23.*	*Exp. No. 24.*	*Exp. No. 25.*	*Exp. No. 26.*
Quantity of water in the boiler, in Bavarian pounds . . .	lbs. 508	lbs. 127	lbs. 254	lbs. 508	lbs. 508
Temperature of the water at the beginning of the experiment	48°	48°	96°	48°	48°
Time required to make the water boil . . .	h. m. 2 4	m. 51	h. m. 1 15	h. m. 2 35	h. m. 3 1
Fuel employed to make the water boil . . .	lbs. 24⅓	lbs. 8¼	lbs. 12¾	lbs. 25	lbs. 24
Time the water continued boiling . .	h. 3	—	—	h. 3	—
Fuel added to keep the water boiling . . .	lbs. 6¹²⁄₃₂	—	—	lbs. 4½	—
Precise Results of the Experiments.					
With the heat generated in the combustion of 1 lb. of the fuel,					
Ice-cold water heated 180°, or made to boil	lbs. 18.87	lbs. 12.74	lbs. 12.69	lbs. 17.48	lbs. 19.01
Or boiling-hot water kept boiling one hour	236.61	—	—	338.66	—

Without stopping to make any observations on the results of these experiments (though they afford matter for several of an interesting nature), I shall proceed to give a brief account of another set of experiments, on a much larger scale, which were made in the copper boiler of a brewery belonging to the Elector.

This boiler, which is rectangular, is 10 feet long, 8 feet wide, and 4 feet deep, *Bavarian measure,** and contains 8176 *Bavarian maasse*, or measures, equal to 1866 gallons wine-measure. On examining this boiler, I found its fire-place was constructed on very bad principles; and on inquiring respecting the quantity of fire-wood consumed in it, I found the waste of fuel to be very great.

This brewery is used for making small *white* beer (as from its pale colour it is called) from malt made of wheat; and as it is worked all the year round, the expense of fuel was very great, and the economy of it an object of considerable importance.

The quantity of fire-wood (pine) that had at an average been consumed daily in this brewery was rather more than four Bavarian *klafters*, or cords. On altering the fire-place of this brewery, and putting a (wooden) cover to the boiler, I reduced this expense to less than 1½ klafters.

In the new fire-place which I caused to be constructed for this boiler, the cavity under the boiler is divided into three flues, by thin brick walls which run in the direction of the length of the boiler. The middle flue, which is twice as wide as one of the side flues, is occupied by the burning fuel, and is furnished with a grate 20 inches wide, and 6 inches long; and the opening by which the fuel is introduced into the fire-place is closed by two iron doors, placed one behind the other, at the distance of 8 inches. The grate, which is placed at the hither end of the fire-place, is horizontal; and it is situated about 20 inches below the bottom of the boiler. The air which serves to feed

* 100 Bavarian inches are equal to 95¾ inches English measure.

the fire is let in under the grate through a register in the ash-pit door.

When the double doors which close the entrance into the fire-place are shut, the flame of the burning fuel first rises perpendicularly against the bottom of the boiler; it then passes along to the farther end of the (middle) flue, which constitutes the fire-place, where it separates, and returns in the two side flues; it then rises up into two horizontal flues (one situated over the other) which go all round the boiler; and, having made the circuit of the boiler, it goes off into separate canals (furnished with dampers) into the chimney.

Though the Figures 17 and 18, Plate III., are not drawings from the fire-place I am now describing, but of another which I shall soon have occasion to describe, yet an inspection of these figures will be found useful in forming an idea of the principles on which the fire-place in question was constructed, and on that account I shall occasionally refer to them.

The burning fuel being confined within a narrow compass, being well supplied with fresh air, and being surrounded on all sides by thin walls of brick (which are non-conductors), the heat of the fire is most intense, and the combustion of the fuel of course very complete. The flame, which is clear and vivid in the highest degree, and perfectly unmixed with smoke, runs rapidly along the bottom of the boiler (which forms the top of the flues), and from the resistance it meets with in its passage, from friction, and from the number of turns it is obliged to make, it is thrown into innumerable eddies and whirlpools, and really affords a most entertaining spectacle.

That I might be able to enjoy at my ease this amusing sight, I caused a glass window to be made in the front wall of the fire-place, through which I could look into the fire when the fire-place doors were shut; and I was well paid for the trouble and the trifling expense I had in getting it executed.

Some may be tempted to smile at what they may think a childish invention; but there are many others, I am confident, and among these many grave philosophers, who would have been very glad to have shared my amusement.

The window of which I am speaking is circular, and only 6 inches in diameter; but as the hole in the wall is conical, and much larger within than without, the field of this window (if I may use the expression) is sufficiently large to afford a good view of what passes in the fire-place.

This conical hole is represented in the Figures 18 and 21 by dotted lines. It is situated on the left hand of the entrance into the fire-place. Into the opening of the hole in the wall, on the outside of it, is fixed a short tube of copper (about 6 inches in diameter, and 4 inches long); and in this tube another short *movable* tube is fitted, one end of which is closed by the circular plate of glass which constitutes the window. As the wall of the fire-place in front is thick, this pane of glass is at a considerable distance from the burning fuel, and, as there is no draught through the hole in the wall, the glass does not grow very hot.

I have been the more particular in my description of this little invention, as I think it may be useful. There are many cases in which it would be very advantageous to know exactly what is going on in a closed fire-place, and this never can be known by opening the door; for the instant the door is opened, the cold air rushing with impetuosity into the fire-place deranges entirely the whole economy of the fire. Besides this, it is frequently very disadvantageous to the process which is going on to open the door of a fire-place, and it is always attended with a certain loss of heat, and consequently should as much as possible be avoided.

I intimated that the window I have been describing afforded me amusement : it did still more, — it afforded me much useful information, it gave me an opportunity of *observing* the various internal motions into which flame may, by proper management of the machinery of a fire-place, be thrown, and of estimating with some degree of precision their different effects. In short, it made me better acquainted with the subject which had so long engaged my attention, — fire; and with regard to *that* subject, nothing surely that is new can be uninteresting. But to return to the brewery. To the top of the boiler was fitted a curb of oak timber. The four straight beams of which this curb was constructed are each about 7 inches thick, and 15 inches wide; and the upper part of the boiler is fastened by large copper nails to the inside of the square frame formed by these four beams. From the top of this curb is raised a wooden building, like the roof of a house with a double slant or bevel, which serves as a cover to the boiler. This building, the sides of which are about 3 feet high inwards, and the top of which is covered in by a very flat roof, slanting on every side from the centre, is constructed of a light frame-work of timber (four-inch deal joists), which is covered within as well as without with thin deal boards, which are rabbeted into each other at their edges, to render the cover which this little edifice forms for the boiler as tight as possible.

From the top of this cover an open wooden tube (*m*, Fig. 17), about 12 inches in diameter, rises up perpendicularly, and going through the roof of the brewhouse ends in the open air. This tube, which is furnished with a wooden damper, is intended to carry off the steam.

On the side of this cover next the mashing-tub, as also on that opposite to it, by which the wort runs off into the coolers, there are large folding wooden doors (*i* and *k*, Fig. 17), which are occasionally lifted up by means of ropes which pass over pulleys fastened to the ceiling of the brewhouse.

There are likewise two glass windows (see Fig. 17) in two opposite sides of the cover, through which, as soon as in consequence of the boiling of the liquid the steam becomes transparent and *invisible* (which happens in a very few minutes after the liquid has begun to boil), the contents of the boiler may be distinctly seen and examined.

Whenever there is occasion during the boiling to open either a door or a window of the cover, it is necessary to begin by opening the damper of the steam-chimney, otherwise the hot steam, rushing out with violence, would expose the by-standers to the danger of being scalded; but when the damper of the steam-chimney is open, no steam comes into the brewhouse, though a door or window of the cover be wide open.

Another similar precaution is sometimes necessary in opening the door of the fire-place, which it may be useful to mention. When the dampers in the canals by which the smoke goes off into the chimney are nearly closed (which must frequently be done to confine and economize the heat), if, without altering the damper, or the register in the ash-pit door, the fire-place door be suddenly opened, it will frequently happen that smoke, and sometimes flame, will rush out of the fire-place by this passage. This accident may be easily and effectually prevented, either by opening the damper, or by closing the register of the ash-pit door, the moment before the fire-place door is opened. This precaution should be attended to in all fire-places of all dimensions, constructed on the principles I have recommended.

To economize the time and the *patience* of my reader as far as it is possible, without suppressing any thing essential relating to the subject under consideration, I shall give him, in a very small compass, the general results of a set of experiments which cost me more labour (or at least more *time*) than it would cost him to read all the Essays I have ever written. I believe I am sometimes too prolix for the taste of the age; but it should be remembered that the subjects I have under-taken to investigate are by no means indifferent to me; that I conceive them to be intimately connected with the comforts and enjoyments of mankind; and that a habit of revolving them in my mind, and reflecting on their extensive usefulness, has awakened my enthu-siasm, and rendered it quite impossible for me to treat them with cold indifference, however indifferent or tire-some they may appear to those who have not been accustomed to view them in the same light.

I have already given an account, in all its various details, of one experiment which was made (on the 15th of April, 1795) with the boiler we have just been describing (see page 66). I shall now recapitulate the general results of that experiment, and compare them with the mean results of two other like experiments made with the same boiler.

	Experiment No. 27.	Experiment No. 28.
Quantity of water in the boiler	12,508 lbs.	12,508 lbs.
Temperature of the water in the boiler at the beginning of the experiment	60°	58°
Time required to make the water boil . . .	3 h. 40 m.	3´h. 48 m.
Fuel employed to make the water boil . .	800 lbs.	825 lbs.
Time the water continued boiling	2 h. 43 m.	——
Fuel added to keep the water boiling . . .	100 lbs.	——
Kind of fuel used	Pine-wood.	Pine-wood.
Precise Results of the Experiments.		
Quantity of *ice-cold water* which might be heated 180°, or made to boil, with the heat generated in the combustion of 1 *lb. of the fuel*	12.06 lbs.	12.70 lbs.
Time in which, according to the result of the experiment, *ice-cold water* might (at Munich) *be made to boil* with the given proportion of fuel	4 h. 20 m.	4 h. 20 m.
Quantity of *boiling hot water kept boiling one hour* with the heat generated in the combustion of 1 lb. of the fuel	339.80 lbs.	——

On comparing the results of these experiments with those made in the boilers of the kitchens of the House of Industry and Military Academy, I was led to imagine that either the boiler or the fire-place of the brewery, or both, were capable of great improvement; for, in some of the experiments with these small kitchen boilers, the economy of fuel had been carried so far that, with the heat generated in the combustion of 1 lb. of pine-wood, it appeared that 20 lbs. of ice-cold water might have been made to boil; but here, though the machinery was on a scale so much larger (and I had concluded, too rashly indeed, as will be shown hereafter, that the larger the boiler, the greater is of course the economy of fuel),

·the results of these experiments indicated that not quite
13 lbs. of ice-cold water could have been made to boil
with the heat furnished in the combustion of 1 lb. of the
wood.

The Experiments No. 22, No. 25, and No. 26, which
were made with the largest of my kitchen boilers, had,
it is true, afforded grounds to suspect that, beyond certain limits, an increase of size in a boiler does not tend
to diminish the expense of fuel in the process of heating
water; yet, as all my other experiments·had tended to
confirm me in the opinion I had at an early period imbibed on that subject, I was disposed to suspect any
other cause than the true one of having been instru-
mental in producing the unexpected appearances I
observed.

I was much disappointed, I confess, at finding that
the brewhouse boiler, notwithstanding all the pains I
had taken to fit up its fire-place in the most perfect
manner, and notwithstanding its enormous dimensions,
when compared with the boilers I had hitherto used in
my experiments, so far from answering my expectations,
actually required considerably more fuel in proportion
to its contents than another boiler fitted up on the
same principles, which was not *one fiftieth* part of its
size.

This unexpected result puzzled me, and I must own
that it vexed me, though I ought perhaps to be ashamed
of my weakness; but it did not discourage me. Finding, on examining the boiler, that its bottom was very
thick, compared with the thickness of the sheet copper
of which my kitchen boilers were constructed, it occurred to me that possibly *that* might be the cause, or
at least *one of the causes*, which had made the consump-

tion of fuel so much greater than I expected; and as
there was another brewhouse in the neighbourhood belonging to the Elector, which, luckily for me, stood in
need of a new boiler, I availed myself of that opportunity to make an experiment, which not only decided
the point in question, but also established a new fact
with regard to heat, which I conceive to be of considerable importance.

Having obtained the Elector's permission to arrange
the second brewhouse as I should think best, I determined to spare no pains to render it as perfect as possible in all respects, and.particularly in every thing relating
to the economy of fuel. As in brewing, in the manner
that business is carried on in Bavaria, where the whole
process, in as far as fire is employed in it, is begun and
finished in the course of a day, *the saving of time* in
heating the water and boiling the wort is an object of
almost as much importance as that of economizing fuel,
and consequently demanded particular attention.

The means I used for the attainment of both these
objects will be evident from the following description of
the boiler and its fire-place, which I caused to be constructed, and which are represented in all their details
in the Plates III., IV., and V.

This boiler is 12 (Bavarian) feet long, 10 feet wide,
and only 2 feet deep. The sheet copper of which it is
made is uncommonly thin for a boiler of such large
dimensions, being at a medium less than *one tenth* of
an English inch in thickness. This boiler, when finished, weighed no more than 674 lbs. Bavarian weight,
equal to 834¾ lbs. avoirdupois, exclusive of 64 lbs. of
copper nails used in riveting the sheets of copper
together.

The top of the boiler is surrounded by a strong curb (*a, b*, Fig. 17) of oak timber, to which it is attached by strong copper nails, and over the boiler is built a roof, or standing cover (see Fig. 17), similar in all respects to that already described. The bottom of the boiler is flat, and reposes horizontally on the top of the thin brick walls by which the fire-place is divided into flues. (See Fig. 18.) These flues do not run in the direction of the length of the boiler, but from one side of it to the other; consequently the door of the fire-place is in the middle of one side of the boiler.

The sheets of copper, of which the bottom of the boiler was constructed, run in the direction of the flues; and they are just so wide that their seams or joinings (where they are united to each other by their sides) repose on the walls of the flues, except only in the middle flue, which, being about twice as wide as the others, one seam was necessarily left unsupported, at least a considerable part of its length. The sheets of copper used in constructing this part of the bottom of the boiler are rather thicker and stronger than the rest: they are just 0.118 of an English inch in thickness.

The fire is made under this boiler in the middle flue, which, as I have just observed, is a little more than twice as wide as one of the other flues. There are five flues under the boiler, namely, one in the middle 44 inches wide, above in the clear (which constitutes the fire-place), and two on each side of it, in which the flame circulates; one 20 inches wide, and the other 19 inches wide.

The side flues are each 14½ inches deep; but as the walls which separate them are much thicker below than above, where the bottom of the boiler reposes on them, the width of these flues below is only 13 inches. The walls of these flues are shown by dotted lines in Fig. 17.

The walls which separate the flues do not run quite from one side of the boiler to the other; an opening being left at one end of each of them, equal to the width of one of the narrow flues, for the passage of the flame from one flue into another, without its going from under the boiler.

The fire being made (on a circular grate) in the middle flue (see Fig. 18), the flame passes on in this flue to its farther end; and then, dividing to the right and left, comes forward in the two adjoining side-flues. Having arrived at the wall which supports the front of the boiler, it turns again to the right and left, and, entering the two outside flues, returns in them to the back of the boiler. Here it went out (before the fire-place was altered) at two openings left for that purpose in the wall which supports the back part of the boiler, and the two currents of flame uniting entered a canal 7 inches wide and 16 inches high, which goes all round the outside of the boiler. (See Fig. 20.) Having made the circuit of the boiler, it went off by a canal (furnished with a damper) into the chimney.

From this description of the fire-place, it appears that the flame and smoke generated in the combustion of the fuel, in passing through those different flues, made a circuit of above 70 feet in contact with the surface of the boiler, before they were permitted to escape into the chimney. This, I thought, must be sufficient to give these hot fluids an opportunity of communicating to the boiler all the heat they could part with, notwithstanding the difficulties which attend their getting rid of it; and I concluded that the communication of their heat to the boiler would be much facilitated and expedited by the various eddies and whirlpools produced in the flame in consequence of the number of abrupt turns and changes of direction it was obliged to make in passing under and round the boiler.

As the experiments which have been made with this boiler were conducted throughout with the utmost care and attention, and as their results are both curious and important in several respects, I have thought them deserving of being made known to the public in all their details.

An Account of three Experiments made at Munich, the 10*th October,* 1796, *with the new Boiler in the Brewery called Neuheusel, belonging to* HIS MOST SERENE HIGHNESS *the* ELECTOR. — *The weather being fair; the barometer standing at* 28 *English inches, and Fahrenheit's thermometer at* 36°.

Dimensions of the boiler, in English } Length . . 11 feet 6.02 inches.
measure, as found by actual ad- } Width . . 9 „ 7.723 „
measurement. } Depth . . 2 „ 0.205 „

Contents of the boiler, when quite full to the brim, 14,163 lbs. *Bavarian weight* of water, at the temperature of 55°, equal to 17.540 lbs. avoirdupois, or 2099 wine-gallons.

The boiler actually contained of water, in the beginning of each of the two following experiments, *in Bavarian weight,* 8120 lbs.. equal to 10,056 lbs. avoirdupois, or nearly 1204 wine-gallons.

The wood used in this and the following experiments was *pine,* which had been moderately seasoned; and the billets were 3 feet 4¼ inches, English measure, in length.

FIRST EXPERIMENT WITH THE NEW BOILER.

Experiment No. 29.

Time.		Quantity of fire-wood put into the fire-place.		Temperature of the water in the boiler.
		No. of billets.	Quantity in weight. lbs.	In degrees of Fahrenheit's therm.
h.	m.			
II	31 A. M.	10	50	50°
	46	15	25	54
12	0	5	25	64
	10 P. M.	5	25	67
	36	—	—	85
	40	4	25	—
	53	5	25	96
I	12	7	25	105
	21	10	50	110
	46	10	50	129
	58	40	50	—
2	17	46	50	156
	29	—	—	164
	34	10	50	—
	41		—	173
	49	—	—	180
	58	40	50	185
3	15	12	50	197
	26	20	25	205
3	35	—	—	The water boiled.

Time employed, 4 h. 4 m.   Wood consumed, 575 lbs.

The boiling water being let off, and it being replaced immediately with cold water, the experiment was repeated as follows : —

Experiment No. 30.

Time.		Quantity of fire-wood put into the fire-place.		Temperature of the water in the boiler.
		No. of billets.	Quantity in weight. lbs.	In degrees of Fahrenheit's therm.
h.	m.			
4	41 P. M.	40	50	60°
	50	40	50	72
5	4	10	50	86
	16	10	50	99½
	29	10	50	114
	42	10	50	126
	56	40	50	142
6	10	40	50	157
	24	40	50	—
	28	—	—	172
	40	40	50	—
	42½	—	—	185½
	53	40	50	—
	55	—	—	198
7	2	—	—	205
7	7	—	—	The water boiled.

Time employed, 2 h. 26 m.   Wood consumed, 550 lbs.

This boiling water being let off, the boiler was again filled (immediately) with cold water; and in this third experiment the quantity of water was increased to 11,368 lbs. *Bavarian weight*, equal to 14,078 lbs. avoirdupois, or 1685 wine-gallons.

The results of this experiment were as follows : —

Experiment No. 31.

Time.		Quantity of fire-wood put into the fire-place.		Temperature of the water in the boiler.
		No. of billets.	Quantity in weight. lbs.	In degrees of Fahrenheit's therm.
h.	m.			
8	51 P. M.	80	100	$65\frac{1}{2}°$
9	7	40	50	79
	21	40	50	90
	44	40	50	107
	57	40	50	118
10	14	40	50	130
	28	40	50	140
	45	40	50	155
11	—	40	50	165
	15	40	50	175
	30	40	50	182
	45	40	50	200
11	58	—	—	The water boiled.

Time employed, 3 h. 7 m.   Wood consumed, 650 lbs.

Experiments Nos. 29, 30, 31.

	No. 29.	No. 30.	No. 31.
Quantity of water in the boiler at the beginning of the experiment, in *Bavarian pounds* . .	8120 lbs.	8120 lbs.	11,368 lbs.
Temperature of the water at the beginning of the experiment .	50°	60°	$65\frac{1}{2}°$
Time employed in making the water boil	4 h. 4 m.	2 h. 26 m.	3 h. 7 m.
Fuel (pine-wood) consumed in making the water boil, in *Bavarian pounds*	575 lbs.	550 lbs.	650 lbs.
Precise Results of the Experiments.			
Quantity of ice-cold water which might have been heated 180°, or made to boil with the heat generated in the combustion of 1 lb. of the fuel	12.54 lbs.	12.28 lbs.	14.59 lbs.
Time in which, according to the result of the experiment, ice-cold water might be made to boil at Munich with the given proportion of fuel	4 h. 31 m.	2 h. 59 m.	3 h. 35 m.

The foregoing table shows the result of these three experiments in a clear and satisfactory manner.

I was surprised, when I compared the results of these experiments with those made in the other brewhouse, to find how little in appearance I had gained by the alterations I had introduced. On a more careful examination of the matter, however, I found that I had gained much more than I at first imagined, both in respect to the economy of fuel and to that of time. The amount of these advantages will appear from the following comparison of the mean result of these two sets of experiments: —

Precise Results of the foregoing Experiments.	Quantity of ice-cold water made to boil with 1 lb. of the fuel.	Time required to make ice-cold water boil, according to the result of the given experiment.	
First Set.	lbs.	h.	m.
In the Experiment No. 27	12.06	4	20
In the Experiment No. 28	12.70	4	20
Sum	24.77	8	40
Means	12.385	4	20
Second Set.			
In the Experiment No. 29	12.54	4	31
In the Experiment No. 30	12.28	2	59
Sum	24.82	7	30
Means	12.41	3	45

The mean results of these two sets of experiments differ very little from each other in appearance; and from this circumstance I shall prove that the new boiler is better adapted for saving fuel than the old.

By comparing the results of the experiments made with the same boiler, but with different quantities of water, we shall constantly find that the expense of fuel was *less* in proportion as the quantity of water was *greater*. In the Experiment No. 23, when 127 lbs. of water were used, the result of the experiment indicated that no more than 12.74 lbs. of ice-cold water could be made to boil with the heat generated in the combustion of 1 lb. of the fuel used; but in the Experiment No. 26, made with the same boiler, but when 4 times as much water was used, or 508 lbs., it appeared from the result of the experiment that 19.01 lbs. of ice-cold water might be made to boil with 1 lb. of the fuel.

Now, in the first set of the experiments we are comparing, as the quantity of water used (12,508 lbs.) was much greater than that used in the second set (8120 lbs.), it is evident that, if the construction of the machinery and the management of the fire had been equally perfect in the two cases, the economy of fuel would have been greatest where the largest quantity of water was used, — that is to say, in the first set of experiments; but, as that was not the case, it is certain that the boiler used in the second set is better adapted to economize fuel than that used in the first.

But we need not go so far to search for proofs of that fact. The result of the Experiment No. 31 is alone sufficient to put the matter beyond doubt. In this experiment, in which the quantity of water (though still considerably short of that used in the former set of experiments) was augmented from 8120 lbs. to 11,368 lbs., the saving of fuel was so much increased as to show in a decisive manner the superiority of the new boiler.

The Precise Results	Quantity of ice-cold water made to boil with 1 lb. of the fuel.	Time required to make ice-cold water boil, according to the result of the experiment.	
	lbs.	h.	m.
Of this Experiment (No. 31) were as follows	14.59	3	37
In the Experiments Nos. 27 and 28, they were, at a medium	12.385	4	20

The difference in the expense of fuel in these experiments with these two boilers is by no means inconsiderable : it amounts to above 14 per cent, and would have amounted to more, if more time had been allowed for heating the water in the experiment with the new boiler; for it is easy to show (what indeed was clearly indicated by all the experiments) that, in causing liquids to boil, the quantity of fuel will be less in proportion as the time employed in that process is long, or, which is the same, as the fire is smaller; and the saving of fuel arising from any given prolongation of the process will be the greater, as the fire-place is more perfect, and as the means used for confining the heat are more effectual.

Though the general results of these two sets of experiments afforded abundant reason to conclude that the alterations I had introduced in arranging the new boiler were real improvements, yet, when I compared the quantity of fuel consumed in the experiments with this new boiler with the much smaller quantities, in proportion to the quantity of water, which were employed in some of my former experiments with kitchen boilers, I was for some time quite at a loss to account for this difference. In all my experiments with boilers of different sizes, from the smallest saucepan up to the largest kitchen boilers, I had invariably found that the *larger* the quantity of water was which was heated, the *less*, in proportion, was the quantity of fuel necessary to be employed in that process; and so entirely had that prejudice taken possession of my mind, that when the strongest reasons for doubt presented themselves, they were overlooked; and it was not till I had searched in vain on every side to discover some other cause to which I could attribute the unexpected appearance that embarrassed me, that I was induced — I may say, forced — to abandon my former opinion, and to be convinced that what I had too hastily considered as a general law does not in fact obtain but within narrow limits; that although in heating *certain quantities* of liquids there is an advantage, in point of the economy of fuel, in performing the process on a larger scale, in preference to a smaller one, yet when the liquid to be heated amounts to a certain quantity this advantage ceases; and, if it exceeds that quantity, it is attended with an expense of fuel proportionally greater than when the quantity is less.

What the size of a boiler must be, in order that the saving of fuel may be a *maximum*, I do not pretend to have determined. I think, however, that there are some reasons for suspecting that it would not be larger than some of the kitchen boilers used in my experiments. But I recollect to have promised my reader that I would not give him my opinion without laying before him at the same time the grounds of those opinions. In the present case they are as follows: —

In an experiment of which I have already given an account (No. 3), $7\frac{15}{16}$ lbs. of water, at the temperature of 58°, were made to boil in a saucepan fitted up in my best manner, in a closed fire-place; and the wood consumed was 1 lb. This gives, for the *precise result* of the experiment, 6.68 lbs. of ice-cold water made to boil with 1 lb. of the fuel.

In another experiment (No. 12) made with one of the small boilers belonging to the kitchen of the Military Academy, fitted up on the same principles, 43.63 lbs. of water, at the temperature of 60°, were made to boil with 3 lbs of wood. This gives 11.93 lbs. of ice-cold water made to boil with 1 lb. of the fuel.

Again, in the Experiment No. 20, which was made with a larger boiler belonging to the same kitchen, and fitted up in the same manner, 187 lbs. of water (equal to about 28 gallons), at the temperature of 55°, were made to boil with the combustion of 8 lbs. of fire-wood. This gives 20.10 lbs. of ice-cold water made to boil with 1 lb. of the wood; and farther than this I have not been able to push the economy of fuel.

In the Experiment No. 26, a boiler was used which had been constructed with the express view to see how far it was possible to carry the economy of fuel in culinary processes; and it was fitted up with the utmost care, and on the most approved principles. As I thought at that time that a large-sized boiler was essential to the economizing of fuel, this boiler was made to contain 106 gallons. In the experiment in question it actually contained 508 Bavarian pounds of water (or about 63 gallons), at the temperature of 48°; and, to make this water boil, 24 lbs. of wood were consumed. This gives 19.01 lbs. of ice-cold water made

to boil with 1 lb. of fuel. Hence it appears that the
expense of fuel was greater in this experiment than in
that last-mentioned.

Again, in the Experiment No. 31, when no less than
11,368 lbs. or 1685 gallons of water were heated and
made to boil in the new brewhouse boiler, the wood
consumed amounted to 650 lbs., which (as the temperature of the water at the beginning of the experiment
was $65\frac{1}{2}°$) gives for the *precise result* of the experiment
14.59 lbs. of ice-cold water made to boil with the heat
generated in the combustion of 1 lb. of the fuel.

As the relative quantities of fuel expended in the
experiments are inversely as the numbers expressing
the quantities of ice-cold water, which, from the result
of each experiment, it appears might have been heated
180°, or made to boil, under the mean pressure of the
atmosphere at the level of the sea, with the heat generated
in the combustion of 1 lb. of the fuel, it is evident that
these numbers measure very accurately the different
degrees to which the economy of fuel was carried in
the different experiments. The economy of fuel in
heating liquids, *depending on the quantity of the liquid*,
as shown by the foregoing experiments, may therefore
be expressed shortly in the following manner : —

	Quantity of water heated in the experiment, *in Bavarian* lbs.	Degrees to which the economy of the fuel was carried.
	lbs.	lbs.
In the Experiment No. 3	7.93	6.68
No. 12	43.63	11.93
No. 16	187	20.10
No. 26	508	19.01
No. 31	11,368	14.59

Before I take my leave of this subject I would just remark that the cause of the appearances observed in the experiments may, I think, be traced to that property of flame from which it has been denominated a non-conductor of heat; for, if the different particles of flame give off their heat only to bodies with which they actually come into contact, the quantity of heat given off by it will be *not as its volume* (and consequently not as the quantity of fuel consumed), but rather *as its surface.* And as the surface of the flame, when fire-places are similar, is proportionally greater in small than in large fire-places, — the surfaces of similar bodies being as the *squares* of their corresponding sides, while their volumes are as the *cubes* of those sides, — it is evident that, on that account, less heat in proportion to the quantity generated in the combustion of the fuel ought to be communicated to the boiler, when the fire-place and boiler are large, than when the process is carried on upon a smaller scale.

There are, however, several other circumstances to be taken into the account in determining the effects *of size* in the machinery necessary for boiling liquids; and one of them, which has great influence, is the heat absorbed by the masonry of the fire-place. This loss will most undoubtedly be the smaller, as the fire-place is larger; but to determine the exact point when, the saving on the one hand being just counterbalanced by the loss on the other, any augmentation or diminution of size in the machinery would be attended with a positive loss of heat is not easy to be ascertained. Provided however that proper attention be paid to the management of the fire, and that as much heat as possible be generated in the combustion of the fuel (which may

always be done in the largest fire-place as well, if not
better, than in smaller ones), as that part of the heat
which goes off in the smoke is indubitably lost, a thermometer placed in the chimney would indicate, with a
considerable degree of precision, the perfections or imperfections of the fire-place.

It is well known that the smoke which rises from the
chimneys of the closed fire-places of very large boilers is
much hotter than that which escapes from smaller fire-
places; and I am surprised that this fact, which has long
been known to me, should not have led me to suspect
that the waste of fuel was proportionally greater in these
large fire-places than in smaller ones.

Besides the experiments of which I have given an account, several others were made with the new brewhouse
boiler; and, among others, four experiments were made
on four succeeding days in brewing beer; and it was
found that considerably less fuel was expended in these
trials than was necessary in brewing the same quantity
of beer in the other brewhouse, in which I first introduced my improvements.   But though the alteration of
form, diminution of the thickness of the metal, etc., which
I had introduced in constructing the new boiler and also
in the manner of fitting it up, had produced a considerable saving of fuel, yet it was not accompanied by a proportional saving of time.   I had flattered myself that by
making the boiler *very thin* and *very shallow*, I should
bring its contents to boil in *a very short time;* but I did
not consider how much time is necessary for the combustion of the fuel necessary for heating so large a quantity of water, otherwise my expectations on this head
would have been less sanguine.   The quantity of heat
generated in any given time being as the quantity of

fuel consumed, it must depend in a great measure on the size of the fire-place; and when it is required to heat a large quantity of water, or of any liquid, in a very short time, either the fire-place must be large, or (what in my opinion would be still better) a number of separate fire-places — two or three, for instance — must be made under the same boiler. The boiler should be made wide and shallow, in order to admit of a great number of flues, in which the flame and smoke of the different fires should be made to circulate separately *under its bottom.*

The combustion of the fuel, and consequently the generation and communication of the heat, may in the same fire-place be considerably accelerated by increasing the draught (as it is called) of the fire; which may be done by increasing the height of the chimney, or by enlarging the canal leading to the chimney, and keeping the damper open, when that passage is too small, or by shortening the length of the flues.

The master brewer having expressed a wish that some contrivance might be used by which the water might be made to boil a little sooner in the new boiler, I made an alteration in its fire-place which completely answered that purpose.

But, besides the desire I had to oblige the master brewer (who only thought how he could contrive to finish as early as possible his day's work), I had another and much more important object in view. Having had reason to suspect that flues which go round on the outside of large boilers do little more than prevent the escape of the heat by their sides, — which, with infinitely less trouble and less expense, may be prevented by other means, — I was desirous of finding out, by a decisive experiment, the real amount of the advantages gained by those flues, or the saving of fuel which they produce. And as I was confident that the suppression of the flue which went round the new boiler would increase the draught of the fire-place, and accelerate the combustion of the fuel, I concluded that, if my opinion was well founded with respect to the smallness of the advantages derived from these *side flues*, the increase of heat arising from the acceleration of the combustion occasioned by the increased draught on closing them up would more than counterbalance the loss of those advantages, and the time employed in heating the water would be found to be actually less than it was before.

The results of the following experiments show how far my suspicions were founded : —

Experiment No. 32. — The flue round the outside of the new brewhouse boiler having been closed up, and two canals (*a* and *b*, Fig. 21) formed from the end of the two outside flues of those situated *under* the boiler, by which two canals (which were both furnished with dampers) the smoke passed off from under the boiler directly into the chimney, the Experiment No. 31, which was made with the same boiler before the outside flues were closed up, was now repeated with the utmost care, in order to ascertain the effects which the closing up of those flues would produce. The quantity of water in the boiler, and its temperature at the beginning of the experiment, were the same ; the wood used as fuel was taken from the same parcel, and it was put into the fire-place in the *same quantities*, and at the *same intervals of time*. In short, every circumstance was the same in the two experiments, excepting only the alterations which had been made in the fire-place. As the length of the flues through which the flame and smoke were obliged to pass to get into the chimney had been diminished more than half (or reduced from 70 to about 30 feet), the strength of the draught of the fire-place was much increased, as was evident not only from the increased violence of the combustion of the fuel, which was very apparent, but also from another circumstance, which I think it my duty to mention. Before the flue round the boiler was closed, if too much fuel was put into the fire-place at once, it not only did not burn with a clear flame, but frequently the smoke, and sometimes the flame, came out of the fire-place door, even when the damper in the chimney was wide open; but, after this flue was closed up, it was found to be hardly possible to overcharge the fire-place, and the fuel always burned with the utmost vivacity.

I ought to inform my reader that, though the entrance into the flue which went round the outside of the boiler was closed, and another and a shorter road opened for the flame and smoke to pass off into the chimney, yet the *cavity* of the flue remained; and, by means of openings (c, c, c, c, c, c, Fig. 21, Plate V.) about 6 inches square in the brick-work which separated this old road (which was now shut up) from the flues *under the boiler*, the flame was permitted to pass into this cavity, and to spread itself round the outside of the boiler. This contrivance (which I would recommend for all boilers) not only prevents the escape of the heat out of the boiler by its sides, but contributes something towards heating it; and, as the openings in the sides of the flues do not sensibly impede the motion of the flame, they can do no harm.

As the two experiments, the results of which I am about

to compare, were made with the greatest care, and as they are on several accounts uncommonly interesting, I shall place them in a conspicuous point of view.

A COMPARATIVE VIEW OF TWO EXPERIMENTS MADE WITH A NEW BREWHOUSE BOILER.

The time is reckoned from the beginning of the Experiment, and was the same in both Experiments.

Quantity of water in the boiler 11,368 *lbs. Bavarian weight.*

Time from the beginning of the experiment.		Fuel put into the fire-place.		Heat of the water in the boiler.	
		Number of billets.	Quantity in weight.	Experiment No. 31 (outside flue open).	Experiment No. 32 (outside flue closed).
h.	m.	No.	lbs.	Degrees.	Degrees.
—	—	80	100	65½	65½
0	16	40	50	79	82
0	30	40	50	90	94
0	53	40	50	107	110
1	6	40	50	118	122
1	53	40	50	130	135
1	37	40	50	140	147
1	54	40	50	155	160
2	9	40	50	165	171
2	24	40	50	175	182
2	39	40	50	182	191
2	54	40	50	200	—
2	59	—	—	—	Boiled.
3	7	—	—	Boiled.	—

Having found, by comparing the results of these two experiments, that I had lost nothing in respect to the economy of fuel by shutting up the outside flue of my boiler, I was now desirous of ascertaining how much I had gained in point of time, or how much the increased draught of the fire-place, in consequence of its flues being shortened, enabled me to abridge the time employed in causing the contents of the boiler to boil, in cases in which it should be advantageous to expedite that process at the expense of a small additional quantity of fuel.

By the following experiment, in which the combustion of the fuel was made as rapid as possible by keeping the fire-place full of wood, and the register in the ash-pit door and the damper in the chimney constantly quite open, may be seen how far I succeeded in the attainment of that object.

Experiment No. 33. — The boiler contained 11,368 lbs. Bavarian weight of water, at the temperature of 47°. The fuel used was pine-wood moderately seasoned, in billets 3 feet 4 inches long, and split into small pieces of about 1 lb. each, that it might burn the more rapidly.

This experiment was made the 29th of November, 1796, the barometer standing at 26 inches 8.7 lines, Paris measure, and Fahrenheit's thermometer at 33°.

Time.		Fuel put into the fire-place.	Temperature of the water in the copper.
h.	m.	lbs.	Degrees.
2	0	100	47
	14	100	58
	34	100	88
	51	100	100
3	9	100	123
	25	100	144
	39	100	151
4	0	100	—
	10	—	200
	17	—	Boiled.
Time employed,	2 17	Wood consumed, 800	

In the Experiment No. 32, the same quantity of water, at the temperature of 65½°, was made to boil in

2 hours 59 minutes, with the consumption of 625 lbs. of the same kind of wood. Had the water in this experiment been as cold as it was in the Experiment No. 33 (namely, at the temperature of 47°), instead of 625 lbs., 705 lbs. of the fuel would have been necessary; and the process, instead of lasting 2 hours and 59 minutes, would have lasted 3 hours and 22 minutes.

Hence we may conclude that to abridge 1 hour and 5 minutes of 3 hours and 22 minutes in the process of boiling 11,368 lbs. of water, this cannot be done at a less additional expense of fuel than that of 95 lbs. of pine-wood; or, to abridge the time *one third*, there must be an additional expense of about *one eighth* more fuel.

In some cases it will be most profitable to save time, in others to economize fuel; and it will always be desirable to be able to do either, as circumstances may render most expedient.

From a comparison of the quantities of fuel consumed, and consequently of heat generated, in the same time, with the quantities of heat actually communicated to the water in the Experiments Nos. 32 and 33 during this time, an idea may be formed of the great quantity of heat that may remain in flame and smoke after they have passed many feet in flues under the thin bottom of a boiler containing cold water; and this shows with how much difficulty these hot vapours part with their heat, and how important it is to be acquainted with that fact in order to take measures with certainty for economizing fuel.

I have been the more particular in my account of these experiments with large boilers, as I believe no experiments of the kind on so large a scale have been yet made; and, as they were all conducted with care, their results have intrinsic value independent of the particular uses to which I have applied them.

As, in the countries where this Essay is likely to be most read, pit-coals are more frequently used as fuel than wood, it will not only be satisfactory, but in many cases may be really useful, to my reader to know the relative quantities of heat producible from coals and from wood, in order to be able to compare the results of experiments in which coals are used as fuel, with those of which I have here given an account; or to determine the quantity of coals necessary in any process which it is known may be performed with a given quantity of wood.

It was my intention to have made a set of experiments on purpose to determine the relative quantities of heat producible from all the various kinds of combustible bodies which are used as fuel; and I made preparations for beginning them, but I have not yet been able to find leisure to attend to the subject.

The most satisfactory account I have been able to procure respecting the matter in question is one for which I am indebted to my friend Mr. Kirwan. By this account, which he tells me is founded on experiments made by M. Lavoisier, it appears that equal quantities of water, under equal surfaces, may be evaporated, and consequently equal heats produced —

In weight,		In measure,
By 403 lbs. of cokes,		By 17 of cokes,
600 „ of pit-coal,		10 of pit-coal,
600 „ of charcoal,		40 of charcoal,
1089 „ of oak;		33 of oak.

I wish I were at liberty to transcribe the ingenious and interesting observations which accompanied this estimate; but, as they make part of a work which I understand is preparing for the press, I dare not antici-pate what Mr. Kirwan will himself soon lay before the public.

According to this estimate it appears that 1089 lbs. of oak produce as much heat in their combustion as 600 lbs. of pit-coal. Now, if we suppose that the pine-wood used in my experiments is capable of producing as much heat *per pound* as oak, — and I have reason to think it does not afford less, — from the quantity of pine-wood used in any of my experiments, it is easy to ascertain how much coal would have been necessary to generate the same quantity of heat; for the weight of the coal which would be required is to the weight of the wood actually consumed, as 600 to 1089.

In one of my experiments (No. 31), 11,368 lbs. of water, at the temperature of $65\frac{1}{2}°$, were made to boil with 650 lbs. of pine-wood. As when the experiment was made the mercury in the barometer stood at about 28 English inches, the temperature of the water when it boiled was only $209\frac{1}{2}°$, consequently its temperature was raised $(209\frac{1}{2} — 65\frac{1}{2})$ 144 degrees. Had the water been boiled in London, or in any other place nearly on a level with the surface of the sea, it must have been heated to 212° to have been made to boil, consequently its temperature must have been raised $146\frac{1}{2}°$; and to have done this, instead of 650 lbs. of wood, $661\frac{1}{2}$ lbs. would have been required (140° is to 650 lbs. as $146\frac{1}{2}°$ to $661\frac{1}{2}$ lbs.).

If pit-coal were used instead of wood, $363\frac{1}{2}$ lbs. of that kind of fuel would have been sufficient; for the quantities in weight of different kinds of fuel required to perform the same process being inversely as the quantities of heat which equal weights of the given kinds of fuel are capable of generating, or directly as the quantities of the kind of fuel in question, which are required to produce the same heat, it is 1089 to 600, as $661\frac{1}{2}$ lbs. of wood to $363\frac{1}{2}$ lbs. of coal, supposing the foregoing estimate to be exact.

Whether it would be possible to cause so large a quantity of water (1681 wine-gallons), at the given temperature ($65\frac{1}{2}°$), to boil, with this small quantity of coal, I leave to those who are conversant in experiments of this kind to determine.

From the result of my 20th Experiment it appeared that $20\frac{1}{10}$ lbs. of ice-cold water might be heated 180 degrees, or made to boil under the mean pressure of the atmosphere at the level of the surface of the ocean, with the heat generated in the combustion of 1 lb. of pine-wood. Computing from the result of this experiment, and from the relative quantities of heat producible from pine-wood and from pit-coal, it appears that the heat generated in the combustion of 1 lb. of pit-coal would make $36\frac{3}{10}$ lbs. of ice-cold water boil.

Hence it appears that pit-coal should heat 36 times its weight of water, from the freezing point to that of boiling; and, as it has been found by experiments made with great care by Mr. Watt that nearly $5\frac{1}{4}$ times as much heat as is sufficient to heat any given quantity of ice-cold water to the boiling-point is required to reduce that same quantity of water, *already boiling-hot*, to steam, — according to this estimation, the heat generated in the combustion of 1 lb. of coal should be sufficient to reduce very nearly 7 lbs. of boiling-hot water to steam.

How far these estimates agree with the experiments that have been made with steam-engines, I know not;

but there seems to be much reason to suspect that the expense of fuel, in working those engines, is considerably greater than it ought to be, or than it would be, were the boilers and fire-places constructed on the best principles, and the fire properly managed.

In attempts to improve, it is always very desirable to know exactly what progress has been made, — to be able to measure the distance we have laid behind us in our advances, and also that which still remains between us and the object in view. The ground which has been gone over is easily measured; but to estimate that which still lies before us is frequently much more difficult.

The advances I have made in my attempts to improve fire-places, for the purpose of economizing fuel, may be estimated by the results of the experiments of which I have given an account in this Essay; but it would be satisfactory, no doubt, to know how much farther it is possible to push the economy of fuel.

In my 4th Experiment, $7\frac{15}{16}$ lbs. of water, at the temperature of 58°, were made to boil, at Munich, with 6 lbs. of wood. If, from the result of this experiment, we compute the quantity of ice-cold water which, with the heat generated in the combustion of 1 lb. of the fuel, might be heated 180°, or made to boil, it will turn out to be only $1\frac{1}{7}$ lb., or more exactly 1.11 lb.

According to the result of the Experiment No. 20, it appeared that no less than $20\frac{1}{10}$ lbs. of ice-cold water might have been made to boil with the heat generated in the combustion of 1 lb. of pine-wood.

It appears, therefore, that about *eighteen times* as much fuel, in proportion to the quantity of water heated, was expended in the Experiment No. 4, as in the No. 20; and hence we may conclude with the utmost certainty, that of the heat generated, or which with proper management might have been generated, in the combustion of the fuel used in the 4th Experiment, less than $\frac{1}{18}$ part was employed in heating the water, — the remainder, amounting to more than $\frac{17}{18}$ of the whole quantity, being dispersed and lost.

I ventured to give it as my opinion, in the beginning of this Essay, that " not less than *seven eighths* of the heat generated, or which with proper management might be generated, from the fuel actually consumed, is carried up into the atmosphere with the smoke, and totally lost." I will leave it to my reader to judge whether this opinion was not founded on good and sufficient grounds.

But though it be proved beyond the possibility of a doubt that the process of heating water was performed in the 20th Experiment with about $\frac{1}{18}$ part of the pro-portion of fuel which was actually expended in the 4th Experiment, yet neither of these experiments, nor any deductions that can be founded on their results, can give us any light with respect to the *real* loss of heat, or how much less fuel would be sufficient were there no loss whatever of heat. The experiments show that the loss of heat must have been at least *eighteen times* greater in one case than in the other; but they do not afford grounds to form even a probable conjecture respecting the amount of the loss of heat in the experi-ment in which the economy of fuel was carried the farthest, or the possibility of any farther improvements in the construction of fire-places. I shall, however, by availing myself of the labours of others, and comparing the results of their experiments with mine, endeavour to throw some light on this abstruse subject.

Dr. Crawford found, by an experiment contrived with much ingenuity, and which appears to have been executed with the utmost care, that the heat generated in the combustion of 30 grains of charcoal raised the temperature of 31 lbs. 7 oz. Troy (= 181,920 grains of water) $1\frac{71}{100}$ degrees of Fahrenheit's thermometer, *when none of the heat generated was suffered to escape.*

But if 30 grains of charcoal are necessary to raise the temperature of 181,920 grains of water $1\frac{71}{100}$ degrees, it would require 3157.9 grains of charcoal to raise the temperature of the same quantity of water 180 degrees, or from the point of freezing to that of boiling; for it is 1.71° to 30 grains, as 180° to 3157.9 grains. Consequently the heat generated in the combustion of 1 lb. of charcoal would be sufficient to heat 57.608 lbs. of ice-cold water 180°, or to make it boil; for 3157.9 grains of charcoal are to 181,920 grains of water as 1 lb. of charcoal to 57.608 lbs. of water.

From the results of M. Lavoisier's experiments, it appeared that the quantities of heat generated in the combustion of equal weights of charcoal and dry oak are as 1089 to 600. Hence we may conclude that equal quantities of heat are generated by 1 lb. of charcoal and 1.815 lbs. of oak; consequently that the heat generated in the combustion of 1.815 lbs. of oak would heat 57.608 lbs. of ice-cold water, — or 1 lb. of oak, 31.74 lbs of ice-cold water 180°, or cause it to boil, *— were no part of the heat generated in the combustion of the fuel lost.*

If now we suppose the quantities of heat producible from equal weights of dry oak and of dry pine-wood to be equal, — and there is reason to believe that this supposition cannot be far from the truth, — we can estimate the real loss of heat in each of the two experiments before mentioned (No. 4 and No. 20), as also in every other case in which the quantity of fuel consumed, and the effects produced by the heat, are known.

Thus, for instance, in the 20th Experiment, as the effects actually indicated that, with *that part* of the heat generated in the combustion of 1 lb. of the fuel which *entered the boiler*, $20\frac{1}{10}$ lbs. of ice-cold water might have been made to boil; as by the above estimate it appears that $31\frac{74}{100}$ lbs. of ice-cold water might be made to boil with *all* the heat generated in the combustion of 1 lb. of the fuel, it is evident that about *one third* of the heat generated was lost, or $\frac{20.1}{31.74}$ of it was saved.

This loss is certainly not greater than might reasonably have been expected, especially when we consider all the various causes which conspire in producing it; and I doubt whether the economy of fuel will ever be carried much farther.

In the Experiment No. 4, as the effects produced by the heat which entered the boiler indicated that no more than 1.14 lb. of ice-cold water could have been made to boil with 1 lb. of the fuel, it appears that in this experiment only about $\frac{1}{25}$th part of the heat generated was saved.

In all the experiments made on a very large scale, with brewhouse boilers, rather more than *one half* of the heat generated found its way up the chimney, and was lost.

CHAPTER VI.

A short Account of a Number of Kitchens, public and private, and Fire-places for various Uses, which have been constructed under the Direction of the Author, in different Places. — Of the Kitchen of the HOUSE *of* INDUSTRY *at* MUNICH; *of that of the* MILITARY ACADEMY; *of that of the* MILITARY MESS-HOUSE; *that of the* FARM-HOUSE, *and those belonging to the* INN *in the* ENGLISH GARDEN *at* MUNICH. — *Of the Kitchens of the Hospitals of* LA PIETÀ *and* LA MISERICORDIA *at* VERONA. — *Of a small Kitchen fitted up as a Model in the House of* SIR JOHN SINCLAIR, *Bart., in* LONDON. — *Of the Kitchen of the* FOUNDLING HOSPITAL *in* LONDON. — *Of a* MILITARY KITCHEN *for the Use of* TROOPS *in* CAMP. — *Of a* PORTABLE BOILER *for the Use of* TROOPS *on a* MARCH. — *Of a large* BOILER *fitted up as a Model for* BLEACHERS *at the* LINEN HALL *in* DUBLIN. — *Of a Fire-place for* COOKING, *and at the same Time* WARMING A LARGE HALL; *and of a* PERPETUAL OVEN, *both fitted up in the* HOUSE *of* INDUSTRY *at* DUBLIN. — *Of the* KITCHEN, LAUNDRY, CHIMNEY FIRE-PLACES, COTTAGE FIRE-PLACES, *and Model of a* LIME-KILN, *fitted up in* IRELAND *in the House of the* DUBLIN SOCIETY.*

MY wish to give the most complete information possible with regard to the *grounds* on which the improvements I propose are founded has induced me to be very particular in my account of my experiments, and of the conclusions and practical inferences

I have thought myself authorized to draw from them; and as these investigations have frequently led me into abstruse philosophical disquisitions, which might not perhaps be very interesting to many of my readers, to whom a simple account of my fire-places, with directions for constructing them, might be really useful; in order to accommodate readers of all descriptions, I have thought it best to divide my subject, and to reserve what I have still to say on the mechanical part of it — the construction of kitchen fire-places — for a separate Essay. In the mean time, for the information of those who may have opportunities of examining any of the kitchens or fire-places, for other purposes, which have already been constructed on my principles, under my direction, I have annexed the following account of them, and of the particular merits and imperfections of each of them. This account, added to what has been said in the foregoing chapters of this Essay on the construction of fire-places, will, I flatter myself, be found sufficient to convey the fullest information respecting the subject under consideration, and enable those who may wish to adopt the proposed improvements to construct fire-places of all kinds on the principles recommended, without any farther assistance.

Those who may not have leisure to enter into these scientific investigations, and who, notwithstanding, may wish to imitate these inventions, will find all the information they can want in my next Essay.

An Account of the Kitchen of the House of Industry at Munich, in its present State.

The large circular copper boiler (which is situated in a small room adjoining to the great kitchen) is fitted up in a very complete manner; its (wooden) cover **is** cheap, simple, and durable, and answers perfectly well for confining the heat; the steam tube (or steam chimney as I have called it) is very useful, as it carries off all the steam generated in cooking, and keeps the air of the kitchen dry and wholesome. To carry off the steam which rises from the hot soup when it is served up, there is a steam-chimney of wood (furnished with a valve), the opening of which is situated at the highest part of the kitchen. To prevent the cold air from coming down by this passage into the kitchen, its damper (which is opened and shut by a cord which goes over a pulley) is, in winter, kept constantly shut, except just when it is necessary to open it for a moment to let out the steam.

The only alteration I would make, were I to fit up this boiler again, would be to leave openings by which the flues might be cleaned occasionally, without lifting the boiler out of its place. This should be done in the fire-places of all large boilers. This boiler, which is used every day, requires to have its flues cleaned, and its bottom and sides scrubbed with a broom, to free them of soot, once in six weeks.

Over against this boiler is a machine for drying potatoes, which has been found to answer perfectly well the end for which it was contrived. Potatoes first moderately boiled, and then skinned and cut into thin slices, and dried in this machine, may be kept good for many years.

The eight iron boilers in the *great kitchen* are fitted up on good principles; and the oven, which is heated by the smoke from the fire-places of two of these boilers, which oven is destined for drying the wood for the use of this kitchen, is deserving of attention.

The wooden covers of these eight boilers, and the horizontal tubes, constructed of wood wound round with canvas and painted with oil colours, by which the steam is carried off, have been found to answer very well the purposes for which they were contrived.

The Kitchen of the Military Academy at Munich.

This kitchen in its present state is so perfect in all its parts, that I do not think it capable of any considerable improvement. The *roaster*, which has been in daily use *seven years*, is still in good condition, and bids fair to last *twenty years* longer. It is large and roomy, and has been found to be extremely useful. Though the different parts of this kitchen are not distributed with so much symmetry as could have been wished, owing to local circumstances, yet it is very complete in its various details, and all the various processes of cookery are performed in it with little labour, and with a very small expense indeed of fuel. Two large boilers and three large saucepans, which are fitted up in a detached mass of brick-work in a corner of the room (on the right hand on going into it), I can recommend as perfect models for imitation. In short, I know of nothing which I could wish to alter in this kitchen. To say the truth, it has already undergone a sufficient number of changes and alterations.

The Kitchen in the Military Hall or Officers' Mess-House in the English Garden at Munich.

This kitchen is much less perfect in its details than that just mentioned. It was built in the spring of the year 1790, and has since undergone only a few trifling alterations. It has three roasters, which are made small on purpose to serve as models for private families; and I have had the pleasure to know that they have often been imitated.

The Kitchen in the Farm-House in the English Garden.

This kitchen is well contrived for the use for which it was designed, and I can recommend it as a very good model for the kitchens of farm-houses, for families consisting of eighteen or twenty persons. One of the boilers, which is destined for warming water for the use of the kitchen and the stables, is in winter heated by the smoke of a German stove, which is situated in an adjoining room, — that inhabited by the overseer of the farm.

The great Kitchen of the Inn in the Garden.

This kitchen, which is adjoining to the farm-house, is contrived almost for the sole purpose of roasting chickens before an open fire, a kind of food of which the Bavarians are extravagantly fond. It has three open fire-places, constructed on the principles recommended in my Essay on Chimney Fire-places, fronting different sides of the kitchen, and all opening into the same chimney, which chimney is built nearly in the middle of the room. This kitchen was built before my roasters were come into use.

The small Kitchen belonging to the Inn.

This kitchen has nothing belonging to it which deserves attention, or which I would recommend for imitation. It was originally designed merely for making coffee, chocolate, etc.

A kitchen which has lately been fitted up on my principles, in the new hospital for the infirm and help-less poor, which is situated on the height called the *Gasteig,* on the side of the river opposite to the town of Munich, is much more interesting, and is a good model for imitation.

The Kitchen of the Hospital of La Pietà at Verona

Is peculiarly interesting, on account of its convenient form and the perfect symmetry of its parts.

The mass of brick-work in which the boilers are fixed occupies the middle of one side of a large high room, which is plastered and white-washed, and neatly paved. The covers of the large boilers are lifted up by ropes which go over pulleys fixed to the ceiling of the top of the room; but were I to build the kitchen again, I should substitute wooden covers with steam-chimneys instead of them, such in all respects as that belonging to the large round copper boiler in the kitchen of the House of Industry at Munich. When the covers are so large that they cannot conveniently be lifted on and off with the hand, they should, in my opinion, always be made of wood, and divided into parts, united by hinges. When they are designed for confining the steam *entirely,* they should be made on a peculiar construction, which will hereafter be de-scribed. The covers for small boilers, and those for saucepans, should always be of tin, and double.

The grates on which the fires are made under the boilers in the kitchen of the Hospital of *La Pietà* are circular; but they are not hollow, or dishing, as that improvement did not occur to me till after that kitchen was finished. The spiral flues under the boilers are also wanting, and for the same reason. In all other respects this kitchen is, I believe, quite perfect.

The Kitchen of the Hospital of La Misericordia at Verona

Is constructed on the same principles as that of *La Pietà*. The only difference between them is in the distribution of the boilers. That of *La Misericordia* is built round two sides of the room. In many cases, this manner of disposing of the boilers will be found more convenient than any other; but in all cases where this method of placing them is preferred, care must be taken to place the largest boilers farthest from the chimney, and the smaller ones nearer to it, and in regular succession as their sizes diminish. This is necessary, in order that in the mass of brick-work in which the boilers are fixed there may be room behind the smaller boilers for the canals which carry off the smoke from the large ones into the chimney.

This circumstance was attended to in constructing the small kitchen which I fitted up last spring in the house of Sir John Sinclair, Bart., President of the Board of Agriculture, Whitehall, London. This kitchen (which was intended to serve as a model, and is open to the public view at all hours) is by no means as perfect as I wished it to be. Having been built during my journey to Ireland, several mistakes were made by the workmen I employed, who, though they have great merit in their different lines of business, had not *then* had sufficient experience in constructing kitchens on my principles, to be able to execute such a job in my absence without committing some faults. Those which were most essential I corrected; but my stay in England, after my return from Ireland, was too short, and my time too much taken up with other matters, to rebuild the kitchen from the foundation, which I was very desirous of doing, and which, with the permission of the proprietor, I shall certainly do when I come to England again. The greatest fault of the kitchen is the want of dampers to the canals by which the smoke is carried off from the closed fire-places of the boilers and saucepans into the chimney. These dampers should never be omitted in any fire-place, however small. They are necessary even in fire-places for the smallest saucepans, and no large boiler should on any account be without one. Some experiments I have lately made (since my return to Bavaria) have showed me how very necessary these dampers are; and I consider it as my duty to the public to lose no time in recommending the general use of them. The flattering attention which has been paid by the public to the various improvements I have taken the liberty to propose, not only demands my warmest gratitude, but lays me under an indispensable obligation to exert myself to the utmost to deserve their esteem, and to merit the distinguished marks of their confidence with which on so many occasions I have been honoured.

But to return to the kitchen in the house of Sir John Sinclair (the place where the meetings of the Board of Agriculture are held, and where of course there is a great concourse of ingenious men from all parts of the kingdom, — of men zealous for the progress of useful improvements). As the room is very small, it was not possible to do more in it than just to fit up a few small boilers and saucepans, and one middling-sized roaster, such as might serve for a small family; which last is a machine so very useful that I cannot help flattering myself that it will soon come into general use. The saving of fuel which it occasions is almost incredible, and the meat roasted in it is remarkably well-tasted and high-flavoured.

One of these roasters, on a large scale, was put up, under my direction, in the kitchen of the Foundling Hospital in London; and though I could not stay in England to see it finished, I have had the satisfaction to learn, since my arrival at Munich, from my friend, Mr. Bernard (who is treasurer to the hospital), that it has answered even beyond his expectations. He informs me, that when 112 lbs. of beef are roasted in it at once, the expense for fuel amounts to no more than *four pence* sterling; and this when the coals are reckoned at an uncommonly high price, namely, at 1s. 4d. the bushel.

In the roaster belonging to the kitchen of the Military Academy at Munich I caused 100 lbs. Bavarian weight (equal to 123.84 lbs. avoirdupois) of veal, in *six large pieces*, to be roasted at once, as an experiment; the fuel consumed was 33 lbs. Bavarian weight of dry pine-wood (equal to 40.86 lbs. avoirdupois), which (at 4½ florins the *klafter*, weighing 2967 lbs. Bavarian weight) cost 3 kreutzers, or about *one penny* sterling.

This experiment was made in the year 1792. Happening to mention the result of it in a large company in London, soon after my arrival there in the autumn of the year 1795, I had the mortification to perceive very plainly by the countenances of my hearers how dangerous it is to promulgate very extraordinary truths. I afterwards grew more cautious, and should not now have ventured to publish this account, had not the results of experiments equally surprising, which have been made with the roaster in the kitchen of the Foundling Hospital, been made known to the public.

Not only the roaster, but the boilers also which have been put up under my direction in the kitchen of the Foundling Hospital, have been found to answer very well; and I am informed that several other great hospitals are about to imitate them. As I left London before the kitchen of the Foundling Hospital was entirely finished, I do not know whether there are dampers to the canals by which the smoke goes off from the fire-places of the boilers, and from that of the roaster to the chimney. If there are not, I could wish they might still be added; and I would strongly recommend it to those who may be engaged in constructing kitchen fire-places on my principles, never to omit them.

Oval grates of cast-iron in the form of a dish, such as I have described in the foregoing chapters of this Essay, were tried in the kitchen of the Foundling Hospital; but the heat was found to be so intense that they were soon melted and destroyed; and we were obliged to have recourse to common flat grates, composed of strong bars of cast-iron. Perhaps the heat generated in the combustion of pit-coal is so intense, when completely confined (as it ought always to be in closed fire-places), that it will not be possible, where coals are used as fuel, to use the hollow dishing grates I have introduced in the public kitchens at Munich, and which have been described and recommended in this Essay.

Since my return to Bavaria, I have made several experiments with grates composed of common bricks, placed edgewise, and I find that they answer for that

use full as well, if not better, than iron bars. By making bricks *on purpose* for this use, of proper forms and
dimensions, and composed of the best clay mixed with
broken crucibles beaten to a coarse powder, kitchen
fire-places might be fitted up with them, which would
be both cheap and durable, and as perfect in all other
respects as any that could possibly be made, even
were the most costly materials to be used in their construction.

To diminish still farther the expense attending the construction of closed kitchen fire-places designed for the use
of poor families, the opening by which fuel is introduced
might be closed with a brick, or with a flat stone; another brick or stone might be made to serve at the same
time as a register and a door to the ash-pit, and a third
as a damper to the chimney or canal for carrying off the
smoke from the fire-place.

I lately had an opportunity of fitting up a kitchen on
these principles, in the construction of which there was
not a particle of iron used, or of any other metal, except
for the boiler. On the approach of the French army
under General Moreau in August last, the Bavarian
troops being assembled at Munich (under my command) for the defence of the capital, the town was so
full of soldiers that several regiments were obliged
to be quartered in public buildings, and encamped on
the ramparts, where they had no conveniences for cooking. For the accommodation of a part of them, four
large oblong square boilers, composed of very thin sheet
coppers well tinned, were fitted up in a mass of brick-
work in the form of a cross; each boiler with its
separate fire-place, communicating by double canals,
furnished with dampers, with one common chimney

which stands in the centre of the cross. The dampers are thin flat tiles; the grates on which the fuel is burned are composed of common bricks, placed edgewise; and the passages leading to the fire-place, and to the ash-pit, are closed by bricks which are made to slide in grooves.

Under the bottom of each boiler, which is quite flat, there are three flues, in the direction of its length; that in the middle, which is as wide as both the others, being occupied by the burning fuel. The opening by which the fuel is introduced is at the end of the boiler *farthest from the chimney;* and the flame, running along the middle flue to the end of it, divides there, and returning in the two side flues to the hither end of the boiler, there rises up into two other flues, in which it passes along the outside of the boiler into the chimney. The boilers are furnished with wooden covers divided into two equal parts, united by hinges. In order that the four boilers may be transported with greater facility from place to place (from one camp to another for instance), they are not all precisely of the same size, but one is so much less than the other, that they may be packed one in the other. The largest of them, which contains the three others, is packed in a wooden chest, which is made just large enough to receive it. In the smallest may be packed a circular tent, sufficiently large to cover them all. In the middle of the tent there must be a hole through which the chimney must pass. The four boilers, together with the tent, and all the apparatus and utensils necessary for a kitchen on this construction for a regiment consisting of 1000 men, might easily be transported from place to place on an Irish car drawn by a single horse.

I have been the more particular in my account of this

portable kitchen, as I think it would be found very useful for troops in camp. The Right Honourable Mr. Thomas Pelham made a trial of one of them last summer for his regiment (the Sussex militia), and found it to be very useful. The saving of fuel was very considerable indeed; and the saving of trouble in cooking not less important. The first experiment we made together in a single boiler, fitted up for the purpose in the open air, in the middle of the court-yard of Lord Pelham's house in London.

I ought, perhaps, to have reserved what I have here said on the subject of these military portable kitchens for my next Essay, where it would more naturally have found its place; but being persuaded of the great advantages that may be derived from them, I am unwilling to lose a moment in recommending them to the attention of those who have it in their power to bring them into use.

Those who wish to know more about them may, I am confident, procure every information they can desire respecting them, by applying to Mr. Pelham, or to any of the officers of the Sussex militia who were in camp with the regiment last summer.

There is one more invention for the use of armies in the field which I wish to recommend, and that is a *portable boiler* of a light and cheap construction, in which victuals may be cooked *on a march*. There are so many occasions when it would be very desirable to be able to give soldiers, harassed and fatigued with severe service, a warm meal, when it is impossible to stop to light fires and boil the pot, that I cannot help flattering myself that a contrivance, by which the pot *actually boiling* may be made to keep pace with the troops as they advance, will be an acceptable present to every humane officer and wise and prudent general. Many a battle has undoubtedly been lost for the want of a good comfortable meal of warm victuals to recruit the strength and raise the spirits of troops fainting with hunger and excessive fatigue.

But to return from this digression. The form of the two principal boilers in the kitchen of the Foundling Hospital is that of an oblong square; that form which, on several accounts, I have reason to think preferable to all others for large boilers, but especially on account of the facility of fitting them up with square bricks, and of cleaning their flues, I first introduced in Ireland in several fire-places designed for different uses, which I fitted up as models, in Dublin, during the visit I made last spring to that country on the invitation of my friend Mr. Secretary Pelham.

The first of these oblong square boilers is that which is fitted up in the court-yard of the Linen-hall at Dublin, *as a model for bleachers.* It is 8 feet wide, 10 feet long, and 2 feet deep; and it is furnished with a wooden cover, which shutting down in a groove in which there is a small quantity of water, the steam is by these means confined in the boiler. This cover is movable on its hinges, which are placed at the end of the boiler farthest from the door of the fire-place; and it is occasionally lifted up by means of a rope, which goes over a compound pulley which is fixed over the boiler at the top or ceiling of the room.

Under this boiler there are five flues which run in the direction of its length, and are arranged and constructed in the same manner as the flues of the new brewhouse boiler which I lately fitted up at Munich. (See Fig. 21, Plate V.) There are no flues round the outside of this boiler; but the brick walls by which they are defended from the cold air are double, and the space between them is filled with charcoal dust.

The fuel burns at the hither end of the middle flue, in an oval dish-grate; and the flame running along in this flue under the middle of the boiler to the farther end of it, there divides, and returns in the two adjoining flues. It then turns to the right and left, and, going back again in the two outside flues to the farther end of the boiler, goes out from under it there in two canals, which, sloping upwards, conduct it to the flues of a *second boiler* of equal dimensions with the first, where it circulates, and warms the water which is designed for refilling the first boiler.

As these boilers are made of exceedingly thin sheet-copper, and *thin boilers* are stronger to resist the effects of the fire, and consequently more durable than very thick ones, they both together cost much less than one single boiler on the common construction; and Mr. Duffin, secretary to the Linen Board, who is a very active, intelligent man, and is himself engaged in a large concern in the bleaching business, showed me a computation, founded on actual experiments which he himself made with this new boiler, by which he proved that the saving of fuel which will result from the general introduction of these boilers in the bleaching trade throughout Ireland will amount to at least fifty thousand pounds sterling a year.

In a laundry which I fitted up in the house belonging to the Dublin Society (and which is designed to serve as a model for laundries for private gentlemen's families), there are also two oblong square boilers, the one heated by the fire, and the other by the smoke; and this smoke, after having circulated in the flues under the second boiler, passes through a long flue (constructed like hot-house flues), which goes round two sides of the *drying-room* (which is adjoining to the *washing-room*), and then, passing through the wall of the drying-room into the ironing-room, it goes off into an open chimney. As the bottom of the second boiler lies on a level with the top of the first, the warm water runs out of the second to refill the first, by a tube furnished with a brass cock, which greatly facilitates the filling of the principal boiler. The wooden covers of these boilers, which are double and movable on hinges, are shut down in grooves in which there is water; and the steam, being by these means confined, is forced to pass off by a wooden tube, which, standing on a part of the cover which is fastened down to the boiler with hooks, carries the steam upwards to the height of seven or eight feet, where it goes off laterally by another (horizontal) wooden tube, through the wall into the drying-room. As soon as this horizontal wooden tube has passed through the wall into the drying-room, it ends in a copper tube, about 3 inches in diameter, which, lying nearly in a horizontal position, conducts the steam through the middle of the drying-room in the direction of its length, and through a hole in a window at the end of the room into the open air.

The steam, in passing through the drying-room in a metallic tube (which is a good conductor of heat), gives off its heat through the sides of the tube to the air of the room, and the water which is condensed runs off through the tube. By sloping the tube *upwards*, instead of downwards, as by accident it was sloped, the condensed water, which is always nearly boiling hot, when it is condensed might be made to return into the boiler, which would be attended with a saving of heat, and consequently of fuel.

The furnace for heating the irons used in smoothing the linen (or ironing, as it is called) is a kind of oven built of bricks and mortar, the bottom of which is a shallow pan of cast iron, 18 inches square and about 3 inches deep, which is nearly filled with fine sand. The fire being made under this pan in a closed fire-place, as the sand defends the upper surface of the pan from the cold air of the atmosphere, the pan is commonly red-hot; and the irons, being shoved down through the sand and placed in contact with this plate of red-hot metal, are heated in a very short time, and at a small expense of fuel.

This contrivance might be used with great success for covering the *hot plates* on which saucepans are made to boil in many private kitchens.

This stove, or oven, for heating the smoothing-irons, projects into the drying-room; but the door by which the irons are introduced, as well as that leading to the fire-place, and that leading to the ash-pit, all open into the ironing-room.

The smoke goes off through the drying-room in an iron tube, and assists in warming the room and in drying the linen.

As it may sometimes be necessary to heat the drying-room when neither the wash-house boilers nor the stove for heating the smoothing-irons are heated, provision is made for that, by constructing a small closed fire-place, designed merely for that purpose, which opens into the flue, by which the smoke from the boilers is carried round the drying-room. This fire-place (which is never used but when it is wanted for drying the linen) is situated just without the drying-room, under the end of the flue where it joins the second boiler. The opening at the top of its fire-place, by which the flame of the burning fuel enters the under part of the flue, is kept closed by a sliding plate of iron, or damper, when this fire-place is not used; and when it is used, the door which closes the opening into the fire-place of the first or principal boiler, and the register in its ash-pit door, are kept shut.

That the top of the principal boiler might not be too high above the pavement of the wash-house for the laundresses to work in the boiler without being obliged to go up steps or stairs, the grate and the bottom of the flues under the boiler are nearly on a level with the pavement, and the ash-pit is sunk into the ground; and, to render the approach to the opening into the fire-place more convenient in introducing the fuel and lighting and managing the fire, there is an area before the fire-place, about 3 feet square and 2 feet deep, sunk in the ground, and walled up on its sides, into which there is a descent by steps. In two of the sides of these vertical walls (those on the right and left when you stand fronting the fire-place) there are vaults for containing fuel, which extend several feet under the pavement. The steps which descend into this area are on the side of it, opposite the fire-place.

Areas of this kind are very necessary for all fire-places for large boilers, otherwise the top of the boiler will necessarily be raised too high above the level of the pavement to be approached with facility and convenience. Steps may be made, it is true, for approaching boilers which are placed higher; but these are always inconvenient, and take up more room, and cost more than the execution of the plan here proposed for rendering them unnecessary.

The areas before the fire-place door of the large boilers in the kitchen of the Foundling Hospital are occasionally closed by trap-doors. As often as this is done there must be a number of small holes bored in the door to permit the air necessary for feeding the fire to descend into the ash-pit; and when the bottom of the passage leading into the fire-place happens to lie above the level of the upper surface of this trap-door, the part of the door immediately under this opening should, to prevent accidents from live coals which may occasionally fall out of the fire-place, be covered with a thin plate of sheet iron.

When large boilers are fitted up in situations where it is not possible to sink an area in front of the fire-place, the mass of brick-work in which the boiler is set must be raised, and steps must be made to approach it. When this is done, the upper step should be made very wide (at least 2 feet), in order that there may be room to stand and work in the boiler; and, for still greater convenience, the steps should be continued round three sides of the boiler, when the boiler stands in a detached mass of brick-work. The bottom of the door of the fire-place should, if possible, be above the upper flat surface of the upper step; and, to preserve the symmetry of the whole, the ash-pit door may be in the front of the upper step, and the passage into the ash-pit (which will be long of course) may descend in a gentle slope. In this manner the kitchen of the Hospital of *La Pietà* at Verona was constructed.

No inconvenience whatever attends the increase of the length of the passage into the ash-pit, except it be

that very trifling one, — which surely does not deserve
to be mentioned, — the increase of labour attending the
removal of the ashes; but the inconvenience would be
very considerable which would unavoidably attend the
discontinuation or breaking off of the steps round the
hither end or front of the boiler, which would be necessary in order to be able to place the ash-pit door *directly*
under the fire-place door, and to make a way to approach it.

The flues under the principal boiler of the laundry
in the house of the Dublin Society are not contrived
so as to divide the flame and cause it to circulate in *two*
currents. They run from side to side under it: the door
of the fire-place is not in the middle, but on one side
of the boiler, and near one end of it. The flame, passing and returning under the boiler twice from its front
to its opposite side, goes off at its end (that farthest from
the fire-place) into a canal furnished with a damper,
which canal, rising upward at an angle of about 45
degrees, leads to the flues under the second boiler.
The bottom of the flues of the principal boiler are just
on a level with the pavement of the wash-house; and
in order that they may easily be cleaned out, and the
bottom of the boiler scrubbed with a broom to free it
from soot, the ends of the flues are, in building the
fire-place, left open, and afterwards, when the boiler is
set, they are closed by temporary (double) walls of dry
bricks. To make these walls tight, the joinings of
the bricks are plastered on the outside with moist
clay.

The sides of the boilers are defended from the cold
air by thin walls of bricks covered with wainscot, and
by filling the space between these walls and the boiler

with pounded charcoal. Were I to fit up these boilers again, I should leave this space void, or filled merely with air, forming several small openings below, through which the flame and hot vapour from the flues might ascend and surround the boiler. In the large boiler fitted up in the Linen-hall as a model for bleachers, this alteration is also necessary to render it complete; and as it might be made in a few hours, and almost without any expense, I cannot help expressing a wish that it might still be done.

The ardent zeal for the prosperity of his country, and indefatigable attention to every thing that tends to promote useful improvement, which so eminently distinguish that enlightened patriot and most respectable statesman, to whom the manufactures and commerce of Ireland, and the linen trade in particular, are so much indebted, encourage me to hope that he will take pleasure in giving his assistance to render the models for improving fire-places and saving fuel, which I have had the satisfaction of leaving in Ireland, as free from faults as they can possibly be made.

Though my stay in Ireland was too short to construct models of all the improvements I wished to have introduced in that delightful and most interesting island, yet the liberality with which my various proposals were received, and the generous assistance I met with from all quarters, enabled me to do more in two months than I probably should have been able to. have effected in as many years in some other older countries, where the progress of wealth and of refinements has rendered it extremely difficult to get people to attend to useful improvements.

I wished much to have been able to have fitted up

the great kitchen in the House of Industry at Dublin, as the expense of fuel is very considerable in that extensive establishment, where more than 1500 persons are fed daily, at an average ; but, not having time to finish so considerable an undertaking, I thought it most prudent not to begin it. I fitted up one large boiler as a model at one end of one of the working-halls ; but this was designed principally to show how a large hall might be heated from a kitchen fire-place, and from the very same fire which is used for cooking.* The smoke from the fire-place is carried along horizontally on one side of the hall from one end of it to the other ; and the boiler being closed by a cover which is steam-tight, the steam from the boiler is also forced along from one end of the hall to the other, in a horizontal leaden pipe, which runs parallel to the flue occupied by the smoke, and lies immediately over it. In warm weather, when the hall does not require to be heated, the smoke and steam go off immediately into the atmosphere by a chimney adjoining the fire-place, without passing through the hall.

To be able to equalize the heat in the hall (which is very long and narrow), or to render it as warm at the end of it which is farthest from the fire-place as at that next the fire, I directed clothing for the steam tube of warm blanketing to be made in lengths of three or four feet, to be occasionally put round it and fastened by buttons.

By clothing or covering the steam tube more or less, as may be found necessary in those parts of the hall

---

* This contrivance might easily be applied to the heating of hothouses, even though the hothouse should happen to be situated at a considerable distance from the kitchen.

where the heat is greatest, the steam, being by this covering prevented from giving off its heat to the air
through the tube, will go on farther and warm those
parts of the hall which otherwise would be not sufficiently heated. The steam tube, which is constructed of
very thin sheet lead, is about 3 inches in diameter, and,
instead of being laid exactly in a horizontal position,
slopes a little upwards, just so much that the water which
results from the condensation of the steam may return
into the boiler.*

The horizontal flue through which the smoke passes
is a round tube of sheet iron, about 7 inches in diameter, divided, for the facility of cleaning it, in lengths
of 12 or 15 feet, fixed nearly horizontally at different
heights from the floor, or, in an interrupted line, in
hollow pilasters or square columns of brick-work. A
common hothouse flue constructed of bricks and mortar would have answered equally well for warming the
hall, but would have taken up too much room, which
is the only reason it was not preferred to these iron
tubes.

---

* I contrived a fire-place for heating one of the principal churches in Dublin on these principles with steam (but without making use of the smoke); and
I promised to give a plan (which, I am ashamed to say, I have not yet been able
to finish) for heating the superb new building destined for the meeting of the
Irish House of Commons.

One of the two chimney fire-places, which I fitted up in the hall in which
the meetings of the Royal Irish Academy are held, will, I imagine, be found to
answer very well for heating high rooms and large halls in private houses. In
this fire-place I have endeavoured, and I believe successfully, to unite the advantages of an open fire with those of a German stove. The grate used in
fitting up this fire-place, and which is of cast iron, and far from being unelegant
in its form, and which cost only *seven shillings and sixpence sterling*, is decidedly
the best adapted for open chimney fire-places, where coals are used as fuel, of
any I have yet seen. By a letter I lately received from a friend in Ireland, I
had the satisfaction to learn that these grates are coming very fast into general
use in that country.

In constructing the boiler (which is of thin sheet iron), I made an experiment which succeeded even beyond my expectation. The flues under the boiler (and there are none round it) are projections from the bottom of the boiler: they are hollow walls of sheet iron, about 9 inches high and an inch and three-quarters thick, into which the liquid in the boiler descends, and which in fact constitute a part of the boiler. By this contrivance the flame is surrounded on all sides, except at the bottom of the flues (where the heat has little or no tendency to pass), by the liquid which is heated, and the fire-place is merely a flat mass of brick-work. The grate is even with the upper surface of this mass of brick-work, and the ash-pit is the only cavity in it.

In constructing the boiler, provision was made, by omitting or interrupting the hollow walls or divisions of the flues, in the proper places, to leave room for introducing the fuel, for the passage of the flame from one flue to another, and from the last flue into the canal by which the smoke goes off into the chimney, or into the iron tubes by which the hall is occasionally warmed.

One principal object which I had in view in this experiment was to see if I could not contrive a boiler, which, being suspended under a wagon or other wheel-carriage, might serve for cooking for troops on a march; or which, being merely set down on the ground, a fire might be immediately kindled under it.

Those who will take the trouble to examine the boiler in question will find that the principle on which it is constructed may easily be applied to the objects here mentioned. But it is not merely for portable boilers that this construction would be found useful: I am convinced that it would be very advantageous for the boilers of steam engines, for distilleries, and for various other purposes. As the escape of heat into the brick-work is almost entirely prevented, and as the surface of the boiler on which the heat is made to act is greatly increased by means of the hollow walls, the liquid in the boiler is heated in a very short time, and with a small quantity of fuel.

There is still another advantage attending this construction, which renders it highly deserving the attention of distillers. By making the tops of the flues arched instead of flat (which may easily be done, and which is actually done in the boiler in question), or in the form of the roof of a house, as the hottest part of the flame will, of course, always occupy the upper part of the flues, and as the thick or viscous part of the liquor in the boiler — that which is in most danger of being burned to the bottom of the boiler, and giving a bad taste to the spirit which comes over — cannot well lie on the convex or sloping surface of these flues, there will be less danger of an accident which distillers have hitherto found it extremely difficult to prevent.

In constructing boilers on these principles for distillers, it will probably be found necessary to increase very much the thickness of the hollow walls of the flues, and perhaps to make them even deeper than the level of the bottom of the flues, in order more effectually to prevent the thick matter which will naturally settle in those cavities from being exposed to too great a heat.

A similar advantage will attend large boilers constructed on these principles for making thick soups for hospitals ; these soups being very apt to burn to the bottoms of the boilers in which they are prepared.

I made another experiment in the House of Industry

in Dublin, which I wished much to have had time to have prosecuted farther. Finding that the expense for wheaten bread for the House was very great (amounting, in the year 1795, to no less than 3841*l.* sterling), I saw that a very considerable saving might be made by furnishing those who were fed at the public expense with oaten cakes (a kind of bread to which they had always been used), instead of rendering them dainty and spoiling them by giving them the best wheaten bread that could be procured, as I found had hitherto been done. But to be able to furnish oaten cakes in sufficient quantities to feed 1500 persons, some more convenient method of baking them than that commonly practised was necessary, and one in which the expense of fuel might be greatly lessened.

With a view to facilitate this important change in the mode of feeding the numerous objects of charity and of *correction*, who were shut up together within the walls of that extensive establishment, I constructed what I would call a *perpetual oven.*

In the centre of a circular, or rather cylindrical mass of brick-work, about 8 feet in diameter, which occupies the middle of a large room on the ground floor, I constructed a small, circular, closed fire-place for burning either wood, peat, turf, or coals. The diameter of the fire-place is about 11 inches, the grate being placed about 10 inches above the floor, and the top of the fire-place is contracted to about 4 inches. Immediately above this narrow throat, six separate canals (each furnished with a damper, by means of which its opening can be contracted more or less, or entirely closed) go off horizontally, by which the flame is conducted into six separate sets of flues, under six large plates of cast iron,

which form the bottoms of six ovens on the same level,
and joining each other by their sides, which are concealed
in the cylindrical mass of the brick-work. Each of these
plates of cast iron being in the form of an equilateral tri-
angle, they all unite in the centre of the cylindrical mass
of brick-work, consequently the two sides of each unite
in a point at the bottom of it, forming an angle of 60
degrees.

The flame, after circulating under the bottoms of these
ovens, rises up in two canals concealed in the front wall
of each oven, and situated on the right and left of its
mouth, and after circulating again in similar flues on the
upper flat surface of another triangular plate of cast iron,
which forms the top of the oven, goes off upwards by a
canal furnished with a damper into a hollow place, situated on the top of the cylindrical mass of the brick-work,
from which it passes off in a horizontal iron tube, about
7 inches in diameter, suspended near the ceiling of the
room, into a chimney situated on one side of the room.

These six ovens which are contiguous to each other in
this mass of brick-work are united by their sides by thin
walls made of tiles, about $1\frac{1}{2}$ inches thick and 10 inches
square, placed edgewise; and each oven having its separate canal, furnished with a register communicating with
the fire-place, any one or more of them may be heated
without heating the others, or the heat may be turned off
from one of them to the other in continual succession;
and, by managing matters properly, the process of baking
may be *uninterrupted*. As soon as the bread is drawn
out of one of the ovens, the fire may immediately be turned
under it to heat it again, while that from under which the
fire is taken is filled with unbaked loaves, and closed up.

A principal object which I had in view in constructing

this oven was to prevent the great loss of heat which is occasioned in large ovens, by keeping the mouth of the oven open for so considerable a length of time as is necessary for putting in and drawing out the bread. As one of these small ovens contains only five large loaves, or cakes, it may be charged, or the bread when baked may be drawn, in a moment; and during this time the other five ovens are kept closed, and consequently are not losing heat; *one* of them is heating, while the other *four* are filled with bread in different stages of the process of baking.

When I constructed this oven, though I had no doubt of its being perfectly well calculated for the use for which it was principally designed, — baking oaten cakes, which are commonly baked on heated iron plates, — yet I was by no means sure it would answer for baking common bread in large thick loaves. I had not made the experiment. And though I could not conceive that any thing more could be necessary in the process of baking than *heat*, — and here I was absolutely master of every degree of it that could possibly be wanted, and could even regulate the succession of different degrees of it at pleasure, — I thought it probable that some particular management might be required in baking bread in these metallic ovens, a knowledge of which could only be acquired by experience.

What served to strengthen these suspicions was a discovery which had accidentally been made by the cook of the Military Academy. In the course of *his* experiments, he found that my roaster is admirably well calculated for baking pies, puddings, and pastry of all kinds: provided, however, that the fire be managed *in a certain way;* for when the fire is managed in the same manner in which it ought to be managed in roasting meat, pies and pastry will absolutely be spoiled. After repeated failures and disappointments, and after having lost all hopes of ever being able to succeed in his attempts, the cook (by mere accident, as he assured me) discovered the important secret; and important he certainly considers it to be, and feels no small degree of satisfaction, not to say pride, in having been so fortunate as to make the discovery. He must pardon me if I take the liberty, even without his permission, to publish it to the world for the good of mankind.

The roaster must be well heated before the pies or pastry are put into it, and the blowers must never be quite closed during the process.

I have lately found that, by using similar precautions, bread may be perfectly well baked in metallic ovens, similar to that in the House of Industry in Dublin.

Thinking it more than probable that means might be devised for managing the heat in such a manner as to perform that process in ovens constructed on these principles, and heated *from without ;* and conceiving that not only a great saving of fuel, but also several other very important advantages, could not fail to be derived from that discovery, on my return to Munich from England, in August last, I immediately set about making experiments, with a view to the investigation of that subject; and I have so far succeeded in them that, for these last four months, my table has been supplied entirely with bread baked in my own house, by my cook, in an oven constructed of thin sheet iron, which is heated (like my roasters) from without; and I will venture to add that I never tasted better bread. All those who have eaten of it have unanimously expressed the same opinion of it. It is very light, most thoroughly baked without being too much dried, and I think remarkably well-tasted. The loaves, which are made small in order that they may have a greater proportion of crust (which, when the bread is baked in this way, is singularly delicate), are placed in the oven on circular plates of thin sheet iron, raised about an inch on slender iron feet. Were the loaf placed on the bottom of the oven, the under crust would presently be burned to a coal, and the bread spoiled. A precaution absolutely necessary in baking bread in the manner here recommended is to leave a passage for the steam generated in the process of baking to escape. This may be done either by constructing a steam chimney for that purpose, furnished with a damper, or simply by making a register in the door of the oven.

As this is not the proper place to enlarge on this subject, I shall leave it for the present; but I cannot help expressing a wish that what I have here advanced may induce others, especially *bakers*, who may find their own advantage in the prosecution of these interesting and important investigations, to turn their attention to them.

How exceedingly useful would my roasters be, and ovens constructed on the principles here recommended, on shipboard! Having served a campaign (as a volunteer) in a large fleet (that commanded by Admiral Sir Charles Hardy, in the year 1779), and having made several long sea voyages, I have had frequent opportunities of seeing how difficult it is in bad weather to cook at sea; and it is easy to imagine how much it would contribute to the comfort of seafaring people, especially at times when they are exposed to the greatest fatigues and hardships, to enable them to have their tables well supplied with warm victuals.

In order that the motion of the vessel might not derange any part of the apparatus used in the process of cooking at sea in my roasters, the form of the roaster should be that of a perfect cylinder; and the dripping-pan in which the meat is placed should be a longitudinal section of another cylinder, less in diameter than the roaster by about an inch, and suspended on two pivots in the axis of the roaster, in such a manner that the dripping-pan may swing freely in the roaster without touching its sides. The roaster should be placed in the brick-work, with its axis in the direction of the length of the ship; and, to prevent the gravy from being thrown out of the dripping-pan when the vessel pitches, its hollow cavity should be divided into a number of compartments, by partitions running across it from side to side.

It remains for me to give some account of the kitchen which I fitted up in the house of the Dublin Society, as a model for private families; and also of a cottage fire-place, and a lime-kiln, which I constructed as models for imitation, in the courtyard of that public building.

With regard to the kitchen, it is necessary that I should remark, at setting out, that it was not intended so much to serve as a complete model of a convenient kitchen for a private family, as to display a variety of useful inventions, all or any of which may at pleasure be easily adopted, in kitchens of all kinds and of all dimensions. I thought this would be more useful than any simple model of a kitchen I could contrive.

It is, however, a very complete kitchen; and though there are some contrivances belonging to it which might have been omitted, yet they will all, I am confident, be found useful for the different purposes for which they were particularly designed, and in a kitchen for a large family would often come into use.

The general disposition of the various parts of this kitchen I consider as being quite perfect. It is the same as that of the Hospital of *La Pietà* at *Verona*, and of a very complete private kitchen which was built about two years ago at Munich, under my direction, in the house of Baron Lerchenfeld, steward of the household to his Most Serene Highness the Elector. In my next Essay, which will treat exclusively of the construction of kitchen fire-places and of kitchen utensils, I shall give a particular detailed account of the manner in which the various boilers — steam-boilers, saucepans, oven, roasters, etc. — are disposed and connected in the mass of brick-work in these kitchens, and shall accompany these descriptions with a sufficient number of Plates to render them perfectly intelligible.

Cottage Fire-place and Iron Pot, for cooking for the Poor.

The cottage fire-place which I fitted up as a model, in the courtyard of the house of the Dublin Society, was not quite finished when I left Ireland; but an idea may be formed from what was done of the general principles on which such fire-places may be constructed. On each side of the open chimney fire-place (which, being small, was built in the middle of one much larger, which was constructed to represent a large open fire-place, such as are now general in cottages) I fitted up an iron pot on a peculiar construction, cast by Mr. Jackson of Dublin, and designed for the use of a poor family in cooking their victuals. This pot is nearly of a cylindrical form, about 16 inches in diameter, and

8 inches deep; and under its bottom, which is quite flat, there is a thin spiral projection, which was cast with the pot, and serves instead of feet to it, the turns of which, when the pot is set down on a flat surface, form a spiral flue in which the flame circulates under the bottom of the pot.   This projection, which is near half an inch thick where it is united with the bottom of the pot, and less than a quarter of an inch below where its lower edge rests on the ground, is about 4 inches wide, or rather deep.   This projection was made tapering, in order to its being more easily cast.   To defend the outside of this pot from the cold air, the pot is enclosed in a cylinder of thin sheet iron, equal in diameter to the extreme width of the pot at its brim, just as high as the depth of the pot and of its spiral flues taken together. The pot is fastened to this cylindrical case by being driven into it with force, a rim in the form of a flat hoop, about an inch and a half deep and a little tapering, being cast on the outside of the pot at its brim, the external surface of which was fitted exactly into the top of this cylinder.   This projection is useful, not only in uniting the pot to its cylindrical case, but also to keep this cylindrical case at some small distance from the sides of the pot, by which means the heat is more effectually confined.

To be able to move about this pot from place to place, it has two handles which are riveted to the outside of its cylindrical case; and it is provided with a wooden cover.

I am sensible that I often expose myself to criticism by anticipating what would more naturally find its place elsewhere.   But what I have here said in regard to this iron pot is intended merely as hints to awaken the curiosity and excite the attention of ingenious men, — of such as take pleasure in exercising their ingenuity in contriving and perfecting useful inventions, and who delight in contemplating the progress of human industry.

Model of a perpetual Lime-kiln.

The particular objects principally had in view in the construction of this lime-kiln (which stands in the court-yard of the Dublin Society) were, *first*, to cause the fuel to burn in such a manner as to consume the smoke, which was done by obliging the smoke to descend and pass through the fire, in order that as much heat as possible might be generated. Secondly, to cause the flame and hot vapour which rise from the fire to come into contact with the limestone by a very large surface, in order to economize the heat and prevent its going off into the atmosphere, which was done by making the body of the kiln in the form of a hollow truncated cone, and very high in proportion to its diameter; and by filling it quite up to the top with limestone, the fire being made to enter near the bottom of the cone. Thirdly, to make the process of burning lime *perpetual*, in order to prevent the waste of heat which unavoidably attends the cooling of the kiln in emptying and filling it, when, to perform that operation, it is necessary to put out the fire. And, fourthly, to contrive matters so that the lime in which the process of burning is *just finished*, and which of course is still *intensely hot*, may, in *cooling*, be made to give off its heat in such a manner as to assist in heating the fresh quantity of cold limestone with which the kiln is replenished as often as a portion of lime is taken out of it.

To effectuate these purposes, the fuel is not mixed with the limestone, but is burned in a closed fire-place, which opens into one side of the kiln, some distance above the bottom of it. For large lime-kilns on these principles there may be several fire-places, all opening into the same cone, and situated on different sides of it; which fire-places may be constructed and regulated like the fire-places of the furnaces used for burning porcelain.

At the bottom of the kiln there is a door, which is occasionally opened to take out the lime.

When, in consequence of a portion of lime being drawn out of the kiln, its contents settle down or subside, the empty space in the upper part of the kiln, which is occasioned by this subtraction of the burned lime, is immediately filled up with fresh limestone.

As soon as a portion of lime is taken away, the door by which it is removed must be immediately shut, and the joinings well closed with moist clay, to prevent a draught of cold air through the kiln. A small opening, however, must be left, for reasons which I shall presently explain.

As the fire enters the kiln at some distance from the bottom of it, and as the flame *rises* as soon as it comes into this cavity, the lower part of the kiln (that below the level of the bottom of the fire-place) is occupied by lime already burned; and as this lime is intensely hot when, on a portion of lime from below being removed, it descends into this part of the kiln, and as the air in the kiln to which it communicates its heat must *rise upwards* in consequence of its being heated, and pass off through the top of the kiln, this lime in cooling is, by this contrivance, made to assist in heating the fresh portion of cold limestone with which the kiln is charged. To facilitate this communication of heat from the red-hot lime just burned to the limestone above in the upper part of the kiln, a gentle draught of air through the kiln from the bottom to the top of it must be established by leaving an opening in the door below, by which the cold air from without may be suffered to enter the kiln. This opening (which should be furnished with some kind of a register) must be very small, otherwise it will occasion too strong a draught of cold air into the kiln, and do more harm than good; and it will probably be found to be best to close it entirely, after the lime in the lower part of the kiln has parted with a certain proportion of its heat.

Conceiving the improvement of lime-kilns to be a matter of very great national importance, especially since the use of lime as manure has become so general, I intend to devote the first leisure time I can spare to a thorough investigation of that subject. In the mean time, I have here thrown out the loose ideas I have formed respecting it, in order that they may be examined, corrected, and improved upon by others who may be engaged in the same pursuits.

The model I caused to be constructed in the courtyard of the Dublin Society is, I am sensible, very imperfect. It was built in a great hurry, being begun and finished the same day, — the day but one before I left Ireland; but I am now engaged in constructing a lime-kiln on the same principles (for the use of the farm in the English Garden at Munich), which I shall take pains to make as perfect as possible; and, should it be found to answer as well as I have reason to hope it will, I shall not fail to give a particular account of it to the public, accompanied with drawings, and all the details that shall be necessary in order to give the most satisfactory account of the result of the experiment.

These investigations will be the more interesting, and their results more generally useful, as the discovery of a mine of pit-coal in the neighbourhood of Munich, which is now worked with success, has put it in my power to use coal as fuel, as well as wood and turf, in the experiments I shall make in burning lime in this kiln.

For the information of those who may be disposed to engage in these pursuits, I have published the annexed sketch of the lime-kiln in question, which is now actually building (see Plate VI.). I thought it right to do this, that we might start fair; and I can assure my competitors in this race, that I shall feel no ill-will on seeing them get before me.

If I do not deceive myself, the laudable exertions of others afford me almost as much pleasure as my own pursuits; at least I am quite certain that when I can flatter myself that I have had any — even the smallest — share in *exciting* those exertions, the satisfaction I feel in contemplating them is inexpressible.

DESCRIPTION OF THE PLATES.

PLATE I.

Fig. 1. A view of a double cover for a boiler or saucepan. In this design the rim is seen which enters the boiler, and the tube by which the steam goes off is seen in part (above), and is in part indicated by dotted lines. (See page 321.)

Fig. 2 shows this cover placed on its boiler. Part of the side of the cover is represented as wanting, in order that the steam tube might be better seen. The height of this cover is represented as being equal to *one half* its diameter; but I have found *one third* of its diameter quite sufficient for its height.

Fig. 3 and Fig. 4 are views of my circular dishing-grates for closed kitchen fire-places. They may be made of any size, from 5 inches to 18 inches in diameter, according to the size of the boiler. The rules I have in general followed, in determining the size proper for the grate for any (circular) boiler, has been to make its diameter equal to half the diameter of the boiler at the brim. (See page 341.)

Fig. 5 is an inverted hollow cone of thin sheet iron, which is placed immediately under the grate, its brim being made to receive the circular rim of the grate. When the fire-place is large, this inverted cone may be made of fire-stone, or constructed of bricks and mortar. For small fire-places it may be made of earthen-ware, which is, perhaps, the very best material for it that can be found. (See page 343.)

Fig. 6, Fig. 7, and Fig. 8, are views and sections of a perforated tile, with its stopper, such as are used for closing the entrance by which the fuel is introduced into closed kitchen fire-places. The diameter of the circular opening, or hole in the tile, may be from 6 to 7 inches. (See page 332.)

PLATE II.

The various figures, from No. 9 to No. 16 of this plate, show the construction of an ash-pit door, with its register. (See page 333.)

Fig. 9 is a front view of the door with its register. The whole is constructed of sheet iron, except the four narrow pieces at the four corners, which hold down in its place the circular plate of the register, and the small circular plate (as large as a half-crown) in the centre of the register, which are made of brass, on account of that metal not being so liable to rust as iron.

Fig. 10 is a side view of the back-side of the door, fixed in its frame, in which the manner of its being shut in its frame is seen; and the iron straps, *a*, *b*, *c*, *d*, are seen, by which the frame is fastened in the brick-work.

Fig. 11 is a horizontal section through the middle of the door and its frame, and through the button which serves for shutting the door.

Fig. 12 is a section of this button, on an enlarged scale, showing the manner in which it is constructed.

Fig. 13 is the plate of sheet iron which forms the front of the door, with the holes in it by which the other parts of the machinery are fixed to it.

Fig. 14 is the circular plate which forms the register. To this plate is fixed a projecting knob, or button (represented in the figure), by which it is turned about.

PLATE I.

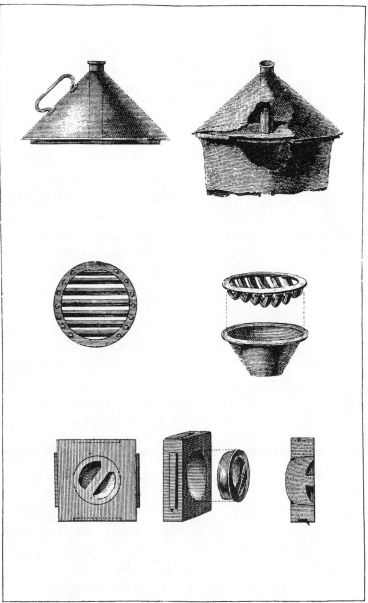

Fig. 15 and Fig. 16 show, on an enlarged scale, one of the four pieces of brass by which the circular plate of the register is kept down in its place.

In constructing these register doors, and in general all iron doors for fire-places, great and small, the door should never shut in a rabbet or groove in the frame, but should merely *shut down on the front edge of the frame*, which edge, by grinding it on the flat surface of a large flat stone, should be made quite level to receive it. If this be done, and if the plate of iron which constitutes the door be made quite flat, and if it be properly fixed on its hinges, the door will always shut with facility and close the opening with precision, notwithstanding the effects of the expansion of the metal by heat; but this cannot be the case when the doors of fire-places are fitted in grooves and rabbets.

Where the heat is very intense, the frame of the door should be made of fire-stone; and that part of the door which is exposed naked to the fire should be covered either with a fit piece of fire-stone, fastened to it with clamps of iron, or a sufficient number of strong nails with long necks and flat heads, or of staples, being driven into that side of the plate of iron which forms the door which is exposed, should be covered with a body about two inches thick of strong clay mixed with a due portion of coarse powder of broken crucibles, which mass will be held in its place by the heads of the nails and by the projecting staples. This mass being put on wet, and gently dried, the cracks being carefully filled up as they appear, and the whole well beaten together into a solid mass, will, when properly burned on by the heat of the fire, form a covering for the door which will effectually defend it from all injury from the fire; and

PLATE II.

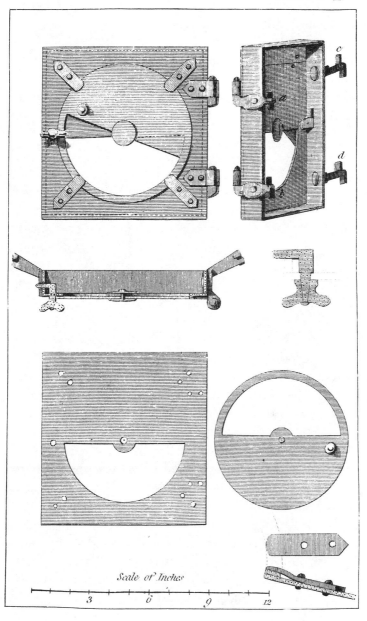

Scale of Inches

3   6   9   12

the door so defended will last ten times longer than it
would last without this defence.

The inside doors of the two brewhouse fire-places
which I have fitted up at Munich are both defended
from the heat in this manner; and the contrivance,
which has answered perfectly all that was expected
from it, has not been found to be attended with any
inconvenience whatever.

PLATE III.

Fig. 17 is a front view of the new boiler of the brewhouse called Neuheusel, or rather of its fire-place and
cover (the boiler being concealed in the brick-work).
The inside door of the fire-place is here represented
shut; and, in order that it might appear, the outside
door is taken off its hinges, and is not shown. The
two vaulted galleries, A, B, in the solid mass of brick-
work, on the right and left of the fire-place (which were
made to save bricks), serve for holding firewood. The
partition walls of the fire-place and the different flues,
as also a section of the boiler, are represented by dotted
lines. The small circular hole on the left of the fire-
place door is the window opening into the fire-place, by
which the burning fuel may be seen.

a, *b*, is the wooden curb of the boiler; *c*, *d*, a platform
on which the men stand when they work in emptying
the boiler, etc.; *e*, *f*, is a platform which serves as a
passage from one side of the boiler to the other. This
platform, which is about 18 inches wide, is 12 inches
higher than the other platforms, in order that the openings *g* and *h*, into the flues, may remain free. These
openings, which are opened only occasionally, — that is
to say, when the flues want cleaning, — are kept closed

PLATE III.

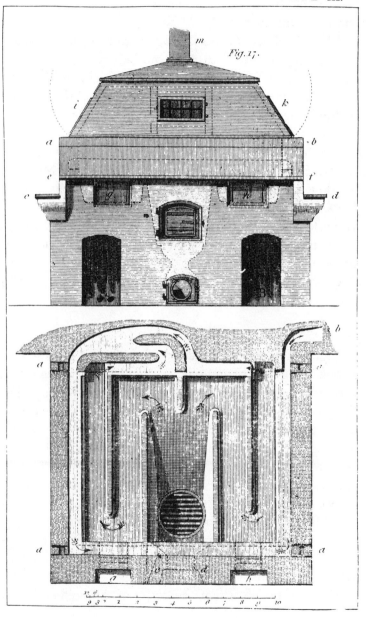

Fig. 17.

by double brick walls. These walls are expressed in the following figure.

Fig. 18. This is a horizontal section of the fire-place at a level with the bottom of the boiler. *a, a, a, a,* are four openings by which the flues which, in the first arrangement of this fire-place, went round the outside of the boiler, were occasionally cleaned; *b* is the canal by which the smoke went off into the chimney.

The entrance into the fire-place, and the conical perforation in the wall of the fire-place which serves as a window for observing the fire, are marked by dotted lines. The position of the inside door of the fire-place is marked by a dotted line, *c, d.* The circular dishing-grate is seen in its place; and the walls of the flues under the boiler are all seen. The crooked arrows in the flues show the direction of the flame. (See page 398.)

PLATE IV.

Fig. 19 is a vertical section of the boiler represented in the foregoing plate (Fig. 17). This section is taken through the middle of the boiler, of the fire-place, and of the cover of the boiler. A is the ash-pit, with a section of its register door; B is the fire-place, and its circular dishing-grate; C is the entrance by which the fuel is introduced, with sections of its two doors; D is a space left void to save bricks; E is the boiler, and F its wooden cover; *m* is the steam chimney, which is furnished with a damper; R, R, is the vertical wall of the house against which the brick-work in which the boiler is fixed is placed; *a, b,* is the curb of timber in which the boiler is set.

The manner in which the cover of the boiler is con-

PLATE IV.

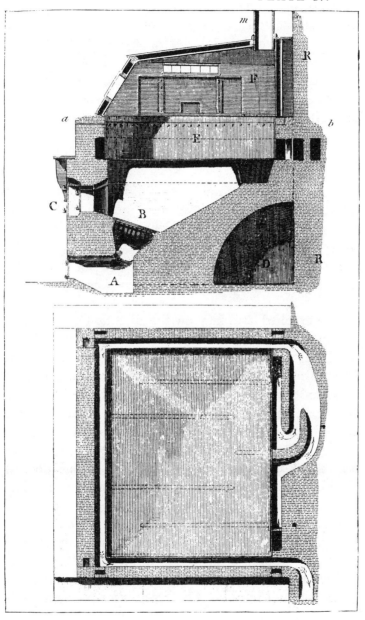

structed, as well as its form, and the door and windows which belong to it, are all seen distinctly in this figure.

Fig. 20 is a horizontal section of this fire-place taken on a level with the bottom of the flue which goes round the outside of the boiler, in which flue, before the fire-place was altered, the flame circulated. The flues under the boiler are, in this figure, indicated by dotted lines.

PLATE V.

Fig. 21 is a horizontal section of the fire-place of the brewhouse boiler, at a level with the top of the flues under the boiler, *after the flue round the outside of the boiler had been stopped up*, or rather the flame prevented from circulating in it. This figure shows the actual state of the fire-place at the present time. (See page 414.)

The crooked arrows show the direction of the flame in the flues; *a, b,* are the two canals (each of which is furnished with a damper) by which the smoke goes off into the chimney; and *c, c, c, c, c, c,* are six small openings communicating with the flues, by which the flame and hot vapour can pass up into the cavity on the outside of the boiler which formerly served as a flue.

Fig. 22 is a front view of the ash-pit door of this brewhouse fire-place, with its register. This door is closed by means of a latch of a particular construction, which is shown in the figure.

Fig. 23 is the door without its register; and

Fig. 24 the circular plate of the register represented alone.

This ash-pit door shuts against the front edge of its frame, and not into it. The reasons for preferring this

PLATE V.

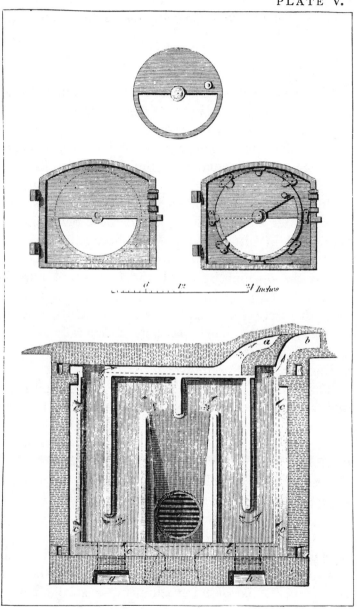

6   12   24 Inches

method of fitting the door to its frame have already been explained. (See descriptions of the Plate II.)

PLATE VI.

Fig. 25 is a section of a small lime-kiln, built, or rather now building, at Munich, for the purpose of making experiments. The height of the kiln is 15 feet; its internal diameter below, 2 feet; and above, 9 inches. In order more effectually to confine the heat, its walls, which are of bricks and very thin, are double, and the cavity between them is filled with dry wood ashes. To give greater strength to the fabric, these two walls are connected in different places by horizontal layers of bricks which unite them firmly.

a is the opening by which the fuel is put into the fire-place. Through this opening the air *descends* which feeds the fire. The fire-place is represented nearly full of coals, and the flame passing off laterally into the cavity of the kiln, by an opening made for that purpose at the bottom of the fire-place.

The opening above, by which the fuel is introduced into the fire-place, is covered by a plate of iron, movable on hinges; which plate, by being lifted up more or less by means of a chain, serves as a register for regulating the fire.

A section of this plate, and of the chain by which it is supported, are shown in the figure.

b is an opening in the front wall of the fire-place, which serves occasionally for cleaning out the fire-place and the opening by which the flame passes from the fire-place into the kiln. This opening, which must never be quite closed, serves likewise for admitting a small quantity

PLATE VI.

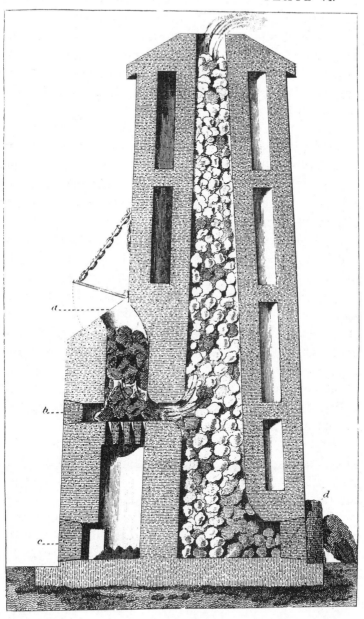

of air to pass horizontally into the fire-place. A small proportion of air admitted in this manner has been found to be useful, and even necessary, in fire-places in which, in order to consume the smoke, the flame is made to descend. Several small holes for this purpose, fitted with conical stoppers, may be made in different parts of the front wall of the fire-place.

The bottom of the fire-place is a grate constructed of bricks placed edgewise, and under this grate there is an ash-pit; but, as no air must be permitted to pass up through this grate into the fire-place, the ash-pit door, *c*, is kept constantly closed, being only opened occasionally to remove the ashes.

d is the opening by which the lime is taken out of the kiln; which opening must be kept well closed, in order to prevent a draught of cold air through the kiln.

As only as much lime must be removed at once as is contained in that part of the kiln which lies below the level of the bottom of the fire-place, to be able to ascertain when the proper quantity is taken away, the lime as it comes out of the kiln may be directed into a pit sunk in the ground in front of the opening by which the lime is removed, this pit being made of proper size to serve as a measure.

While the lime is removing from the bottom of the kiln, fresh limestone should. be put into it above; and during this operation the fire may be damped by closing the top of the fire-place with its iron plate.

Should it be found necessary, the fire and the distribution of the heat may, in burning the lime, be farther regulated by closing more or less the opening at the top of the lime-kiln with a flat piece of fire-stone, or a plate of cast iron.

The double walls of the kiln, and the void space between them, as also the horizontal layers of bricks by which they are united, are clearly and distinctly expressed in the figure. The kiln is represented as being nearly filled with small round stones, such as are used at Munich in burning lime. These stones are brought down from the calcareous mountains on our frontiers, by the river (the Isar), and are rounded by rubbing against each other as they are rolled along by the impetuosity of the torrent.

EXPERIMENTS AND OBSERVATIONS

ON THE

ADHESION OF THE PARTICLES OF WATER TO EACH OTHER.

W E often see small bodies of a specific gravity much exceeding that of water float upon the surface of that fluid. Su:h, for example, are very small grains of sand, fine filings of the metals, and even small sewing-needles.

So extraordinary a phenomenon has not failed to excite the attention of philosophers. It formed a subject of discussion at the last sitting of the Class, and as this remarkable fact is intimately connected with a subject of research upon which I have been long employed, I shall here give an account of some experiments I have made to elucidate the same, which have afforded results of considerable interest.

Suspecting that the presence of air adhering to these small floating bodies, which is generally considered as the cause of their supension, is not indispensably necessary for the success of the experiment, I made the following experiments.

Experiment No. 1.— Having half filled with water a wine-glass one inch and a half in diameter at its edge, I poured on the surface of the water a stratum of sulphuric ether, one inch and a half in thickness; and when the whole was perfectly still, I took a very small sewing-needle with a pair of pincers, which I introduced below the ether, where, holding it horizontally at a small distance from the surface of the water, I let it fall. The needle descended to the water, and there floated on its surface.

Experiment No. 2. — Having melted some tin, I poured it into a spherical wooden box, and, shaking it strongly, the metal in cooling was reduced to powder, which was then sifted.

On examining this powder with a magnifier, it appeared composed of small spherules of different sizes; but these spherules were too small to be distinguished by the naked eye.

I took up on the point of a spatula a very small quantity of this metallic powder, and poured it gently from the height of a quarter of an inch on to the surface of the ether which rested upon the water in the glass.

The powder descended wholly through the ether, and when it arrived at the surface of the water, it remained floating.

Experiment No. 3. — Having poured a large drop of mercury into a china plate, I broke it into a great number of small spherules.

In order to take up and convey these small spherules one by one, I made a small tool or shovel out of a piece of brass wire, five inches long, and about one twentieth of an inch in diameter, bent to a right angle at one of its extremities. This bent part was about a quarter of an inch long, and was hammered flat, sharpened, and made a little concave.

By means of this tool I took up a small spherule of mercury, about one sixtieth of an inch in diameter, which I carefully conveyed into the stratum of ether to the

distance of about one twentieth of an inch from the surface of the water beneath; and there, by a little inclination of the instrument, I caused the spherule of mercury
to roll gently on to the surface of the water.

The spherule descended to that surface, and there remained floating.

When the eye was placed lower than the surface of
the water, and the spherule was observed by looking upwards through the glass, it appeared suspended in a kind
of bag, a little below the level of the surface.

Having placed a second spherule of mercury on the
surface of the water, it immediately moved towards the
former, and, approaching it with an accelerated motion,
fell down into the same cavity, which then became longer; but the two spherules did not unite.

Having placed a third spherule on the surface of the
water, it joined the two others; but the weight of these
three spherules together being too great to be supported
by the kind of pellicle which is formed at the surface of
the water, the bag was broken, and the spherules descended through the water to the bottom of the vessel.

When the experiment was made with a spherule of
mercury a little larger, namely, about the fortieth or fiftieth of an inch, it never failed to break the pellicle of
the water, and to descend through that liquid to the
bottom of the glass. But when the viscidity of the
water was increased by dissolving a small quantity of
gum-arabic in it, still larger spherules of mercury were
supported at the surface of the liquid.

A spherule of mercury of a proper size to be supported by water at its surface, if placed gently there,
would not fail to make its way through the pellicle of
the water, if let fall from too great an height.

All the preceding experiments were repeated with a stratum of essential oil of turpentine, and afterwards with one of oil of olives, placed on the water contained in the glass instead of the ether, and the results were in all respects similar. I thought, however, that the spherules of mercury which were suspended upon the water were rather larger when the surface of the water was covered with oil than with ether; and in the experiments made with the powder of tin poured on the oil, the finest parts of the powder in very small quantity floated on the surface of the oil.

Experiment No. 4. — Having found means to place a stratum of alcohol on the water contained in the glass, so that the two liquids appeared as distinct from each other as when the upper stratum was oil, I poured from a very small height a small quantity of the very fine powder of tin upon the alcohol.

This powder totally descended through the alcohol and the water, without giving the smallest indication of its having been subjected to any resistance at the surface of the latter fluid.

Though this last surface appeared very distinctly to the eye, yet, judging from the manner in which the metallic powder descended to the bottom of the glass, I am disposed to think that it had no existence; and, in fact, it is probable that it was destroyed by the chemical action of the alcohol in contact with the water.

In order to examine more accurately the kind of film which is formed at the surface of the water, I made the following experiment.

Experiment No. 5. — In a cylindrical glass with a solid foot, the diameter of which was fourteen lines, or about an inch and a half English, and ten inches in height, I poured very limpid water to the height of nine inches, and on the water I placed a stratum of ether, three lines or twelfths of an inch in thickness. I then placed on the surface of the water a number of small solid bodies, which remained suspended, such as a small spherule of mercury, some pieces of extremely fine silver wire, two or three lines in length, and a little of the powder of tin. When the whole was perfectly tranquil, I took the glass in both hands, and carefully raising it, I turned it three or four times round its axis with considerable rapidity, keeping it in a vertical position. All the small bodies suspended at the surface of the water turned round along with the glass and stopped when it was stopped; but the liquid water below the surface did not at first begin to turn along with the glass, and its motion of rotation did not cease all at once upon stopping that of the vessel. In fact, all the appearances showed that there was a real pellicle at the surface of the water, and that this pellicle was strongly attached to the sides of the glass so as to move along with it.

Upon examining with a good magnifier, through the stratum of ether, the small bodies which were supported at the surface of the water, the existence of this pellicle could no longer be doubted; more particularly when it was touched with the point of a needle. For in this case all the small bodies were observed to tremble at the same time.

Having left this small apparatus at repose in a quiet chamber until the stratum of ether was entirely evaporated, i examined it again with a magnifier. The surface of the water was precisely in the same state; the small solid bodies were still there, in the same situation, and at the same distances from each other.

When this experiment was made with a cylindrical glass of much larger diameter, the effects of the adhesion of the pellicle of the water to the sides of the vessel were much less sensible with regard to those parts of the same which were situated near the axis. It was dif ficult to prevent the small bodies which floated on the surface of the water from uniting, and when united they often formed masses too heavy to continue to be supported; and, having broken the pellicle of the water, they fell to the bottom of the vessel.

If the particles of water adhere strongly to each other, it appears to me to be a necessary consequence that a kind of pellicle will be formed at the upper surface of the liquid, and even at all its surfaces, whatever may be in other respects the mobility of these particles, or 'rather of the small liquid masses composed of a great number of them, when they are remote from the surface and possess their fluidity without impediment.

When a small solid body, placed on the surface of water, becomes wetted, it immediately descends beneath the pellicle, which no longer opposes its resistance. At this period the viscidity of the water begins to manifest itself in a very different manner, but with infinitely less effect than when it acts at the confines of the liquid. But it is not yet time to inquire into this part of our subject.

With a view to render sensible the resistance which the pellicle of the inferior surface of a stratum of water opposes to a solid body which passes through that stratum by falling freely downwards, I made the following experiment.

Experiment No. 6. — Having filled a small wine-glass to about half its height with very pure mercury, I poured a stratum of water of three lines in thickness upon the mercury, and upon that a stratum of ether of two lines.

When the whole was at rest, I took with the small tool before described a spherule of mercury of about one third of a line in diameter, and let it fall through the stratum of ether.

This spherule, being too heavy to be supported by the pellicle at the superior surface of the water, broke it, and descended through that fluid; but upon its arrival at the inferior surface it was stopped, and remained there, preserving its spherical form.

I moved this spherule with the extremity of a feather, and even compressed it; but it always preserved its form without mixing with the mass of mercury on which it appeared to rest.

It was no doubt the pellicle of the inferior surface of the stratum of water which prevented this contact, and as this pellicle was supported by the mercury on which it rested, I was not at all surprised to find that it could support, without being broken, a spherule of mercury much larger than the pellicle of the superior surface could support.

In order to satisfy myself that the viscidity of the water was the cause of the suspension of this mercurial globule at the bottom of that fluid, I repeated the experiment and varied it by substituting water containing a certain quantity of gum-arabic, in solution, in the place of pure water; and I found, in fact, that much larger spherules were supported when the viscidity of the water was thus augmented.

To prove this fact in another manner, I again varied the experiment, by placing a stratum of ether immediately upon the mercury. The particles of this liquid appear to have very little adhesion to each other; for which reason I imagined that the kind of film that would be formed at its surface must have very little force. The results of my experiment fully confirmed this conjecture.

The very smallest spherules of mercury which I let fall through this liquid seldom failed to mix immediately with the mass of mercury on arriving at its surface, where they entirely disappeared; and I have never succeeded in causing either a spherule of mercury, or the smallest metallic particle, or any other body of greater specific gravity than ether, to swim upon its surface.

The results of the experiment were not perceptibly different when alcohol was substituted in the place of ether.

It is known that ether evaporates very rapidly. Is not this another proof that the particles of this liquid adhere to each other with much less force than those of water? But the following experiment proves this fact in a decisive manner.

Experiment No. 7. — Having half filled a small cylindrical glass with mercury, I placed on the mercury a stratum of ether four lines in thickness, and blew upon the ether with a pair of common bellows.

In less than one minute the ether had disappeared.

The same experiment being made with water, no sensible quantity of this fluid had disappeared in one minute.

The objects which are before our eyes from the earliest periods of our lives seldom employ our meditation, and not often our attention. We see, without surprise, immense masses of dust raised by the winds and carried to great distances; and at the same time we know that every particle of this powder is really a stone, almost three times as heavy as water, and of a size so considerable that its form may be perfectly seen by means of a good microscope.

And we see also, without surprise, that water, which is much lighter than dust, and is composed of particles incomparably smaller, is not carried off by the wind in the same manner.

In order to convince ourselves that the particles of water do strongly adhere to each other, and that they require to do so in order to prevent the greatest confusion in the universe, we need only figure to ourselves the inevitable consequences that would result from the want of such an adhesion.

The particles of water would be raised and carried off by the winds with infinitely more facility than the finest and lightest dust. Every strong breeze setting in from the ocean would bring with it a great inundation. Navigation would be impossible, and the banks of all the seas, lakes, and large rivers would be uninhabitable.

The adhesion of the particles of water to each other is the cause of the preservation of that liquid in masses. It covers the surface with a very strong pellicle, which defends and prevents it from being dispersed by the winds. Without this adhesion, water would be more volatile than ether, and more fugitive than dust.

But the adhesion is also the cause of other phenomena, which are of the greatest importance in the phenomena of nature.

The viscidity which results from the mutual adhesion of the particles of water renders this fluid proper to hold all kinds of bodies in solution, as well the most

heavy as the lightest, provided always that they be reduced to very minute particles.

I have found, by a calculation founded on facts which
appear to me to be decisive, that a solid spherule of
pure gold, of the diameter of $\frac{1}{300,000}$ of an inch,
would be suspended in water by the effect of its viscidity, even though this small body should be completely
wetted and submerged in a tranquil mass of the fluid.

This viscidity, or want of perfect fluidity, which
causes it to hold every kind of substance in solution,
renders it eminently proper to become the vehicle of
nourishment to plants and animals; and we accordingly
see that it is exclusively employed in this office.

If the adhesion of the particles of water to each other
were to cease, and the fluidity of this body were to
become perfect, every living being would perish by
inanition.

May I be permitted to remark the simplicity of the
means employed by Nature in all her operations!

May I be permitted to express my profound admiration and adoration of the Author of so many wonders!

CONTINUATION

EXPERIMENTS AND OBSERVATIONS

ON THE

ADHESION OF THE PARTICLES OF LIQUIDS TO EACH OTHER.

B EFORE proceeding with the account of my experiments, I shall take the liberty of going back
to a distant period, and of describing to the Class an
occurrence which first fixed my attention on this subject
and led me to engage in these researches.

Being occupied in the year 1786 with a series of
experiments on the oxygen gas which is disengaged from
water when this liquid mixed with various solid substances is exposed to the action of the sun's rays, among
the substances employed in my investigation was a
quantity of raw silk, wound from the cocoon on purpose for this experiment, in a single thread, just as it is
produced by the silkworm.

It being necessary for completing my calculations that
I should determine with precision the amount of the
surface of this thread, which was almost two leagues in
length, and which weighed in the air only about 20
grains Troy, and having no means of measuring directly the exact diameter of the thread, I undertook to
calculate it from the known length of the thread and
the specific gravity of the substance.

It was in weighing this substance in water to ascertain
its specific gravity, that I encountered difficulties which

for a long time seemed to me insurmountable, but by the exercise of patience and due precaution, I succeeded, after somewhat long and difficult labour, in accomplishing my object.

Those who are in the habit of making delicate experiments with the hydrostatic balance will conjecture immediately, before I have time to say it, that it was the air which remained obstinately attached to the surface of the silk when I weighed it in water, which rendered this operation so difficult.

I do not wish to abuse the patience of the Class by giving it a detailed account of all the means I was obliged to try before finding an efficient remedy for this inconvenience; it will suffice to say, that the silk was weighed finally in water, and with precision, and I will here add in passing, that the specific gravity of this substance was found to be to that of water as 1734 is to 1000. The following phenomenon, however, which I noticed while weighing the silk in water, struck me forcibly.

The silk being in the form of a skein about 6 inches long, and tied loosely in order to allow the water readily to enter among all the threads, it was hung from one of the arms of an excellent hydrostatic balance in a large mass of distilled water which had previously been freed from air by long boiling.

The weight of the silk in this situation having been determined, it was then placed, by means of silver pincers, and without taking it from the water, into a small glass vessel of oval form, about 2 inches in diameter and 3 inches long, and weighed again.

The weight of the silk when weighed in the small glass vessel was sensibly greater than when it was weighed out of the vessel in the same large amount of water, and on repeating the experiment several times, the result was always the same.

The following appeared to me to be a satisfactory explanation of this phenomenon.

As silk is one of those substances which can be wet by water, it is evident that the particles of the liquid which were in immediate contact with the surface of the thread must have remained attached to it. These particles, having become thus fixed and immovable, were in contact with other particles which still enjoyed their freedom of motion, and these particles again were in contact with others farther from the silk, and so on. Now, as the fluidity of various liquids is evidently very different, it is more than probable that no liquid possesses perfect fluidity; consequently water does not: and if any force whatever is needed to separate its particles and make them move on each other, it is evident that, in this case, if a solid body specifically heavier than water were plunged into a quiet mass of this liquid, there should be an apparent loss of weight on account of the viscosity of the liquid, and this loss of weight would be in proportion to the extent of the surface of the body.

If, for example, the body is suspended by a thread, the thread will not support all the excess of the weight of the body over the weight of a mass of the liquid equal to the volume of the solid body; for a part of this excess would be supported by the adhesion which exists among those particles of the liquid which are in contact with the particles attached to the surface of the body.

This appeared to me too evident to need demonstration or even further explanation.

In one of the experiments in question, the silk being suspended freely in the water, it was in contact with this liquid by a very great surface (about 550 square inches), and its loss of weight on account of the viscosity of the liquid was very sensible; but when it was weighed in a small vessel which had been previously counterpoised very exactly in the water, the arm of the balance supported all the excess of the weight of the silk over that of an equal bulk of water without any diminution.

The results of these experiments have furnished data for calculating, with sufficient exactness, the degree of force with which particles of water adhere to each other, when it is a question of causing them to move one upon the other at a temperature of about 60° F., and I found it to be such that a solid body specifically heavier than water, having a surface equal to 368 square inches (English), when submerged in water, ought to lose in weight, on account of the viscosity of the water, an amount equal to 1 grain Troy.

The discovery of this fact has put me in position, not only to prove that all bodies in nature, the heaviest as well as the lightest, can be suspended and supported in still water, on account of its viscosity, provided they are reduced to a sufficiently small size, but also to determine by calculation that a solid spherule of gold about $\frac{1}{300,000}$ of an inch in diameter would remain suspended in this manner, as I have already announced to the Class in the memoir read at the session held on the 16th of June, the past year (1806).*

Having announced facts as remarkable as these, I refrained from entering into more minute details. I did not even think it necessary to observe that, even if I should have deceived myself somewhat in my estimation of the force of cohesion of the particles of water, still, if

* This calculation will be found in a note at the end of this paper (page 503).

it be only granted (and this cannot be called into question) that the fluidity of this liquid is not absolutely perfect, but that a certain amount of force is necessary, no matter how small it may be, to separate the particles from each other, this alone will be sufficient to establish all that I have asserted with regard to the necessary consequences of the adhesion of the particles of liquids to each other.

It would only be a question in each case of supposing that a solid body immersed in any liquid be reduced to a sufficiently small size, and it could be proved that it must necessarily remain suspended there. But it is easy to see that the greater the force of cohesion between the particles of a liquid, the more capable this liquid becomes of holding in suspension foreign bodies of all sorts.

Water appeared to me to possess this quality to a remarkable degree; and it is certain that if there had been need of a vehicle for the nourishment of plants and animals, one capable of holding in suspension and of transporting from one place to another all sorts of substances, very different in weight and size, *without affecting them chemically*, it would never have been possible to find one more fitted for this purpose than water.

Is it not probable that this is one of the principal designs of the existence of this liquid in the economy of Nature?

Being accustomed to see traces of great wisdom and of admirable simplicity in all those dispositions of Nature which I have been able to comprehend, I have been perhaps too much inclined, in my ordinary meditations, to admit this conclusion. I must, however, confess that the facts which have seemed to me to render it probable have made a deep impression upon me.

Having found that the adhesion of the particles of water to each other is so considerable, I was not slow to perceive that this adhesion ought to manifest itself in a very peculiar and sensible manner at the surface of the liquid ; and it was then that I saw clearly that it might be possible to explain in a satisfactory manner several phenomena which have always been regarded as difficult of explanation ; as, for example, the suspension of heavy bodies of small size which appear to float on the surface of the water ; the concave form taken on by the surface of water when confined in a small vessel ; the change of this form into convex when, the vessel having been filled to the brim, more liquid is added ; the suspension of liquids in capillary tubes, etc.

I wrote, in the winter of 1800, a memoir on this subject, which I afterwards showed to several persons, among others to Professor Pictet, of Geneva, when he was in London in 1801 ; also to Sir Charles Blagden. The reason for not publishing it at that time was that I needed the assistance of profound analysis in order to finish it.

When I arrived at Paris in the spring of 1802, I took advantage of this occasion to consult the greatest geometers of the century on the embarrassing question which stood in my way. Four persons now present in this Assembly can remember the circumstance. I desired to know the form which the vertical middle section of a drop of water, or other liquid substance, would take if placed on a plane horizontal surface, supposing that the liquid was restrained solely by the resistance of a pellicle exerting a given force on its surface.

The problem appeared very simple, but its solution is extremely difficult. I did not know at that time

that Segner had attempted to solve it.   I had no knowledge of the memoir which he published on this subject
more than forty years ago in the first volume of the
memoirs of the Royal Society of Göttingen.

If I recall these facts, it is simply to prove that I have
not taken the liberty of occupying the attention of the
Class with a subject as difficult as a research on the
adhesion of the particles of liquids, and the various phenomena dependent upon it, without previous meditation; and to prove that the opinions which I have ventured to bring before it were adopted a long time since,
and have been often examined before being announced.

I have most certainly nothing more at heart than to
preserve the esteem and deserve the confidence of every
member of this illustrious Assembly.   The favour which
they have shown me in giving me the right to sit among
them, which I regard in the light of a very distinguished
honour, as well as my respect for their talents, makes
me hold it as a sacred duty never to abuse their attention with trifles, or crude ideas, or opinions formed in
haste and ill-digested.

If I ventured to speak of the *pellicle* of the water, it is
because I really believed in its existence; and I believe
in it still, and more firmly than ever.

Allow me to recall to the Class the phenomena which
have seemed to me to indicate its existence.

When I have seen little steel needles float on the surface of this liquid without sinking into it, and even
without being wet; when I have seen little globules of
mercury roll about on the surface of the water, then, coming to rest, and sinking to a certain depth in the liquid
without, however, being wet by it, remain as though
suspended in a small pocket; when I have seen diminu-

tive spider-like insects, with long legs, run about over
the water without their feet sinking into the liquid, or
even being wet by it; when I have seen several minute
bodies at a distance from each other, resting upon the
surface of the water contained in a small vessel, tremble
every time that the surface of the water was touched
with the point of a needle, — I have been unable to
doubt the existence of a resisting surface, a sort of pellicle on the surface of the liquid.

There is another phenomenon which seems to me to
furnish a demonstrative proof of the existence of this resisting surface. When water is heated in any vessel,
as soon as the liquid begins to become warm a considerable quantity of air is disengaged in the form of spherical bubbles, larger or smaller, which, passing through the
liquid from below upwards, escape into the air. Now
it very often happens that these little bubbles, after
having traversed the liquid with great rapidity, are
stopped all of a sudden when they have nearly reached
the surface.

What is it that stops these bubbles if not a resisting
pellicle at the surface of the liquid?

I endeavoured, but in vain, to explain these facts, by
calling to my aid the atmospheric air. I saw clearly,
as I observed the little globule of mercury situated in
its little pocket, which sank sensibly lower than the
level surface of the water, and which was scarcely large
enough to hold the globule, — I saw, I say, that the film
of air, which we might suppose still attached to the surface of the globule (if such a film really existed), could
not be thick enough to buoy up this heavy body and
make it float, hydrostatically, on the surface of the water.
But when to the testimony of these experiments and to

that of several others of the same sort, which can readily be performed, is added the evidence furnished by the certain knowledge which we have of a strong adhesion which exists among the particles of water, and of the effect which this adhesion must necessarily produce at the surface of the liquid, it seems to me impossible to call into doubt the existence of a resisting layer extremely thin at the surface of the water.

In announcing the existence of a sort of pellicle at the surface of liquids, I was far from thinking that it was a new idea. I am aware that several philosophers, and among others one of our celebrated colleagues, M. Monge, had suspected it before I did, but I think that I was the first to devise and perform decisive experiments which have established the fact beyond doubt ; and it is certain that the observations that I have published on the effect which the adhesion of the particles of liquids to each other must have in the economy of Nature, have been borrowed from no one.

If the existence of a resisting film at the surfaces of liquids has just been confirmed by the results of the learned analytical researches of one of our celebrated colleagues, I ought, without doubt, to regard this event as a proof very flattering to me, that my conjectures on this subject were not ill founded.

I know that there are some persons who imagine that the results of the calculations of the illustrious author of the *Mécanique Céleste* on the rising of liquids in capillary tubes are opposed to the opinions which I have published on the adhesion of the particles of liquids to each other ; but, as far as I have been able to understand the data on which these calculations are founded, it seems evident to me that the *attraction* with which M. La Place supposes the particles of the liquid to be endowed, does not differ essentially from the force which I have designated by the name *adhesion;* and with regard to the pellicle, of which I have often spoken, since the calculation of this learned geometrician and philosopher is founded on the supposition that the mutual attractions of the particles of the liquid situated a certain distance below the surface of the liquid do not contribute in any way to the rising of the liquid in a capillary tube, nor to any other similar effects which he has considered, it seems to me that the calculations of M. La Place simply relate to the force of cohesion of the layer of particles at the surface, or, in other words, to the pellicle in question.

I must, however, confess that I am not sufficiently well versed in the higher geometry to understand fully the calculations of M. La Place on this subject ; and I shall take good care not to pass judgment on them. One must have, without doubt, a very profound acquaintance with analytical methods to feel the force of his demonstrations ; but I have such a high opinion of the talents of this man, learned and worthy of esteem both as a geometrician and as a natural-philosopher, that I am always inclined to receive his opinions in matters of science (as well as on every other subject) with the greatest deference.

The researches to which I have sought to call the attention of philosophers would be, no doubt, of less importance if it was merely a question of the explanation of a few facts, isolated and of little utility in their applications ; but the adhesion of the particles of liquids to each other is probably the cause of a great variety of phenomena which affect us intimately ; and for this reason the subject must be regarded as very interesting. It seems to me that it is to this adhesion, and to the changes of its intensity, arising from different circumstances, that we must look for the proximate cause of the growth of plants and of animals.

I have already observed that the strong force of adhesion existing among the particles of water renders this liquid peculiarly fitted to serve as the vehicle for conveying nourishment to all living beings; and I think that I can show that this force of adhesion can be very much decreased, that this actually happens very often, and that one of the necessary consequences of such a diminution would be the deposition or precipitation of foreign matters which this liquid holds in suspension on account of its viscosity.

If water ascends as sap in the capillary tubes of trees as far as the leaves, it is possible that it there undergoes some change, or that it there receives some addition, which diminishes its viscosity, and disposes it in this way to deposit matters which it holds in suspension and which contribute to the growth of the plant.

If, during the digestion of food which takes place in the stomach, water, aided perhaps by the gastric juice, seizes at first upon nutritive particles of every sort which are there found and holds them in suspension, is it not possible that this liquid thus loaded, being mixed subsequently with a portion of bile, at its entrance into the intestinal canal, is by this means rendered less viscous and consequently better fitted to pass easily through the lacteal veins, and more disposed to yield up the nutritive particles as it enters into circulation?

In case this conjecture be well founded, we ought, undoubtedly, to find that a mixture of bile with water would diminish, to a sensible extent, the viscosity of this latter liquid ; and I actually found by experiments which I shall have the honour of laying before the Class at some future time in detail, that mixing 1 part of bile with 1000 parts of water diminishes the adhesion of the particles of water to each other nearly one third, that is to say, in the ratio of 23 to 16 ; and that if 1 part only of bile be mixed with 30,000 parts of water, the diminution is still very apparent. In a mixture of 1 part of bile with 300 parts of water, the adhesion in question is reduced almost one half.

Milk is a liquid which seems to be already elaborated and fitted to serve as nourishment for animals ; now I have found, by decisive experiments, that the adhesion among the particles of this animal fluid is less than that among the particles of water in the proportion of 13 to $19\frac{3}{4}$.

The adhesion of the particles of urine to each other varies considerably. I have found it from $13\frac{1}{4}$ to 16, that of water being $19\frac{3}{4}$.

Many persons have endeavoured to discover the nature of diseases by the examination of the urine ; no one, however, as far as I know, has ever proposed to measure the force of adhesion of its particles to each other, a thing as easy to determine as it is useful to ascertain.

How interesting it would be to know the force of adhesion of the particles of the gastric juice, of the pancreatic juice, of the lymph, and of the blood, both in health and in the various diseases ! Of how great importance would a knowledge of these facts be to the physiologist and to the physician !

How useful it would be for those who study vegetable physiology to know the adhesive force of the parti-

cles of the sap when rising and when descending, and
that too in the various seasons !

How much light would be thrown on all chemical
operations taking place in the wet way, if we could estimate exactly the force of adhesion existing among the
particles of the various liquid agents which there come
into play !

How many wonderful reactions there are which seem
to depend on such a simple thing as the imperfect fluidity of liquids !

It seems to me that the facts which I have just announced are of such a character as to excite all our curiosity, and I hasten to make them public in order to
induce all those who cultivate the sciences to assist me
in these interesting researches.

I feel deeply that all that a single individual can effect
by his own labours during the course of his short life,
in extending the vast domain of science, is unfortunately
a very small matter. It is only by the simultaneous
efforts of a large number of men with good heads and
skilled hands that we can hope to see a sensible advance
of this great enterprise, of which men will never see the
completion ; and for this reason those who with true
love for science take more delight in seeing its progress
than in obtaining the pleasures of gratified vanity, ought
rather to seek to associate with themselves a great number of zealous and skilful co-labourers than to endeavour to do everything themselves.

Happily for the progress of this new branch of research, the apparatus to be employed is portable, and of
great simplicity, and the experiments are as easy to perform as their results are decisive and satisfactory. In
general, the only thing needed will be an inverted si-

phon, with one of its capillary arms provided with a scale for measuring the height of the liquid in this arm above the level of the top of the column in the large arm ; for it is now well established, by the results of conclusive experiments, that the heights to which various liquids rise in the same capillary tube are in proportion to the degrees of force with which the particles of the several liquids adhere to each other.

The experiments for determining the diminution produced in this force by a given increase of temperature demand more complicated apparatus and special care. The apparatus which I have used in this research is before the Class. Since, in the present state of the physical sciences, we can hardly flatter ourselves that we are able to take a single step in advance, except with the aid of instruments devised with care and executed with the utmost precision, I always regard it as a duty to afford the Class an opportunity of judging, by its own observation, of the excellence of those used by me in such new experiments as I have the honour of describing to the Class.

I will show, presently, the way in which this apparatus is used, and I will give to the Class, at a subsequent sitting, the account of the results of the experiments in which it has been employed.

I will conclude this memoir with some observations on a very important point, which should, perhaps, be still further elucidated.

I have shown how I have proceeded in measuring that sort of adhesion of the particles of water to each other which produces the viscosity of this liquid, that is, the force which must be exerted in order to cause those particles to move on each other ; but we must by no means suppose that the same force will suffice to separate these same particles from each other, when two of them, which are in contact, are drawn in opposite directions along the line passing through their centres.

The very considerable weight of a drop of water which remains suspended from a solid body shows evidently that this latter force is incomparably greater than I have found the former to be. Now, when a solid body rests on the surface of a liquid, it cannot penetrate into it without breaking the layer of particles which are at this surface, and which may be considered as forming a sort of pellicle ; in order to break this pellicle, it is evidently necessary to separate the particles which compose it, by compelling them to withdraw from each other directly or nearly in lines passing through their centres, and it is for this reason that small solid bodies specifically heavier than water remain on the surface of this liquid without penetrating into it.

Likewise, when water issuing from the upper extremity of the shortened capillary tube of an inverted siphon forms a small hemispherical mass, resting on the end of the tube and attached to its walls, the convex surface of this small mass of liquid is formed by a layer of particles which resist, with all the force of their attraction for each other, every effort tending to separate them ; and it is the resistance of this single layer of particles, or of several layers resting immediately one upon another, and together forming a sort of very thin pellicle, which sustains the entire weight of the column of water in the other arm of the siphon, which is situated above the level of the surface of this small mass of liquid.

I have recently established this fact by means of an experiment, which I regard as decisive.

Having found a way of placing in the middle of this small hemispherical mass of water little isolated solid bodies which displaced a great part of the liquid without being wet by that which remained, this arrangement produced no change, either in the exterior form or in the dimensions of the little hemisphere, or in the force displayed in resisting the pressure of the more elevated column of water in the other arm of the siphon.

NOTE.

(See page 491.) The following calculation, which is neither long nor difficult to follow, may be of service in understanding what has just been advanced.

A cubic inch of water, English measure, weighs 253.175 grains Troy ; consequently a spherical mass of this liquid 10.8233 inches in diameter, and which would have a surface of 368 square inches, would weigh 168,060 grains. And since the specific gravity of gold is to that of water as 192,581 is to 10,000, a sphere of gold of the same diameter would weigh 3,236,525 grains in vacuo.

Now, a similar sphere weighed in water would lose of its weight in vacuo an amount equal to the weight of a mass of water of a volume equal to that of the sphere. It would weigh, therefore, 3,236,525 — 168,060 = 3,068,465 grains, a deduction being made for the slight amount of its weight which it would lose on account of the viscosity of the liquid.

Since the surface of the globe is equal to 368 square inches, we see, from the result of the experiment of which we have just given an account, that this decrease of weight must be exactly one grain. Consequently the sphere suspended in water will weigh on the beam of the balance only 3,068,464 grains, and it will lose $\frac{1}{3,068,465}$ of its weight on account of the viscosity of the liquid.

Let us suppose, now, that the diameter of this sphere were 10 times as small, or 1.08233 inches, and let us see according to what law the effect produced on the viscosity of the liquid will be increased by this diminution of volume.

The volumes and consequently the weights of spheres of different diameters being as the cubes of those diameters, while their surfaces are as the squares of the same lines, it is evident that the weight of the small sphere mentioned above must be 1000 times less than the weight of the large sphere (the cube of 10 being 1000 and its square 100), consequently the smaller sphere ought to weigh in water only 3068.465 grains (deduction being made for the effect of the viscosity of the liquid), and its surface would be 3.68 square inches.

Since the diminution of weight which was due to the viscosity of the liquid was only 1 grain when the surface of the sphere was 368 square inches, it is evident that this diminution ought to be 100 times smaller, or $\frac{1}{100}$ of a grain, in the case of the smaller globe which has 100 times less surface ; now $\frac{1}{100}$ of a grain in the case of a body which weighs only 3068.465 grains is $\frac{1}{306,846.5}$ of the real weight of the body in water, and by this amount the weight will be diminished on account of the viscosity of the liquid. This quantity is precisely 100 times more considerable, relatively to the weight of the body in water, than we have found it to be in the case of a body 10 times as large.

Hence we may conclude (and this can also easily be shown by a rigorous demonstration) that when a solid sphere heavier than water is submerged in this liquid, the decrease of weight due to the viscosity of the liquid is inversely proportional to the diameter of the sphere.

For example, if, when the diameter of the sphere was 10.8233 inches, the decrease of its weight in water due to the viscosity of that liquid is to its weight in the same liquid in the ratio of 1 to 3068465 ; —

When the diameter is reduced to	The diminution of weight due to the viscosity of the liquid will be to the weight of the body in water
1.08233 of an inch	as 1 to 306846.5
0.108233	" 1 " 30684.65
0.0108233	" 1 " 3068.465
0.00108233	" 1 " 306.8465
0.000108233	" 1 " 30.68465
0.0000108233	" 1 " 3.068465
0.00000108233	" 1 " 0.3068465

And in this last case it is evident that the minute body must of necessity remain suspended in the liquid.

I know very well that these long numerical calculations must seem superfluous to geometers accustomed to algebraic calculation ; but many persons who are unacquainted with algebra desire to have brought within their reach satisfactory proofs of the truth of a conclusion which is given out to them as certain, especially when it is to serve as the foundation of a theory which is applied to very interesting phenomena.

As soon as it is shown that the diminution of weight which a sphere plunged into water experiences as a result of the viscosity of this liquid is inversely proportional to the diameter of the sphere, and when we know the amount of this diminution in a particular case, it is easy to determine, by a very simple calculation, what will be the diminution taking place in another case.

For example, we can determine what will be the diameter of the largest sphere of gold which will remain suspended in water on account of the viscosity of this liquid. Proceeding in this manner, I found this diameter equal to $\frac{1}{283,505}$ of an inch.

REFERENCES TO RUMFORD'S OWN WORKS

(References are to the present edition)

1. "An Inquiry concerning the Nature of Heat and the Mode of Its Communication." *Collected Works of Count Rumford,* Vol. I, p. 323.
2. "Experiments on Cooling Bodies (Experimental Investigations concerning Heat, Section II)," Vol. II, p. 1.
3. Vol. I, pp. 161–163.
4. "Propagation of Heat in Fluids," Vol. I, p. 119.
5. "An Account of a Curious Phenomenon Observed on the Glaciers of Chamouny," Vol. II, p. 31.
6. "Experiments tending to show that Heat is communicated through Solid Bodies, by a Law which is the same as that which would ensue from Radiation between the Particles," Vol. II, p. 9.
7. "Of the Slow Progress of the Spontaneous Mixture of Liquids Disposed to Unite Chemically with each Other," Vol. II, p. 74.
8. "The Specific Gravity of Silk," II Vol. III.
9. "Account of some New Experiments on Wood and Charcoal," Vol. II, p. 162.
10. "Researches upon the Heat Developed in Combustion and in the Condensation of Vapours," Vol. II, p. 80.
11. "Chimney Fireplaces," Vol. II, p. 221.
12. "An Experimental Inquiry concerning the Source of the Heat which is Excited by Friction," Vol. I, p. 1.
13. "On the Propagation of Heat in Various Substances," Vol. I, p. 53.
14. "An Inquiry concerning the Weight Ascribed to Heat," Vol. I, p. 27.
15. "On the Propagation of Heat in Various Substances," Part II, Vol. I, p. 85.

FACTS

OF

PUBLICATION

EXPERIMENTS ON COOLING BODIES (EXPERIMENTAL
INVESTIGATIONS CONCERNING HEAT, SECTION II)

Read at the Institut de France, 10 Floréal, An 12 (April 30, 1804).

Mémoires de la classe des Sciences, Mathématiques et Physiques de l'Institut de France, (Paris: Baudouin, Imprimeur de l'Institut de France, January 1806), VI, 97–105.

A Journal of Natural Philosophy, Chemistry and the Arts: Illustrated with Engravings, edited by William Nicholson (London, 1805), XII, 70–75.

The Complete Works of Count Rumford (Boston: American Academy of Arts and Sciences), II (1873), 137–144.

EXPERIMENTS TENDING TO SHOW THAT HEAT IS COMMUNICATED THROUGH SOLID BODIES, BY A LAW WHICH IS THE SAME AS THAT WHICH WOULD ENSUE FROM RADIATION BETWEEN THE PARTICLES (EXPERI-MENTAL INVESTIGATIONS CONCERNING HEAT, SECTION III)

Read at the Institut de France, 17 Floréal, An 12 (May 7, 1804).

Mémoires de la classe des Sciences, Mathématiques et Physiques de l'Institut de France (Paris: Baudouin, Imprimeur de l'Institut de France, January 1806), VI, 108–122.

A Journal of Natural Philosophy, Chemistry and the Arts: Illustrated with Engravings, edited by William Nicholson (London, 1805), XII, 154–164.

The Complete Works of Count Rumford (Boston: American Academy of Arts and Sciences), II (1873), 144–157.

DESCRIPTION OF A NEW INSTRUMENT OF PHYSICS

Read at the Institut de France, 28 Ventôse, An 12 (March 19, 1804).

Mémoires de la classe des Sciences, Mathématiques et Physiques de l'Institut de France (Paris: Baudouin, Imprimeur de l'Institut de France, 1806), VI, 71–78.

Neues Allegemeines Journal der Chemie, edited by Adolf Ferdinand Gehlen (Berlin, 1804), II, 657–663.

AN ACCOUNT OF A CURIOUS PHENOMENON OBSERVED ON THE GLACIERS OF CHAMOUNY; TOGETHER WITH SOME OCCASIONAL OBSERVATIONS CONCERNING THE PROPAGATION OF HEAT IN FLUIDS

Read before the Royal Society, December 15, 1803.

Philosophical Transactions of the Royal Society of London, 94 (London, 1804), 23–29.

Bibliothèque Britannique (Science et Arts), edited by Auguste Pictet, Charles Pictet, and F. G. Maurice (Geneva, 1804), XXVI, 3–13; remarks by P. Prevost, pp. 13–28.

A Journal of Natural Philosophy, Chemistry and the Arts: Illustrated with Engravings, edited by William Nicholson (London, 1804), IX, 207–212.

Annalen der Physik, begun by F. A. C. Gren, continued by L. W. Gilbert (Halle, 1804), XVIII, 361–369.

Le Comte de Rumford, *Mémoires sur la Chaleur* (Paris: Chez Firmin Didot, 1804), pp. 156–166 (translated by Auguste Pictet).

The Complete Works of Count Rumford (Boston: American Academy of Arts and Sciences), II (1873), 251–257.

AN ACCOUNT OF SOME NEW EXPERIMENTS ON THE
TEMPERATURE OF WATER AT ITS MAXIMUM DENSITY

Read at the Institut de France, July 15, 1805.

*Mémoires de la classe des Sciences, Mathématiques et Physiques
de l'Institut de France* (Paris: Baudouin, Imprimeur de l'Institut
de France, 1806), VII, 78–97.

*A Journal of Natural Philosophy, Chemistry and the Arts:
Illustrated with Engravings*, edited by William Nicholson
(London, 1805), XI, 225–235.

Annalen der Physik, begun by F. A. C. Gren, continued by
L. W. Gilbert (Halle, 1805), XX, 369–383.

Bibliothèque Britannique (Science et Arts), edited by Auguste
Pictet, Charles Pictet, and F. G. Maurice (Geneva, 1806),
XXXIV, 113–120; observations by Auguste Pictet.

The Complete Works of Count Rumford (Boston: American
Academy of Arts and Sciences), II (1873), 258–273.

INQUIRES CONCERNING THE MODE OF THE PROPAGA-
TION OF HEAT IN LIQUIDS

Read at the Institut de France, June 9, 1806.

Bibliothèque Britannique (Science et Arts), edited by Auguste
Pictet, Charles Pictet, and F. G. Maurice (Geneva, 1806),
XXXII, 123–141.

*A Journal of Natural Philosophy, Chemistry and the Arts:
Illustrated with Engravings*, edited by William Nicholson
(London, 1806), XIV, 353–363; translated from the French by
W. Cadell.

The Complete Works of Count Rumford (Boston: American
Academy of Arts and Sciences), II (1873), 274–289.

OF THE SLOW PROGRESS OF THE SPONTANEOUS MIX-
TURES OF LIQUIDS DISPOSED TO UNITE CHEMICALLY
WITH EACH OTHER

Read at the Institut de France, March 29, 1807.

Mémoires de la classe des Sciences, Mathématiques et Physiques de l'Institut de France (Paris: Baudouin, Imprimeur de l'Institut de France, 1807), VIII, ii, 100–115.

The Complete Works of Count Rumford (Boston: American Academy of Arts and Sciences), II (1873), 318–323.

DESCRIPTION OF A NEW CALORIMETER (RESEARCHES UPON THE HEAT DEVELOPED IN COMBUSTION AND IN THE CONDENSATION OF VAPOURS, SECTION I)

Read before the Institut de France, February 24, 1812.

Bibliothèque Britannique (*Science et Arts*), edited by Auguste Pictet, Charles Pictet, and F. G. Maurice (Geneva, 1812), LI, 3–17, 97–116.

A Journal of Natural Philosophy, Chemistry and the Arts: Illustrated with Engravings, edited by William Nicholson (London, 1812), XXXII, 105–125.

The Philosophical Magazine, comprehending the Various Branches of Science, the Liberal and Fine Arts, Agriculture, Manufactures and Commerce, edited by Alexander Tilloch (London: T. Cadell, Jr. and W. Davies, 1813), XLI, 285–297, 434–439.

Recherches sur la Chaleur développée dans la Combustion et dans la Condensation des Vapeurs (Paris: Chez Éverat, 1812), pp. 3–29; 8ᵛᵒ, 7½ sheets (1000 copies).

The Complete Works of Count Rumford (Boston: American Academy of Arts and Sciences), II (1873), 370–387.

EXPERIMENTS MADE WITH SPIRIT OF WINE, ALCOHOL, AND SULPHURIC ETHER (RESEARCHES UPON THE HEAT DEVELOPED IN COMBUSTION AND IN THE CONDENSATION OF VAPOURS, SECTION III)

Read at the Institut de France, November 30, 1812.

The Philosophical Magazine, comprehending the Various Branches of Science, the Liberal and Fine Arts, Agriculture, Manufactures and Commerce, edited by Alexander Tilloch (London: T. Cadell, Jr. and W. Davies, 1813), XLI, 439–444; XLII, 296–307.

Recherches sur la Chaleur développée dans la Combustion et dans la Condensation des Vapeurs (Paris: Chez Éverat, 1813), pp. 30–73.

The Complete works of Count Rumford (Boston: American Academy of Arts and Sciences), II (1873), 387–417.

ON THE QUANTITY OF HEAT DEVELOPED IN THE CONDENSATION OF THE VAPOUR OF WATER (RESEARCHES UPON THE HEAT DEVELOPED IN COMBUSTION AND IN THE CONDENSATION OF VAPOURS, SECTION X)

The Philosophical Magazine, comprehending the Various Branches of Science, the Liberal and Fine Arts, Agriculture, Manufactures and Commerce, edited by Alexander Tilloch (London: T. Cadell, Jr. and W. Davies, 1814), XLIII, 64–69 (where this paper is said to have been read as a supplement to "Experiments made with Spirit of Wine . . .").

Recherches sur la Chaleur développée dans la Combustion et dans la Condensation des Vapeurs (Paris: Chez Éverat, 1813), 74–83.

The Complete Works of Count Rumford (Boston: American Academy of Arts and Sciences), II (1873), 417–424.

EXPERIMENTS AND OBSERVATIONS ON THE COOLING OF LIQUIDS IN VESSELS OF PORCELAIN, GILDED AND NOT GILDED

Read at the Institut de France, August 10, 1807.

Mémoires de la classe des Sciences, Mathématiques et Physiques de l'Institut de France (Paris: Baudouin, Imprimeur de l'Institut de France, 1807), VIII, i, 249–260.

The Complete Works of Count Rumford (Boston: American Academy of Arts and Sciences), II (1873), 241–250.

ON THE CAPACITY FOR HEAT OR CALORIFIC POWER OF VARIOUS LIQUIDS

The Philosophical Magazine, comprehending the Various Branches of Science, the Liberal and Fine Arts, Agriculture, Manufactures and Commerce, edited by Alexander Tilloch (London: T. Cadell, Jr. and W. Davies, 1814), XLIII, 212–218 (where this paper is said to have been read as a supplement to that on "Experiments made with Spirit of Wine . . .").

Recherches sur la Chaleur développée dans la Combustion et dans la Condensation des Vapeurs (Paris: Chez Éverat, 1813), 84–98.

The Complete Works of Count Rumford (Boston: American Academy of Arts and Sciences), II (1873), 425–434.

OBSERVATIONS RELATIVE TO THE MEANS OF INCREASING THE QUANTITIES OF HEAT OBTAINED IN THE COMBUSTION OF FUEL

Journal of the Royal Institution of Great Britain (London: T. Cadell, Jr. and W. Davies), I (1802), 28–33.

Bibliothèque Britannique (Science et Arts), edited by Auguste Pictet, Charles Pictet, and F. G. Maurice (Geneva, 1801), XVIII, 313–316.

The Annual Register, or a View of History, Politics and Literature, (London: Baldwin, Cradock and Joy, 1802), XLIII, 467–470.

The Complete Works of Count Rumford (Boston: American Academy of Arts and Sciences), II (1873), 345–351.

ACCOUNT OF SOME NEW EXPERIMENTS ON WOOD AND CHARCOAL

Read before the First Class of the French Institute, December 30, 1811.

A Journal of Natural Philosophy, Chemistry and the Arts: Illustrated with Engravings, edited by William Nicholson (London, 1812), XXXII, 100–105.

Bibliothèque Britannique (Science et Arts), edited by Auguste Pictet, Charles Pictet, and F. G. Maurice (Geneva, 1812), LI, 220–232.

Annalen der Physik, begun by F. A. C. Gren, continued by L. W. Gilbert (Halle, 1813), XLV, 142–149.

Published separately under the title *Recherches sur les Bois et le Charbon* (Paris, 4to, 8 sheets, 1812, 125 copies; 8vo, 8 sheets, 1813, 1000 copies).

The Complete Works of Count Rumford (Boston: American Academy of Arts and Sciences), II (1873), 362–369.

INQUIRIES RELATIVE TO THE STRUCTURE OF WOOD, THE SPECIFIC GRAVITY OF ITS SOLID PARTS, AND THE QUANTITY OF LIQUIDS AND ELASTIC FLUIDS CONTAINED IN IT UNDER VARIOUS CIRCUMSTANCES; THE QUANTITY OF CHARCOAL TO BE OBTAINED FROM IT; AND THE QUANTITY OF HEAT PRODUCED BY ITS COMBUSTION

A Journal of Natural Philosophy, Chemistry and the Arts: Illustrated with Engravings, edited by William Nicholson (London, 1813), XXXIV (supplement), 319–335 (where this paper is said to have been read before the First Class of the French Institute, September 28 and October 5, 1812); XXXV, 95–117.

Bibliothèque Britannique (Science et Arts), edited by Auguste Pictet, Charles Pictet, and F. G. Maurice (Geneva, 1812), LI, 299–329; (1813), LII, 35–53.

Annalen der Physik, begun by F. A. C. Gren, continued by L. W. Gilbert (Halle, 1813,) XLV, 1–41.

The Complete Works of Count Rumford (Boston: American Academy of Arts and Sciences), II (1873), 435–483.

CHIMNEY FIREPLACES, WITH PROPOSALS FOR IM-
PROVING THEM TO SAVE FUEL; TO RENDER DWELLING-
HOUSES MORE COMFORTABLE AND SALUBRIOUS, AND
EFFECTUALLY TO PREVENT CHIMNEYS FROM SMOKING

Bibliothèque Britannique (Science et Arts), edited by Auguste
Pictet, Charles Pictet, and F. G. Maurice (Geneva, 1796), III,
213–271; translated into French from the original English and
written in the third person.

Sir Benhamin Thompson, Count of Rumford, *Essays,
Political, Economical and Philosophical*, (London: T. Cadell,
Jr. and W. Davies), I (1796), 305–387.

Published separately (London: T. Cadell, Jr. and W. Davies,
1796); (Geneva, 1801), 8vo.

The Complete Works of Count Rumford (Boston: American
Academy of Arts and Sciences), II (1873), 484–557.

SUPPLEMENTARY OBSERVATIONS CONCERNING CHIM-
NEY FIREPLACES

Sir Benjamin Thompson, Count of Rumford, *Essays, Political,
Economical and Philosophical* (London: T. Cadell, Jr. and W.
Davies), III (1802), 387–400.

The Complete Works of Count Rumford (Boston: American
Academy of Arts and Sciences), II (1873), 559–570.

OF THE MANAGEMENT OF FIRE AND THE ECONOMY OF
FUEL

Sir Benjamin Thompson, Count of Rumford, *Essays, Political,
Economical and Philosophical* (London: T. Cadell, Jr. and W.
Davies), II (1798), 1–196.

Published separately (London: T. Cadell, Jr. and W. Davies,
1797).

The Complete Works of Count Rumford (Boston: American
Academy of Arts and Sciences), III (1874), 1–165.

EXPERIMENTS AND OBSERVATIONS ON THE ADHESION
OF THE PARTICLES OF WATER TO EACH OTHER

Read at the Institut de France, June 16, 1806.

Mémoires de la classe des Sciences, Mathématiques et Physiques de l'Institut de France (Paris: Baudouin, Imprimeur de l'Institut de France, 1807), VIII, ii, 97–108.

Bibliothèque Britannique (Science et Arts), edited by Auguste Pictet, Charles Pictet, and F. G. Maurice (Geneva, 1806), XXXIII, 3–16.

A Journal of Natural Philosophy, Chemistry and the Arts: Illustrated with Engravings, edited by William Nicholson (London 1806), XV, 52–56, 157–159, 173–175.

Annalen der Physik, begun by F. A. C. Gren, continued by L. W. Gilbert (Halle, 1807), XXV, 121–132.

Le Moniteur Universel ou Gazette Nationale (Paris, July 17, 1806), 914–915 (where the paper is said to have been read July 7, 1806).

Nuova Scelta d'Opuscoli interessanti sulle Scienze e sulle Arti, edited by Carlo Amoretti (Milan, 1804), I, vi, 393–399.

The Complete Works of Count Rumford (Boston: American Academy of Arts and Sciences), II (1873), 290–299.

CONTINUATION OF EXPERIMENTS AND OBSERVATIONS
ON THE ADHESION OF THE PARTICLES OF LIQUIDS TO
EACH OTHER

Read at the Institut de France, March 9, 1807.

Bibliothèque Britannique (Science et Arts), edited by Auguste Pictet, Charles Pictet, and F. G. Maurice (Geneva, 1807), XXXIV, 301–313; XXXV, 3–16.

The Complete Works of Count Rumford (Boston: American Academy of Arts and Sciences), II (1873), 300–317.

INDEX

Acoustical theory of heat, 138
Adhesion in liquids
  body fluids, 499
  for growth of plants and animals, 498
  milk, 499
  sap, 498
  water, 478, 491
Air
  to increase heat, 338
  pollution, 280
Alcohol
  heat of combustion, 97
  heat of condensation,
  specific heat, 154
Argand, Francois Pierre Aimé (1755–1803), 339

Banks, Sir Joseph (1743–1820), 225
Bernard, 434
Besborough, Earl of, 225
Blagden, Sir Charles (1748–1820), 493
Body fluids, 499
Boilers
  experiments, 369
  laundry, 440;
  portable, 438

Caloric, 3, 5, 138, 336
  radiant, 7, 141
Calorimeter, combustion, 80
Charcoal, 172
  combustion, 95, 113
  experiment on, 162;
  heat loss in making, 217;

maximum temperature, 123;
  production, 171, 197
Chemical reaction, 74
Chimney
  draught, 233
  pots, 278;
  smoking, 221, 244, 276;
  sweep, 231, 252
Churches, heating, 448
Coal, burning, 238
Combustion
  calorimeter, 80
  heat of, 80; of alcohol, 97; of charcoal, 113; fuel, 155; maximum, 117; of naphtha, 111; of sulphuric ether, 97, 103; of tallow, 112; of wax, 90; of wood, 115, 171, 204
Condensation, heat of
  alcohol, 129;
  vapors, 80;
  water, 127
Conduction
  poor, 58
  glass, 1
  good, 58
  solids, 12
Conductors, heat, 348
Convection, 49, 57, 354
  demonstration, 273
  insulating liquids, 59
Cooking, 314
  bread and cake, 454
  pies, puddings and pastry, 454
  pots, 457, double covers, 321
  soup, 327

Cooling, 1
  of liquids, 135
  gilded vessels, 135
Crawford, Adair (1748–1795), 94,
  97, 105, 114, 116, 119, 215,
  424
Cruickshanks, 107

Damper
  chimney, 340
  fireplace, 333
Density, maximum
  in lakes, 33
  of water, 39
Drying
  wood, 163; with smoke, 325
  rooms, 441
Dublin
  House of Industry, 454
  Linen-hall, 439
  Society, 440, 457, 459
Duffin, 440

Elector Palatine, 313
English Garden, Munich, 329,
  461
  kitchen, 430
Essay
  Propagation of Heat in Fluids,
    35
  An Inquiry concerning the Na-
    ture of Heat and the Mode of
    Its Communication, 33
Ether
  evaporation, 485
  sulphuric, heat of combustion,
    97, 103; specific heat, 154
Evaporation, ether, 485
Eves and Sutton, 225
Experiments, wood and charcoal,
  162

Fire-balls, 157, 280
Fireplace
  Air-tube, 304
  Argand lamp, 339;

breast, 245;
construction, 156, 221, 224,
  230, 244, 296, 304, 339,
  faults, 296; flues, 413
cottage, 457
covings, 248;
experiments with boilers, 369,
  glass-windowed furnace, 392;
improvements, 159
radiation from, 240;
smoking, 221;
throat, 245;
whitewash, 242
Flame as insulator, 361
Fluids, elastic, in wood, 171
de Fontana, 360
Form of drop of water, 493
Foundling Hospital, kitchen, 434
Fuel
  consumed in kitchens, 317
  cooking, 328;
  economy of, 309;
  heat of combustion, 156;
  waste of, 221, 287, 310
Furnace, 339;
  construction, ash pits, 343

Garden, English, 329, 430
Gay-Lussac, Nicolas F. (1778–
  1850), 89, 95, 164, 201, 209,
  215
Gilded and nongilded vessels,
  cooling, 135
Glaciers, Chamouny, 31, 39
Grew, 171, 177

Hardy, Admiral Sir Charles (1716–
  1780), 455
Heat
  capacity of liquids for, 145, 153,
    154
  of combustion, 80; of alcohol,
    97; of charcoal, 113; of fuel,
    155; maximum, 117; of naph-
    tha, 111; of sulphuric ether,

97, 103; of tallow, 112; of wood, 115, 171, 204
of condensation: of alcohol, 129; of vapors, 80; of water, 127
conductors, 348
insulators, 348
intensity increased by air flow, 338
loss, experiments, 1; in making charcoal, 217; opaque bodies, 5
megaphone, 30;
nature of, 7;
propagation in fluids, 31;
radiant, 240, 345;
radiation cooling, 3, 4;
specific: of alcohol, 154; of naphtha, 153; of olive oil, 153; of sulphuric ether, 154; of turpentine, 154
theory, acoustical, 138
transmission: by convection, 59; in liquids, 57; in solids, 9, 57; in solid metals, 46; in vacuum, 356
vibratory theory of, 30
Heating
churches, 448
hothouses, 447;
large halls, 447;
water with steam, 324
Hempel, Mrs., 265
Hopkins, 265
Hospital of La Misericordia, Verona, kitchen, 432
Hospital of La Pieta, Verona, kitchen, 431, 444, 457
House of Industry, Munich, 313, 318, 327, 384
kitchen, 375, 426, 427
Hothouses, 447
Hydrogen, heat of combustion, 94

Industrial lime-kiln, 459
Industrial processes

bleaching, 439
brewing, 398
Ingen-housz, Dr. Jan (1730–1799), 350
Instruments
combustion calorimeter, 80
experiments on heat, 24;
glass balloons, 78;
Little Sentinel, 76;
thermoscope, 24;
visible convection currents, 36;
Wedgwood pyrometer, 120
Insulator, heat, 348
adjustable, 447
flame, 361
steam, 358;
water, 20

Jackson, 457

Kindling balls, 280
Kirwan, Richard (1733–1812), 419
Kitchen
construction, 426
English Garden, Munich, 430
Foundling Hospital, London, 434;
Hospital of La Misericordia, Verona, 432;
Hospital of La Pieta, Verona, 431, 444, 457;
House of Industry, Munich, 375, 426, 427;
Military Academy, Munich, 381, 429;
Military Hall, Munich, 429

La Misericordia Hospital, Verona, 432
de Laplace, Pierre S. (1749–1822), 497
La Pieta Hospital, Verona, 431
Lavoisier, Antoine Lauren (1743–1794), 89, 91, 93, 95, 97, 115, 419, 424

Lavoisier, Marie Anne Pierrette (Paulze) (1758–1836), 31
Lerchenfeld, Baron von, 457
Lime-kiln, perpetual, 459
Linseed oil, specific heat, 153
Liquid
  in wood, 171, 188
  character of surface, 494
Liquids
  mixing of, 54
  spontaneous mixture, 74
London, Foundling Hospital, 434
Lowitz, 101
de Luc, Jean André (1727–1817), 39

Malpighi, Marcello (1628–1694), 171, 177
Management of fire, 309
Mer de Glace, 31, 39
Military cooking, 449
Military stoves, 438
Military Academy, Munich, 314, 329, 381, 429
Military Hall, Munich, 429
Milk, adhesion, 499
Monge (1746–1818), 496
Montagu, Mrs., 225
Moreau, General Jean Victor (1763–1813), 436
Motion, continual, in liquids, 78
Motivation to work, 323
Munich, kitchen
  in English Garden, 430
  in House of Industry, 375, 427
  in Military Acadeny, 381, 429
  in Military Hall, 429

Naphtha
  heat of combustion, 111
  specific heat, 153
Nast, 136

Olive oil
  combustion, 93
  specific heat, 153

Oven
  baking bread, 454
  construction, register, 302, ship, 455
  perpetual, 451
Oxygen from silk, 488

Palmerston, Lord (1739–1802), 225, 360
Patent, 265
Pelham, Thomas, 438
Pictet, M. Auguste (1752–1825), 31, 493
Proust, Louis Joseph (1754–1826), 203
Pryometer, Wedgwood, 120

Radiation
  absorption, 347
  cooling of vessels, 135;
  from fireplaces, 241;
  heat, 345;
  theory, 137
Rape oil, combustion, 93
Roaster, 429, 434, 453; ship, 456
Robbery, 370
Royal Institution of Great Britain, 161
Royal Irish Academy, fireplaces, 448

St.Paul's churchyard, 370
Salisbury, Marquis of, 225
Sap, 177, 181, 498
de Saussure, H. B. (1740–1799), 89, 97, 103, 105, 131
de Schwachheim, Baron, 315
Sea, currents, 40
Segner, Johann (1704–1777), 494
Silk, oxygen from, 488
Sinclair, Sir John (1754–1835), 222, 432
Smoke
  ascent of, 271
  for drying wood, 325

Soup, 327
Specific gravity, wood, 173
Specific heat
  alcohol, 154
  naphtha, 153;
  olive oil, 153;
  sulphuric ether, 154;
  turpentine, 154
Spencer, Countess Dowager, 225
Steam for hot water, 324
Stoves
  German, 228
  for heating irons, 442;
  portable, 126, 438
Sudley, Lord, 225
Surface tension, 478, 494, 495
Suspension of gold in water, 70

Tallow, heat of combustion, 112
Temperature of different countries, 40
Templeton, Lady, 225
Thénard, Baron L., Joseph (1777–1857), 89, 95, 164, 201, 209, 215
Thermoscope, 24
Thompson, Dr. (Edinburgh), 36
Tree-sap, 177, 181
Turpentine, specific heat, 154

Vacuum, heat transmission in, 356

Vapor, heat of condensation of, 80
Vauquelin, 104, 111
Verona
  kitchen of Hospital of La Misericordia, 432
  kitchen of Hospital of La Pieta, 431, 444, 457
Viscosity, 491

Water
  absorption in wood, 191
  adhesion, 478, 491;
  heat of condensation, 127;
  heating with steam, 387;
  viscosity, 491
Watt, James (1736–1819), 129, 421
Wax, combustion, 90
Wedgwood, Josiah (1730–1795), 120
Winds, polar regions, 40
Windows, double, 348, 352
Wood
  drying, 163
  experiment on, 162;
  heat of combustion, 115, 171, 204;
  liquid in, 171, 188;
  specific gravity, 173;
  structure of, 171;
  water absorption, 191